17
Topics in Organometallic Chemistry

AF786248

Topics in Organometallic Chemistry

Recently Published and Forthcoming Volumes

Organometallic Oxidations Catalysis
Volume Editors: F. Meyer, C. Limberg
Vol. 22, 2006

N-Heterocyclic Carbenes (NHC) in Transition Metal Catalysis
Volume Editor: F. Glorius
Vol. 21, 2006

Dendrimer Catalysis
Volume Editor: L. H. Gade
Vol. 20, 2006

Metal Catalyzed Cascade Reactions
Volume Editor: T. J. J. Müller
Vol. 19, 2006

Catalytic Carbonylation Reactions
Volume Editor: M. Beller
Vol. 18, 2006

Bioorganometallic Chemistry
Volume Editor: G. Simonneaux
Vol. 17, 2006

Surface and Interfacial Organometallic Chemistry and Catalysis
Volume Editors: C. Copéret, B. Chaudret
Vol. 16, 2005

Chiral Diazaligands for Asymmetric Synthesis
Volume Editors: M. Lemaire, P. Mangeney
Vol. 15, 2005

Palladium in Organic Synthesis
Volume Editor: J. Tsuji
Vol. 14, 2005

Metal Carbenes in Organic Synthesis
Volume Editor: K. H. Dötz
Vol. 13, 2004

Theoretical Aspects of Transition Metal Catalysis
Volume Editor: G. Frenking
Vol. 12, 2005

Ruthenium Catalysts and Fine Chemistry
Volume Editors: C. Bruneau, P. H. Dixneuf
Vol. 11, 2004

New Aspects of Zirconium Containing Organic Compounds
Volume Editor: I. Marek
Vol. 10, 2004

Precursor Chemistry of Advanced Materials
CVD, ALD and Nanoparticles
Volume Editor: R. Fischer
Vol. 9, 2005

Metallocenes in Stereoselective Synthesis
Volume Editor: T. Takahashi
Vol. 8, 2004

Transition Metal Arene π-Complexes in Organic Synthesis and Catalysis
Volume Editor: E. P. Kündig
Vol. 7, 2004

Organometallics in Process Chemistry
Volume Editor: R. D. Larsen
Vol. 6, 2004

Organolithiums in Enantioselective Synthesis
Volume Editor: D. M. Hodgson
Vol. 5, 2003

Organometallic Bonding and Reactivity: Fundamental Studies
Volume Editor: J. M. Brown, P. Hofmann
Vol. 4, 1999

Activation of Unreactive Bonds and Organic Synthesis
Volume Editor: S. Murai
Vol. 3, 1999

Bioorganometallic Chemistry

Volume Editor: Gérard Simonneaux

With contributions by

C. S. Allardyce · P. A. Butler · P. J. Dyson · J. C. Fontecilla-Camps
B. Kräutler · T. Hirao · P. Le Maux · T. Moriuchi · K. Severin
G. Simonneaux · A. Volbeda

Springer

The series *Topics in Organometallic Chemistry* presents critical overviews of research results in organometallic chemistry. As our understanding of organometallic structure, properties and mechanisms increases, new ways are opened for the design of organometallic compounds and reactions tailored to the needs of such diverse areas as organic synthesis, medical research, biology and materials science. Thus the scope of coverage includes a broad range of topics of pure and applied organometallic chemistry, where new breakthroughs are being achieved that are of significance to a larger scientific audience.
The individual volumes of *Topics in Organometallic Chemistry* are thematic. Review articles are generally invited by the volume editors.
In references *Topics in Organometallic Chemistry* is abbreviated *Top Organomet Chem* and is cited as a journal.

Springer WWW home page: springer.com
Visit the TOMC content at springerlink.com

ISBN 978-3-642-06958-1 e-ISBN 978-3-540-33049-3

ISSN 1436-6002

DOI 10.1007/b97910

Springer is a part of Springer Science+Business Media

springer.com

Softcover reprint of the hardcover 1st edition 2006

Cover design: *Design & Production* GmbH, Heidelberg

Printed on acid-free paper 02/3141 YL – 5 4 3 2 1 0

Volume Editor

Gérard Simonneaux

DR CNRS (UMR 6226)
Bât 10C pièce 126
Campus de Beaulieu
35042 Rennes Cedex, France
gerard.simonneaux@univ-rennes1.fr

Editorial Board

Topics in Organometallic Chemistry Also Available Electronically

For all customers who have a standing order to Topics in Organometallic Chemistry, we offer the electronic version via SpringerLink free of charge. Please contact your librarian who can receive a password or free access to the full articles by registering at:

springerlink.com

If you do not have a subscription, you can still view the tables of contents of the volumes and the abstract of each article by going to the SpringerLink Homepage, clicking on "Browse by Online Libraries", then "Chemical Sciences", and finally choose Topics in Organometallic Chemistry.

You will find information about the

- Editorial Board
- Aims and Scope
- Instructions for Authors
- Sample Contribution

at springer.com using the search function.

Preface

One of the more recently developing areas of organometallic chemistry, indeed in all of chemistry, is that of its applications to living systems. The name *bioorganometallics* applied to compounds with metal–carbon bonds and organometallic reactions have found a place in living systems. In this book, you will find general overviews of selected in vivo processes presented from the viewpoint of the biochemist, and the study of organometallic complexes of biological and medical interest.

Naturally occurring bioorganometallic complexes, such as vitamin B12, are first considered. The B_{12}-coenzymes are the organometallic cofactors in various important enzymatic reactions and are particularly relevant in the metabolism of archaea and (other) anaerobic microorganisms. Surprising characteristics of recently discovered iron and nickel hydrogenases, including a possible role in the geochemical theory of the origin of life, are highlighted. The possible formation of carbene complexes of cytochrome P450 enzymes in various metabolisms of xenobiotics is also discussed.

Bioorganometallic chemistry is envisioned to provide not only organometallic receptors such as polynuclear organometallic macrocycles for biologically interesting molecules but also ferrocene–peptide bioconjugates giving a peptidomimetic basis for protein folding. These chapters illustrate the usefulness of organometallic complexes in water as a molecular scaffold, a sensitive probe, a chromophore, an NMR shift reagent, a redox-active site and a chemosensor.

One of the other major areas of bioorganometallic compounds originated in a possible use of these complexes as therapeutic drugs. Thus the medicinal properties of organometallic compounds are reviewed, with notable applications in the treatment and diagnosis of cancer and in the treatment of viral, fungal, bacterial and parasitic infections.

This monograph is not intended to provide a comprehensive view of all explored fields of research activity in bioorganometallic chemistry. However, the reader will get a balanced view of this rapidly developing and promising area. I hope this book will stimulate its readers to enter the exciting field of bioorganometallic chemistry.

Rennes, France, April 2006 Gérard Simonneaux

Contents

Top Organomet Chem (2006) 17: 1–55
DOI 10.1007/3418_004

Published online: 30 March 2006

Biological Organometallic Chemistry of B_{12}

Philip A. Butler · Bernhard Kräutler (✉)

Institute of Organic Chemistry & Center of Molecular Biosciences (CMBI),
University of Innsbruck, 6020 Innsbruck, Austria
bernhard.kraeutler@uibk.ac.at

Abstract Vitamin B_{12}, the "antipernicious anemia factor" required for human and animal metabolism, was discovered in the late 1940s. B_{12}-derivatives are cobalt complexes of the unique and remarkably complex corrin ligand, that belongs to the class of the natural tetrapyrroles. The B_{12}-coenzymes are the organometallic cofactors in various important enzymatic reactions and are particularly relevant in the metabolism of some archaea and (other) anaerobic microorganisms. Indeed, the microorganisms are the only natural sources of the B_{12}-derivatives, while (with the exception of the higher plants) most living organisms depend on these cobalt-corrinoids. Vitamin B_{12}-derivatives thus hold an important position in the life sciences and have attracted particular interest from medicine, biology, chemistry and physics.

Keywords Cobalt · Coenzyme B_{12} · Methyl group transfer · Radical reaction · Vitamin B_{12}

Abbreviations

Ado	Adenosyl
AdoCbl	5′-Deoxy-5′-adenosylcobalamin, adenosylcobalamin, coenzyme B_{12}
Cbl	Cobalamin (DMB-cobamide)
CNCbl	Cyanocobalamin
DCE	Dichloroethene
DD	Diol dehydratase
DMB	5,6-Dimethylbenzimidazole
EAL	Ethanolamine ammonia lyase
GD	Glycerol dehydratase
GM	Glutamate mutase
H_2OCbl	Aquocobalamin, B_{12a}
HOCbl	Hydroxocobalamin
ICM	Isobutyryl-CoA mutase
LAM	Lysine aminomutase
MeCbl	Methylcobalamin
MGM	Methyleneglutarate mutase
MMCM	Methylmalonyl-CoA mutase
NHE	Normal hydrogen electrode
NMR	Nuclear magnetic resonance
NOE	Nuclear Overhauser effect
PLP	Pyridoxal-phosphate
PCE	Tetrachloroethene
SAM	*S*-adenosylmethionine (= AdoMet)
SCE	Saturated calomel electrode
TCE	Trichloroethene
UV	Ultraviolet spectrum
UV-vis	Ultraviolet visible absorbance spectrum

1 Introduction

B_{12}-coenzymes are conceivably Nature's most complex and physiologically most broadly relevant organometallic cofactors. B_{12}-derivatives, therefore, (co)catalyze unique enzymatic reactions that directly depend upon the reactivity of the cobalt coordinated organic ligands and they hold an exceptional position in the area of bioorganometallic chemistry. The metabolism of most living organisms depends on catalysis by B_{12}-dependent enzymes [1].

Nearly 60 years ago, the red cyanide-containing cobalt-complex vitamin B_{12} (**1**, cyanocob(III)alamin, CNCbl) was discovered and isolated as the (extrinsic) antipernicious anemia factor [2, 3]. Vitamin B_{12} (**1**, CNCbl) crystallizes readily and is a relatively inert Co(III)-complex. It is the most important

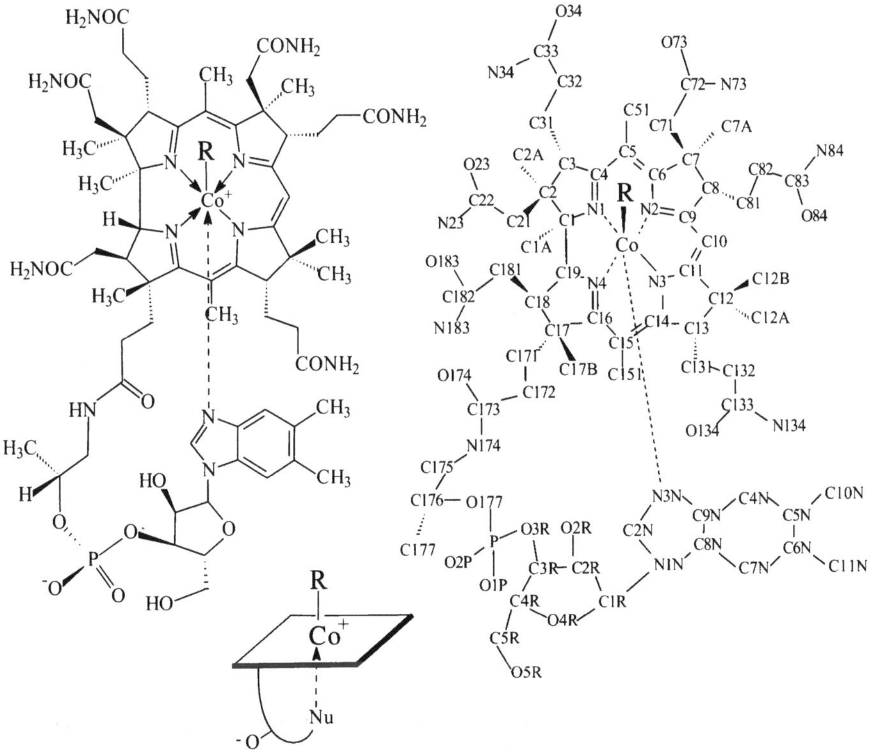

Fig. 1 Structural formulae of selected cobalamins (DMB-cobamides, Cbl = cobalamin, ado = adenosyl, *left*), atom numbering used (*right*) [29] and symbol used (*bottom*): vitamin B_{12} **1** (CNCbl, R = CN); coenzyme B_{12} **2** (R = 5′-deoxy-5′-ado); methylcobalamin **3** (MeCbl, R = CH_3); aquocobalamin **4**$^+$ (R = H_2O^+); hydroxocobalamin **5** (HOCbl, R = HO); chlorocobalamin **12** (R = Cl); sulfitocobalamin **13** (R = SO_3^-); nitritocobalamin **15** (R = NO_2); thiocyanato-Cbl **16** (R = NCS); selenocyanato-Cbl **17** (R = NCSe); thiosulfato-Cbl **18** (R = S_2O_3); cob(II)alamin **23** (B_{12r}, R = e^-); α-adenosyl-Cbl **25** (R = 5′-deoxy-5′-α-ado); adeninylpropyl-Cbl **26** (R = 3-adenosyl-propyl); homocoenzyme B_{12} **27** (R = 5′-deoxy-5′-ado-methyl); 2,3-dihydroxypropyl-Cbl **28** (R = 2,3-dihydroxy-propyl); trifluoromethyl-Cbl **29** (R = CF_3); difluoromethyl-Cbl **30** (R = CHF_2); vinylcobalamin **32** (R = CH = CH_2); *cis*-chlorovinyl-Cbl **33** (R = CH = CHCl); bishomocoenzyme B_{12} **36** (R = 2-[5′-deoxy-5′-ado-ethyl); 2′-deoxycoenzyme B_{12} **50** (R = 2′,5′-dideoxy-5′-ado); 2′,3′-dideoxycoenzyme B_{12} **51** (R = 2′,3′,5′-trideoxy-5′-ado); 5-adeninyl-pentyl-Cbl **53** (mR = 5-adeninyl-pentyl); 3-aminopropyl-Cbl **54** (R = 3-aminopropyl)

commercially available form of the naturally occurring B_{12}-derivatives, but it appears to have no physiological function itself [4]. The physiologically relevant vitamin B_{12}-derivatives are the highly light-sensitive and chemically more labile organometallic coenzymes, coenzyme B_{12} (**2**, 5′-deoxy-5′-adenosylcobalamin, AdoCbl) and methylcobalamin (**3**, MeCbl), as well as the "inorganic" and easily reducible B_{12}-derivatives aquocob(III)alamin (**4**$^+$, H_2OCbl^+) and hydroxocob(III)alamin (**5**, HOCbl).

During the last five decades, the remarkable scientific advances towards the solution of some of the major "B_{12} mysteries" have been reported in a series of European symposia on "Vitamin B_{12} and B_{12}-Proteins", the first of which was in 1956 in Hamburg (Germany), followed by, again, Hamburg (1961), Zürich (Switzerland, 1979) [5], Innsbruck (Austria, 1996) [6] and Marburg (Germany, 2000). Some of the top achievements in this field concern the elucidation of the structure of vitamin B_{12} [7, 8] and of coenzyme B_{12} [9], the synthetic conquest of the vitamin B_{12} structure [10–12], the biosynthetic pathways to B_{12} [13–16], as well as the first x-ray crystal structures of B_{12}-binding proteins [17–21].

Several concise books on the subject have been written, with earlier ones by Pratt [22] and by Friedrich [23] describing the chemistry of B_{12}. The more recent ones on B_{12} [24], vitamin B_{12} and B_{12}-proteins [6], on chemistry and biochemistry of B_{12} [25] and a review [26], provide more systematic information around the cofactor role of the B_{12}-coenzymes.

2
B_{12}: Structure and Reactivity

2.1
Crystallographic Structural Studies

The structures of vitamin B_{12} (**1**; see Fig. 1) and of coenzyme B_{12} (**2**) were established through the pioneering x-ray crystallographic studies of Hodgkin et al. [7–9], which discovered the composition of the corrin core of **1** and the nature of the organometallic ligand of **2**. Since these landmark analyses, work in this field has turned away from the initial constitutional and stereochemical questions concerning B_{12}-molecules. Studies towards more accurate structural data of B_{12}-derivatives, have become of interest, as presented in recent reviews [27, 28].

Vitamin B_{12} (**1**, CNCbl), and other B_{12}-derivatives, where the cyanide ligand of **1** is replaced by a different "upper" β-ligand are 5′, 6′-dimethylbenzimidazolyl-cobamides and are the most commonly discussed B_{12}-derivatives. Only "base-on" cobalamins, where the nucleotide functionality coordinates in an intramolecular mode, have been analyzed by x-ray crystallography [27, 28]. In this present chapter a systematic atomic numbering is used for vitamin

B_{12}-derivatives [29], which builds on the convention that atom numbers of the heavy atoms of a substituent reflect the number of the points of attachment to the corrin ligand and are indexed consecutively [30].

The "old" structure of vitamin B_{12} [7] has been reanalyzed using modern cryocrystallography techniques [31, 32], which showed the molecular geometry of the B_{12} moiety to agree within experimental error of Hodgkin's original result. Cyano-13-epicobalamin (neovitamin B_{12} **6**; see Fig. 2), a derivative of vitamin B_{12} where the propionamide chain at the C13 position is in the β-configuration was also studied [33, 34]. A notable difference between the two structures is an increased "nonplanarity" in the corrin ring of the neoderivative (expressed as a 6° larger fold angle, 23.7° vs. 17.9°). The C8 epimer of vitamin B_{12}, cyano-8-epicobalamin (**7**), has an even larger fold angle of the corrin core (23.8°) [35].

Due to the discovery of the replacement of the cobalt coordinated 5,6-dimethylimidazole (DMB) base by a protein-derived imidazole in several B_{12}-dependent enzymes (see later), the analysis of Co_β-cyano-imidazolylcobamide (**8**) [31] was of particular interest. The less bulky and more nucleophilic imidazole base of **8** caused a number of structural differences. The corrin ring fold angle of **8** decreased to 11.3° and the axial Co – N bond shortened (from 2.011 Å in **1** to 1.968 Å in **8**). In addition, the "base tilt" of **8** (i.e., half the difference between the two Co – N – C angles to the coordinating base) decreased to practically zero, within experimental error. In all crystal structures of 5′,6′-dimethylbenzimidazoyl-cobamides a tilt of about 5° is found [36], which appears to be an inherent property of the cobalt-coordinated DMB.

Norpseudovitamin B_{12} (Co_β-cyano-7″-[2-methyl]adeninyl-176-norcobamide) (**9**) represents the first example of a complete B_{12}-derivative that lacks one of the methyl groups (of C176) of the cobamide moiety. X-ray crystal structures were determined for **9** whose nucleotide base is different from DMB and the previously known analogues pseudovitamin B_{12} (**10**) and factor A (Co_β-cyano-7″-[2-methyl]adeninylcobamide) (**11**) [37]. These first accurate crystal structures of complete corrinoids with an adeninyl pseudonucleotide confirmed the expected coordination properties around Co and corroborated the close conformational similarity of the nucleotide moieties of **9** and its two homologues. Originally the axial Co – N bond of **11** was reported as 2.118 Å and with a fold angle of 15.7° [38]. With the new analysis these have been determined again to be a more feasible Co – N bond of 2.026 Å and a fold angle of 16.9° [37]. For **9** and **10** the axial Co – N bonds were 2.035 Å and 2.021 Å, respectively, and both had a fold angle of 19.6°. The data reported showed that in cyano-Co(III) cobamides the structural consequences of a replacement of 5,6-dimethylbenzimidazole by adenine or 2-methyladenine were of hardly any significance.

Crystal structures of a cobalamin complex with a central Co – CN – Re feature have been formed where the cyanide ligand in vitamin B_{12} acts as

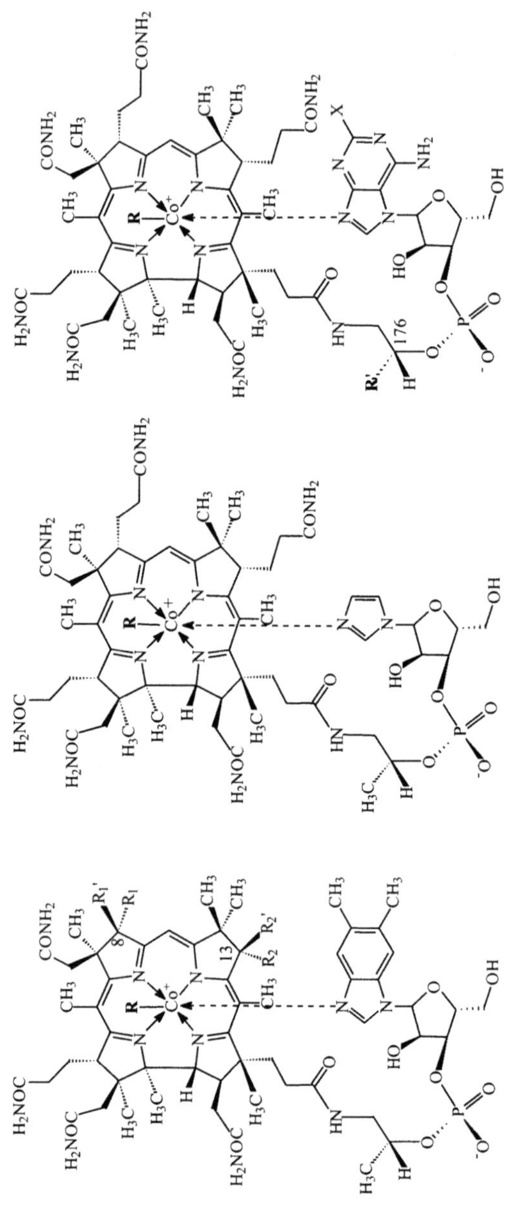

Fig. 2 Structural formulae of complete-cobamides: *left*: neovitamin B_{12} (**6**, cyano-13-epicobalamin, R = CN, $R_1' = R_2 = H$, $R_1 = R_2' =$ propionamide), cyano-8-epi-cobalamin (**7**, R = CN $R_1 = R_2' = H$, $R_1' = R_2 =$ propionamide), neocoenzyme B_{12} (**39**, R = 5′-deoxy-5′-adenosyl, $R_1' = R_2 = H$, $R_1 = R_2' =$ propionamide); *center*: Co_β-cyano-imidazolylcobamide (**8**, R = CN), Co_β-methyl-imidazolylcobamide (**31**, R = CN); *right*: norpseudovitamin B_{12} (**9**, R = CN, R′ = H; X = H), pseudovitamin B_{12} (**10**, R = CN, R′ = CH_3; X = H), Factor A (Co_β-cyano-2′-methyladeninyl-cobamide, **11**, R = CN, R′ = CH_3; X = CH_3), pseudocoenzyme B_{12} (**37**, R = 5′-deoxy-5′-adenosyl, R′ = CH_3; X = H), adenosyl-factor A (**38**, R = 5′-deoxy-5′-adenosyl, R′ = CH_3; X = CH_3)

bridging ligand between the rhenium carbonyl compounds [39]. This concept paves the way for radiolabeling of vitamin B_{12} or metal-mediated coupling of bioactive molecules.

For the crystal structure of aquocobalamin perchlorate (**4**$^+$-ClO_4^-) [40] the shortest known axial $Co_\alpha - N$ bond of a vitamin B_{12} derivative was observed (1.925 Å). Together with the large upwards folding angle of 18.7°, the conclusion stated was, that steric repulsion between the DMB-base and corrin core led to a flexing of the corrin ring [36]. The short axial $Co_\alpha - N$ for the **4**$^+$-ion was consistent with the weak donor ability of the *trans*-axial aquo ligand. Crystal structures of numerous other inorganic B_{12}-derivatives have been solved and previously reviewed elsewhere [1,28]. More recent structures though that have been analyzed include chlorocobalamin (**12**) [36], [(SO_3)Cbl]NH_4 (**13**), [(thiourea)Cbl](PF_6) (**14**) [41], NO_2 – Cbl (**15**), NCS – Cbl (**16**), NCSe-Cbl (**17**) [42, 43] and $S_2O_3Cbl^-$ (**18**) [43].

For incomplete cobamides the earlier x-ray investigations have been reviewed by Glusker [27]. This includes the important α-cyano-β-aquo cobyric acid (**19**; see Fig. 3) structure solved by Hodgkin and coworkers [44]. Cobyric acid is the natural nucleus of the B_{12} vitamins and the initial target for Woodward and Eschenmoser for their total synthesis of vitamin B_{12} (as it had already been shown how **19** could be converted to vitamin B_{12}) [11, 12]. Since then work has focused on dicyano-heptamethyl cobyrinate (**20**, "cobester") and its analogues. Structures analyzed include cobester (**20**) [45–47], cobester-b-monoacid (**21**) [48] and 15-norcobester (**22**) [49, 50] as reviewed in [1].

Information on cob(II)alamin (**23**, B_{12r}) was of particular interest as it is the product of Co – C bond homolysis of coenzyme B_{12} (**2**), which occurs during the catalytic cycle of coenzyme B_{12}-dependent enzymes. The crystal structure of cob(II)alamin showed that the corrin moiety of **23** and **2** is very similar [51]. The fold angle of the corrin ring in **23** is 16.3° compared to 13.3° in **2**. Surprisingly, the axial cobalt-nitrogen bond is slightly shorter in the five coordinated cob(II)alamin (2.13 Å) than in the six coordinated coenzyme B_{12} (2.24 Å). However, the distance between the corrin ring and the coordinated DMB-base is almost the same, due to a "downward" displacement of the cobalt atom from the plane of the corrin ligand in **23**. It was expected that the reduced Co(II) would have a longer bond than the Co(III) species. In view of this result, in **2** and related organocobalamins, the "structural *trans* effect" of the organic ligand appears to increase the axial Co(III) – N bond, which compensates for the larger covalent radius of Co(II) compared to Co(III). From these observations the conclusion was made that the interactions (apoenzyme/coenzyme) at the corrin moiety of the coenzyme appear to be inadequate to provide the major means for a protein-induced activation of the bound coenzyme toward homolysis of its Co – C bond. Instead, the organometallic bond may be labilized by way of apoenzyme (and substrate) induced separation of the homolysis fragments,

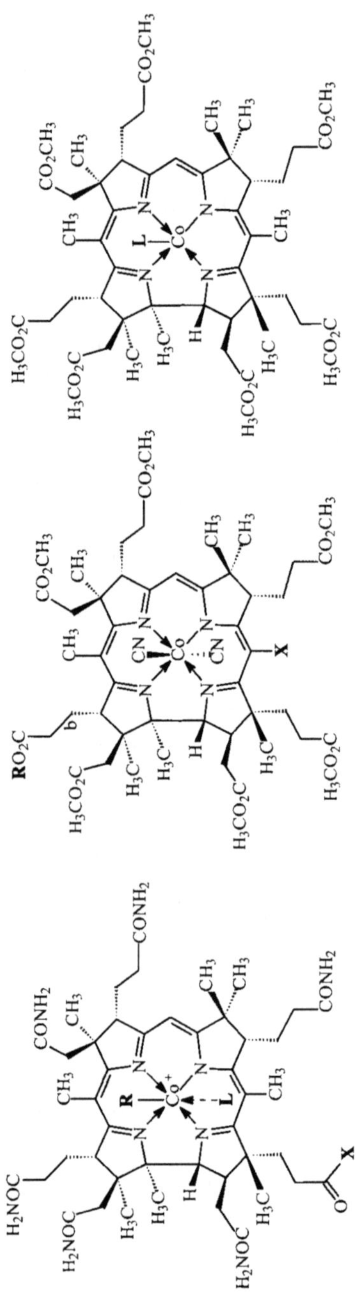

Fig. 3 Structural formulae of *left*: "incomplete" cobamides: α-cyano-β-aquo cobyric acid (**19**, $R = H_2O$, $L = CN$, $X = OH$); Co_β-5′-deoxy-5′-adenosylcobinamide (**35**, R = 5′-deoxy-5′-ado, $L = H_2O$, $X = NH - CH_2 - CHOH - CH_3$); *center*: cobester (**20**, $R = CH_3$, $X = CH_3$), cobester-b-monoacid (**21**, R = H, $X = CH_3$), 15-nor-cobester (**22**, $R = CH_3$, X = H); *right*: perchlorato-heptamethylcob(II)yrinate (**24**, $L = ClO_4$), heptamethylcob(I)yrinate (**34**, L = absent)

made possible by strong binding of the separated components to the protein [51]. Cob(II)alamin (**23**) has also been recently studied using neutron Laue diffraction studies, which came to the same conclusions regarding its structure [52].

The crystal structure of heptamethyl-cob(II)yrinate (**24**) revealed a preference for incomplete Co(II)-corrins to coordinate the axial ligand at the sterically less hindered upper β-face, in contrast with the result obtained from the complete cob(II)alamin. Analysis of complex **24** revealed a five-coordinated Co(II)-center to which a perchlorate ligand was coordinated at the β-face, a long axial cobalt-oxygen bond (2.31 Å) and a 6° fold angle of the corrin ligand [53].

The original crystal structure of the organometallic coenzyme B_{12} (**2**) [9, 54] has been confirmed by more extensive studies by x-ray and neutron crystallography [55–57]. The findings show both axial Co – C (2.030 Å) and Co – N (2.237 Å) bonds are relatively long [27, 28]. The organometallic adenosyl moiety is bound in an anticonformation and the adenine ring is found to be above ring C of the corrin ligand. A large Co – C – C bond angle of 125.4° is observed for the organometallic group [54].

In α-adenosylcobalamin (**25**) the organometallic adenine base is attached at the ribose moiety in an α-configuration and is thus a stereoisomer of AdoCbl (**2**). The crystal structure of **25** showed the lengths of the axial Co – C (2.02 Å) and Co – N (2.24 Å) bonds to be similar to **2** but the corrin ring was flatter (fold angle = 11.7° vs. 13.3° in **2**) [58]. The adenosyl ligand, as in **2**, was placed over the southeast quadrant (ring C), but the position of the adenine moiety relative to the ribose unit of the organometallic ligand was disordered due to the different conformations of the adenine heterocycle.

Adeninylalkylcobalamins, where a methylene chain connects the adenine with the cobalt center [59], inhibit various AdoCbl-dependent enzymes depending upon the length of the alkyl chain [60]. Adeninylpropylcobalamin (**26**) has been studied in its crystalline form, as well as in solution [61]. The structure of the corrin ring and the lower nucleotide loop closely resembled that of **2**. However, the adenine group of **26** is oriented almost parallel to the corrin plane and is positioned over ring D of the corrin ligand around 120° clockwise from its position in coenzyme B_{12}.

The homologue of coenzyme B_{12} "homocoenzyme B_{12}" (**27**, Co_β-(5′-deoxy-5′-adenosylmethyl)-cob(III)alamin) has been recently examined as it has been suggested to function as a covalent structural mimic of the hypothetical enzyme bound "activated" state of the B_{12}-cofactor [62]. In the crystal structure of **27** the cobalt center was observed to be at a distance of 2.99 Å from C5′ of the homoadenosine moiety and the latter to be in the unusual *syn*-conformation. The crucial distance from the corrin-bound cobalt center to the C5′ of the homoadenosine moiety in **27** is, thus, roughly the same as found in one of the two activated forms of coenzyme B_{12} in the crystal structure of glutamate mutase [63].

To investigate if the large Co – C – C bond angle of AdoCbl (**2**) is typical for organocobalamins the crystal structures of the 2,3-dihydroxypropylcobalamins (the diasteromeric *R*- and *S*-isomers **28R** and **28S**) were examined earlier [64]. The Co – C distances (2.00 and 2.08 Å for **28R** and **28S**, respectively) were similar to AdoCbl (2.03 Å) but the bond angles were smaller (119.6° for **28R** and 113.6° for **28S**). The value for **28S** should be considered the "normal" angle, with little interactions between the corrin ring and β-substituent.

The crystal structure of the simplest organometallic B_{12}-derivatives methylcobalamin (**3**) was solved in 1985 by Rossi et al. [65]. The structure has been further investigated [32] to provide a more accurate structure. The structures confirmed the folding of the corrin core of **3** to be similar to that of coenzyme B_{12} (**2**) (fold angle in **3** = 14.7° [32]). This proved that the bulkiness of the 5′-deoxyadenosyl ligand in **2** was not a main contributor to the conformation of the corrin ligand of AdoCbl. The lengths of the axial Co – C (1.979 Å) and Co – N (2.162 Å) are slightly shorter in **3** when compared to **2**. The shorter axial bond to the DMB-base is consistent with the stronger nucleotide coordination in **3**. The structures of the fluorinated analogues trifluoromethyl-cobalamin (**29**) [66] and difluoromethyl-cobalamin (**30**) have been elucidated and compared to methylcobalamin [67].

Co_β-methyl-imidazolylcobamide (**31**) was prepared as a model for organometallic B_{12}-cofactors in a "base-off/His-on" form and its crystal structure was analyzed [68]. The substitution by a less bulky and more nucleophilic imidazole base had the expected structural effects. The axial Co – C (1.97 Å) and Co – N (2.09 Å) are shorter in **31** than in methylcobalamin and the fold angle of the corrin ligand was reduced by over 2° to 12.5°.

Finally the first crystal structure analyses of organocobalamins with sp^2-hybridized carbon ligands have been reported, vinylcobalamin (**32**) [69] and of *cis*-chlorovinylcobalamin (**33**) [70], the latter is a putative intermediate in the reductive degradation of chlorinated ethylenes. As expected for a vinyl ligand the Co – C bond length (1.912 Å for **32** and 1.952 Å for **33**) is shorter than in adenosylcobalamin (**2**) and methylcobalamin (**3**). The Co – C bond in **32** is significantly shorter than that in **33**, presumably because of steric repulsion between the chlorine of **33** and the β-substituents on the corrin ring. The axial Co – N bond lengths of 2.166 Å for **32** and 2.144 Å for **33** are also shorter than in **2** and **3** and provide a good example of the "inverse" *trans* effect.

The "thermodynamic" and "structural" *trans*-effects of B_{12}-derivatives are the effect of one cobalt-coordinated axial ligand on chemical equilibria and coordination properties of an axial ligand *trans* to the first one [71]. An increasing σ-donor power of the Co_β-ligand X was found to correlate with the size of the thermodynamic *trans*-effect in B_{12}-derivatives. The length of the axial Co_α – N bond to the DMB-base in cobalamins generally increases with the σ-donor property of the Co_β-ligand [27, 28]. In the same sequence, the σ-ligand influences the base-on/base-off equilibria. A linear correlation thus exists between free enthalpy of the base-on/base-off equilibria in aqueous solution and the length of the Co – N bond [34]. However in B_{12}-derivatives, both axial bonds lengthen simultaneously with increasing σ-donor character of the axial ligands [72].

The saturated and direct *trans*-junction between two of its four five-membered rings is the main cause of the nonplanar nature of the corrin core in B_{12}-derivatives. The characteristic "ligand-folding" is a main factor to the variability in the conformation of the corrin ligand [73]. The fold has always been found as "upwards" (towards the β-face), about the C10 – Co axis, and the "fold angle" is defined as the angle between the best planes through N1 – C4 – C5 – C6 – N2 – C9 – C10 and C10 – C11 – N3 – C14 – C15 – C16 – N4 [27]. Fold angles are usually smaller in incomplete corrinoids (with a minimal observed value of 1.9° in Co_α-aquo-Co_β-cyano-8-dehydro-cob(III)yrinic acid c lactone [27]) when compared to complete corrinoids, where a value of 23.8° has been found in cyano-8-epicobalamin (**7**) [35].

The bulky DMB-base was therefore suggested to be a relevant contributor to the upwards folding of corrins [27]. This possible effect of the intramolecular coordination of the DMB-base on the folding of the corrin in Cob(III)alamins has been examined in detail [31, 32, 36, 40]. Both inorganic and organometallic cob(III)alamins have been compared and the conclusion is that longer Co_α – N bonds correlates with smaller fold angles (and vice versa) [28]. For example aquocobalamin perchlorate (**4**$^+$-ClO_4^-, Co_α – N = 1.925 Å, fold angle = 18.7°) and coenzyme B_{12} (**2**, Co_α – N = 2.237 Å, fold angle = 13.3°). In contrast, the folding of the corrin ligand in Co_β-cyano-imidazolylcobamide (**8**) (11.3°) is less than half of that of vitamin B_{12} (**1**) regardless of the shorter Co_α – N bond (1.97 Å vs. 2.01 Å) [31]. Accordingly folding is more apparent in cob(III)alamins with short Co_α – N bonds (near 2.0 Å or less), to which the known inorganic B_{12}-derivatives belong to. In organocobalamins (such as methylcobalamin and coenzyme B_{12}) the length of the Co_α – N bond is close to or greater than 2.2 Å, so there is less steric interaction of the nucleotide base with corrin ligand.

2.2 Structural Studies of B_{12}-Derivatives by Nuclear Magnetic Resonance Spectroscopy

Nuclear magnetic resonance (NMR) spectroscopy has had a strong influence in the development of B_{12} chemistry. The early NMR spectroscopic studies established the nature of many noncrystalline B_{12}-derivatives, mostly in their Co_β-cyano forms, using one-dimensional analyses. These studies were based on the ^1H- and ^{13}C-chemical shift values from spectra of several already well-characterized B_{12}-derivatives and used to identify and describe the structure of synthetic and natural analogues of vitamin B_{12} [74]. The natural corrinoids from a range of bacteria were first characterized by NMR [75, 76].

Earlier assignment problems regarding B_{12}-derivatives in aqueous or non-aqueous solutions have now been eliminated by the use of heteronuclear NMR spectroscopy [74, 77]. Following on the pioneering studies of coenzyme B_{12} (**2**) [78, 79], Co(I)-heptamethyl-cobyrinate (**34**) [80] and the non-crystalline B_{12}-derivative Co_{β}-5′-deoxy-5′-adenosylcobinamide (**35**) [81], the newer NMR studies have begun to compliment (and in certain aspects rival) x-ray analytical studies of B_{12}-derivatives in the solid state.

By applying a selection of now well-established homo- and heteronuclear 2D experiments, the assignment of signals in ^{1}H-, ^{13}C- and ^{15}N-spectra provide a reliable basis for detailed structure and dynamic information of B_{12}-derivatives. Techniques for suppression of the solvent (water) signal allow the recording of spectra from an aqueous solution with little or no loss of information [77]. Characteristic chemical shift values from ^{1}H-, ^{13}C-, ^{15}N- and ^{31}P-spectra provide important information on the constitution and conformation of complete B_{12}-derivatives. The coordination of the DMB-base, in base-on compounds, induces a high-field shift of the ^{1}H-NMR signal of HC10, due mainly to an increase in the electron density of the corrin ligand by the axial coordination of the base. This characteristic has been used to determine the temperature dependent base-on/base-off equilibria (in aqueous solutions) of organometallic B_{12}-derivatives (e.g., methylcobalamin **3**) [74]. In the ^{1}H-NMR spectrum of, e.g., **3** , the anisotropic shielding effect of the coordinated DMB-base also induces high-field shift of protons located nearby, such as methyl group H_3C1A and methylene groups H_2C81 and H_2C82 [82]. Shielding by the cobalt-corrin in the axial direction leads to high-field shifts of the DMB-protons closest to the cobalt-corrin, HC2N and HC4N. Likewise protons of organometallic ligands are characteristically up-field as seen in the ^{1}H-NMR spectra of homocoenzyme B_{12} (**27**) and bishomocoenzyme B_{12} (**36**) [62]. Significant conformational differences between the solution and crystal structure were revealed in some cases, such as in the studies of AdoCbl (**2**) [78] and MeCbl (**3**) [82].

A major factor in the importance of heteronuclear NMR spectroscopy is that the structures (in solution) of noncrystalline B_{12} derivatives can be characterized. One of the main examples of this is the natural complete but base-off protonated form of coenzyme B_{12} (**2**-H^{+}) [79]. More recently, the solution structures of the organometallic derivatives pseudocoenzyme B_{12} (**37**), adenosyl-factor A (**38**) [76] and neocoenzyme B_{12} (**39**) [83] could be analyzed in great detail. The structures and the base-on/base-off equilibria of a range of protonated base-off cobamides have also been investigated in aqueous solution [74, 77].

From NOE measurements the first reliable assignment (upper/Co_{β} or lower/Co_{α}) of the cobalt-bound methyl group in noncrystalline methyl-cob(III)yrinates was determined [84]. Also from NOE data and three-bond coupling constants, detailed and important information on the conformational properties of the nucleotide moiety, the organometallic group, and of other peripheral side chains was extracted [77]. Such studies resulted in the detection of significant conformational dynamics of the organometallic 5′-deoxy-5′-adenosyl moiety in the pioneering study of coenzyme B_{12} (**2**) [79]. In a related context extensive conformational dynamics of the organometallic adenosyl ligand and the unusual *syn*-orientation of the adenine heterocycle were observed in a series of coenzyme B_{12} analogues, such as homo- and bishomocoenzyme B_{12} (**27** and **36**) [62] pseudocoenzyme B_{12} (**37**) [76], neocoenzyme B_{12} (**39**) [81], and other adenosyl-cobamides [77]. Conformational flexibility of the organometallic ligand was also discovered in the solution structures of adeninyl-alkyl-cobamides [58, 61].

The use of 2D-NMR spectroscopy has proven to be a versatile method in the detection of intra- and intermolecular H-bonding. The water ligand of aquocobalamin perchlorate (4^+-ClO_4^-), which from the crystal structure forms an H-bond to an acetamide side chain, was shown by NMR to still form a similar H-bond in aqueous solution [40]. Pseudointramolecular H-bonding of a specific "external" water molecule to the nucleotide portion of methylcobalamin (**3**) [82] (and some other organometallic cobamides), which is accompanied by a remarkable adjustment in the conformation of the nucleotide moiety, was characterized by NMR spectroscopy [77]. In this way the first contributions to the hydration behavior of B_{12}-derivatives in aqueous solutions have been identified. Further exploratory studies have been undertaken to investigate in greater detail B_{12}-derivatives in their solvent environment [77] and these complement other recent results obtained from studies on the structure of the water networks in crystals of B_{12}-derivatives [55, 56, 85]. The aqueous solution environment of **3** has been investigated in such a way, by measuring NOEs between the solvent and the protonated base-off form (**3**-H^+) of **3**, initial results show a water molecule to be the Co_α-axial ligand. This would be the first experimental evidence, for the (solution) structure of an organometallic "yellow" cobyrinic acid derivative to be a hexacoordinated cobalt-corrin (personal communication with Fieber and Konrat, 2005).

3
B_{12}-Electrochemistry

Under physiological conditions vitamin B_{12}-derivatives have been observed in three different oxidation states, Co(III), Co(II), and Co(I), each possessing different coordination properties and qualitatively differing reactivities [22, 75]. Oxidation–reduction processes are therefore of key importance in the chemistry and biology of B_{12}. Electrochemical methods have been applied in the synthesis of organometallic B_{12}-derivatives [86, 87], as well as for the purpose of generating reduced forms of protein bound B_{12}-derivatives [88] and electrode-bound B_{12}-derivatives for analytical applications [89].

Axial coordination to the corrin-bound cobalt center depends on the formal oxidation state of the cobalt ion [28] and, as a rule, the number of axial ligands decreases with the cobalt oxidation state. In the thermodynamically predominating forms of cobalt-corrins, the diamagnetic Co(III) (coordination number 6) has two axial ligands bound, the paramagnetic (low spin) Co(II) (coordination number 5) has one axial ligand bound and for the diamagnetic Co(I) (coordination number 4) no axial ligands are bound, or only very weakly [22, 90]. Electron transfer reactions involving B_{12}-derivatives are, therefore, accompanied by a change in the number of axial ligands. The nature of the (potential) axial ligands heavily influences the thermodynamic and kinetic features of the electrochemistry of cobalt corrins [87, 90].

In Co(III)-corrins, such as vitamin B_{12} (**1**), coenzyme B_{12} (**2**) and hydroxocobalamin (**5**), the corrin-bound cobalt center binds two kinetically rather labile axial ligands. In the case of base-on cobalamins one of the axial ligands is the DMB-base. In contrast, the metal center in Co(I)-corrins, such as Cob(I)alamin (**40**$^-$, B_{12s}), is highly nucleophilic [91] with very low basicity [90, 92]. The intermediate oxidation state of Co(II)-corrins, such as in cob(II)alamin (**23**, B_{12r}), provides a highly reactive metal-centered radicaloid species [51, 93]. The use of electrochemistry thus provides an excellent means for generating, under controlled conditions, B_{12}-derivatives of specific reactivity, as well as investigating the redox processes in their interconversion between oxidation states as reviewed by Lexa and Savéant [90].

3.1
Thermodynamic Redox Properties of Cobamides

The electrochemistry of the B_{12} derivative aquocobalamin (**4**$^+$) has been particularly well studied [90, 94–99] where the one-electron reduction of **4**$^+$ gives B_{12r} (**23**) and then B_{12s} (**40**$^-$) (see Fig. 4). Typically, electrochemical studies of aquocobalamin (**4**$^+$) were carried out in aqueous solution.

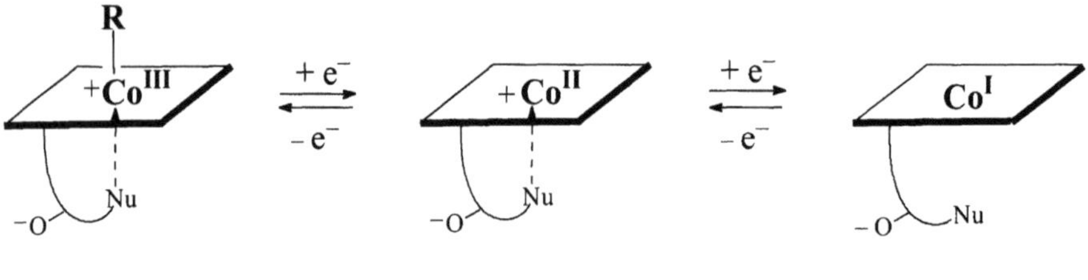

vitamin B_{12} (**1**, R = CN) cob(II)a amin (**23**) cob(I)alamin (**40**$^-$)
aquocobalamin (**4**$^+$, R = H_2O^+)

Fig. 4 Outline of the redox transitions between the cob(III)alamins 1 or **4**$^+$, cob(II)alamin (**23**) and cob(I)alamin (**40**$^-$)

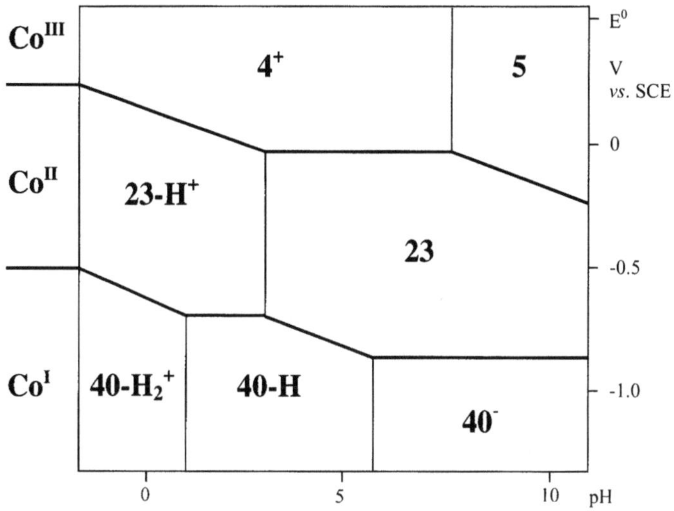

Fig. 5 The dependence of standard potentials of the redox system Co(III)-/Co(II)-/Co(I)-corrin (B_{12a}, $\mathbf{4}^+$/B_{12r}, **23**/B_{12s}, $\mathbf{40}^-$) upon pH in aqueous solution (at 22° C), adapted from Lexa and Saveant [90]; electrochemical potentials are referenced to the saturated aqueous calomel electrode (SCE), which is at 0.242 V vs. normal hydrogen electrode (NHE) [100]

A standard potential vs. pH diagram correlates the thermodynamics of the aquocobalamin ($\mathbf{4}^+$)-B_{12r} (**23**)-B_{12s} ($\mathbf{40}^-$) system (see Fig. 5). The interconversion between the different oxidation states of B_{12}-derivatives can usually be monitored effectively by UV-vis spectroscopy, and the relevant data were obtained from potentiostatic measurements, which were followed by UV-vis spectroscopy [90, 94]. Within the pH range – 1 to 11 and applied potentials $E = 0.5$ V and – 1.2 V vs. SCE, seven solution cobalamins are thermodynamically predominant spanning a range of the three formal oxidation states of B_{12} [90].

Aquocobalamin ($\mathbf{4}^+$) and HOCbl (**5**) differ by protonation of the upper (β) axial ligand with pK_a ($\mathbf{4}^+$) = 7.8 [90]. The Co(II)-corrin B_{12r} (**23**) represents the base-on form of the Co(II) oxidation level (i.e., the nucleotide loop is coordinated intramolecularly), this is converted into the base-off (**23**-H^+) by protonation of the DMB-base, with pK_a (**23**-H^+) = 2.9 [90]. At the Co(I)-level, cob(I)alamin B_{12s} ($\mathbf{40}^-$) is first protonated at the nucleotide base to give **40**-H. For the pK_a of **40**-H, an original value of 4.7 was determined [90, 94, 96], but more recently this has been estimated to be 5.6 [101, 102]. A second protonation then occurs at the Co(I)center to give the "Co(III)-hydride" [92] **40**-H_2^+, with pK_a (**40**-H_2^+) = 1 [90, 94, 103].

In the pH range 2.9 to 7.8, $\mathbf{4}^+$ and (base-on) B_{12r} (**23**) represent the predominant Co(III)-/Co(II)-redox couple, with a standard potential of – 0.04 V (see Fig. 5). For the Co(II)-/Co(I)-redox system there are two pH-independent standard potentials [90]: at a pH less than 5.6 the Co(II)-/Co(I)-couple (base-off) **23**-H^+/**40**-H predominates at a standard potential of – 0.74 V, but for the redox couple (base-on) B_{12r} (**23**)/B_{12s} (**40**$^-$) a more negative standard potential of – 0.85 V [90] (– 0.88 V [102]) is required.

This shift by about 110–140 mV to a more negative potential for the reduction, of (base-on) B_{12r} (**23**) when compared to that of the protonated base-off form **23**-H^+, reflects the selective stabilization of the Co(II)-corrin **23** by intramolecular nucleotide coordination [75, 90]. A dependence of the standard potentials of the Co(III)-/Co(II)-redox couples occurs at approximately 60 mV per pH unit, at pH > 7.8 for HOCbl (**5**)/B_{12r} (**23**) and at pH < 2.9 for **4**$^+$/**23**-H^+ and this reflects the effect of the removal by protonation of one axial ligand. An analogous dependence of the potential occurs between pH 2.9 and ca. 5.6 for **23**/**40**-H as well as below pH 1 for the Co(II)-/Co(I)-redox couple **23**-H^+/**40**-H_2^+. At all pH-values the disproportionation of Co(II)-corrins to Co(III)- and Co(I)-corrins is thermodynamically disfavored (the disproportionation equilibrium constant is below 10^{-10}) [90].

A complex interplay between the thermodynamic and kinetic factors of electron transfer reactions occur in the analogous studies of vitamin B_{12} (**1**), due to the strongly coordinating cyano ligand [87, 90]. Coordination of (one or two) cyanide ligands to the Co(III)-center stabilizes it against reduction and the Co(III)-/Co(II)-standard potentials are shifted to more negative values [90, 104]. Cyanide ions transform **1** into (base-off) dicyano-cob(III)alamin (**1**-CN^-) with an equilibrium constant of about 10^4 M^{-1} [22, 104].

Electrochemical studies of the incomplete diaquocobinamide (**41**$^{2+}$) (Fig. 6) gave a standard electrochemical potential for the diaquo-cob(III)inamide (**41**$^{2+}$)/aquo-cob(II)inamide (**42**$^+$) couple of + 0.27 V [90, 105]. This corresponds to the extrapolated value for the highly acidic protonated base-off form (**4**-H^{2+}) of aquocob(III)alamin (**4**$^+$), with pK_a (**4**-H^{2+}) = ca. – 2.4 [90]. The potential of the corresponding aquo-cob(II)inamide (**42**$^+$)/cob(I)inamide (**43**) couple was determined as – 0.73 V [90]. The stan-

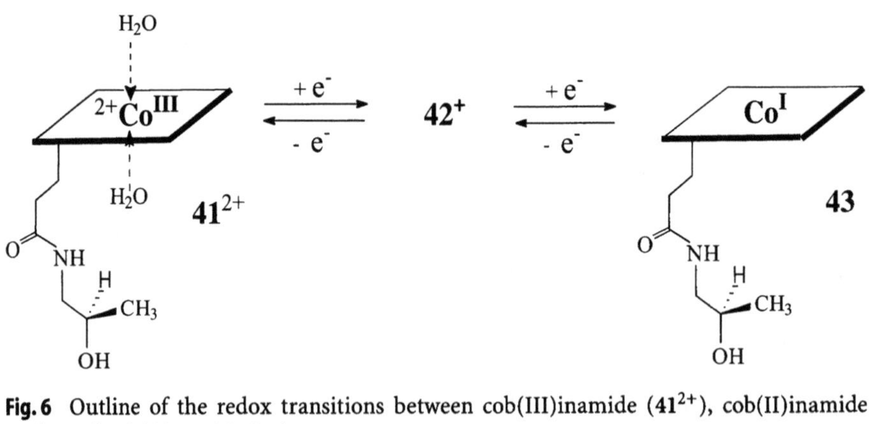

Fig. 6 Outline of the redox transitions between cob(III)inamide (**41**$^{2+}$), cob(II)inamide (**42**$^+$) and cob(I)inamide (**43**)

dard potential of the redox couple between **42**$^+$ and **43** is thus indistinguishable from that of the base-off cobalamins B_{12r}-H$^+$/ B_{12s}-H.

The electrochemical studies of organometallic studies of B_{12}-derivatives are complicated due to the rapid and irreversible loss of the organic ligand upon reduction [90]. Low temperature conditions are therefore required to obtain pertinent thermodynamic of organometallic B_{12}-derivatives [106]. The standard potential (at – 30 °C) for the methylcob(III)alamin (**3**)/methylcob(II)alamin (**44**$^-$) redox couple was estimated as – 1.60 V vs. SCE [90, 106] similar to the value obtained for the coenzyme B_{12}/5′-deoxy-5′-adenosylcob(II)-alamin pair [102]. With one-electron reduction of **3** a decoordination of the nucleotide base occurs to give **44**$^-$, which rapidly decomposes into cob(I)alamin (**40**$^-$) and a methyl radical. The standard potential of the typical Co(III)-/Co(II)-redox pair of organometallic B_{12}-derivatives is significantly more negative than that of B_{12r} (**23**)/B_{12s} (**40**$^-$) and out of the reach of biological reductants.

The thermodynamic trends of B_{12}-redox systems can be summarized as:

1. Intramolecular coordination of the nucleotide base or strong coordinating or nucleophilic ligands (such as cyanide ions) stabilize the corrin-bound cobalt center against one-electron reduction and shift the Co(III)-/Co(II)-redox couples to more negative potentials.
2. The one-electron reduction of organometallic Co(III)-corrins typically occurs at more negative potentials than the Co(II)-/Co(I)-redox couple B_{12r}/B_{12s} [90]. Exceptions to this are provided by organometallic B_{12}-derivatives with electron withdrawing substituents on the organometallic group, such as methoxycarbonylmethyl-cob(III)alamin [107].

3.2
Kinetic Redox Properties of Cobamides

With a one-electron transfer reaction of a cobalt-corrinoid complex cleavage or formation of a bond to an axial ligand occurs. A reduction is accompanied by an expulsion, and an oxidation by the coordination, of the ligand [90]. The electron transfer step accordingly takes place either in a concerted fashion or in a rapid associated step with coordination or dissociation of the axial ligand.

Electron transfer in the protonated Co(II)-/Co(I)-couple B_{12r}-H$^+$ (**23**-H$^+$)/B_{12s}-H (**40**-H) is fast in aqueous solution ($k_s^{app} > 0.1$ cm s^{-1}) as the presumed axial water ligand is only kinetically weakly bound in the base-off Co(II)-corrin **23**-H$^+$ [90, 95]. However, when the aquo ligand in **23**-H$^+$ is substituted by a stronger axial ligand, e.g., by the nucleotide base as in base-on B_{12}, the electron transfer is slowed down sufficiently so that its kinetics can be conveniently measured by cyclic voltammetry [90, 97, 98]. For example in the Co(II)-/Co(I)-redox couple **23**/**40**$^-$ $k_s^{app} = 0.0002$ cm s^{-1} [90], the electron transfer is at least a thousand times slower than in the base-off forms **23**-H$^+$/**40**-H.

The trend in kinetics for Co(III)-/Co(II)-couples follows the same trend as those for the corresponding Co(II)-/Co(I)-couples, albeit much slower. The Co(III)-Co(II)-couple aquocob(III)alamin (4^+)/B_{12r} (**23**) has a rate constant for heterogeneous electron transfer of about 10^{-5} cm s^{-1} [90]. The electron transfer steps for the cyano-cob(III)- and cyano-cob(II)alamins **1**-CN^- and **23**-CN^- are slower still [90, 104].

There is an approximate linear correlation between the equilibrium constant for the coordination of the axial ligand and the standard apparent rate constant for electron transfer [90]. This correlation has been rationalized by a model, in which stretching of the bond between the cobalt ion and the axial ligand represents the main factor of the kinetics of the electron transfer. As a consequence, kinetic and thermodynamic dependence of the electron transfer on the strength of the complexation of the axial ligands both add up, resulting in more negative reduction potentials as the strength of the ligand increases.

Organocobalamins, such as coenzyme B_{12} (**2**) and MeCbl (**3**), have a different kinetic behavior from CNCbl (**1**) and other Co(III)-corrins with strong axial ligands. Whereas the Co(III)-Co(II)-reduction potentials are quite negative the kinetics of electron transfer are fast [90]. The one-electron reduction of **3** to the unstable methylcob(II)alamin anion (**44**$^-$) was estimated to have a rate constant of 1200 s^{-1} at – 30 °C. However, the product of the one-electron reduction of methylcobinamide (**45**$^+$), methylcob(II)inamide (**46**), has a half-life of only about 0.1 s at – 20 °C before decomposing into a methyl radical and cob(I)inamide (**43**) (see Fig. 7). An Arrhenius plot of the kinetics of the decomposition of **46** gave the activation energy to be 19 kcal/mol and a pre-exponential factor $A = 10^{17.6}$ s^{-1} [90, 106]. From the values of the (Co – C)-bond dissociation energy (37 kcal/mol) of MeCbl [108, 109] and the kinetics of the decomposition of the intermediate **44**$^-$, the one-electron reduction is suggested to reduce the strength of the (Co – C)-bond of **3** (by about 12 kcal/mol) to "half" of its value [90, 106].

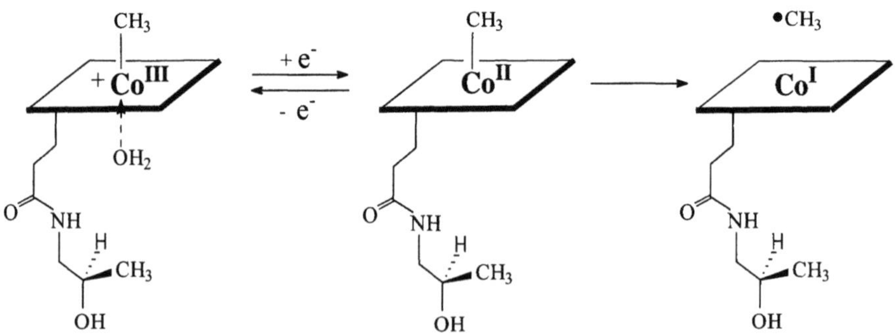

Fig. 7 One-electron reduction of methylcob(III)inamide (**45**$^+$) presumably occurs with loss of a water ligand and gives methyl-cob(II)inamide (**46**), which rapidly decomposes into cob(I)inamide (**43**) and a methyl radical

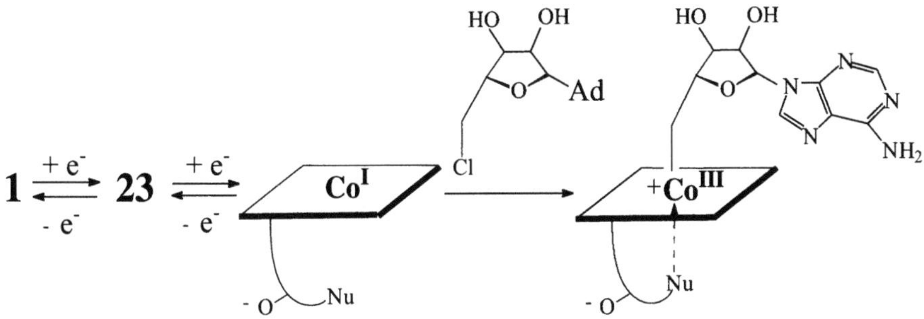

40⁻ 2

Fig. 8 Preparation of coenzyme B_{12} (**2**) from CNCbl (**1**) by electrochemical reduction to cob(I)alamin (**40**⁻) and alkylation with 5′-deoxy-5′-chloroadenine [110]

3.3
Organometallic Electrochemical Synthesis

Electrochemistry is an excellent method for the selective and controlled production of reduced B_{12} forms under potentiostatic control. As alkyl halides or alkyl tosylates react quickly and efficiently with Co(I)-corrins [22, 91], which are cleanly generated at controlled electrode potentials near that of Co(II)-/Co(I)-couples, electrochemistry provides a suitable method for the synthesis of organometallic B_{12} derivatives [87].

Using electrolysis at a controlled potential of -1.1 V vs. SCE, coenzyme B_{12} (**2**) was prepared in 95% yield from vitamin B_{12} (**1**) or from aquocobalamin (**4**$^+$) by alkylating cob(I)alamin (**40**$^-$) with 5′-chloro-5′-deoxyadenosine (see Fig. 8) [110]. Other organometallic B_{12}-derivatives produced in an analogous method were, e.g., pseudocoenzyme B_{12} (**37**) (78% yield from pseudovitamin B_{12}) [76], neocoenzyme B_{12} (**39**) (89% from neovitamin B_{12}) [83] and homocoenzyme B_{12} (**27**) (99% from **4**$^+$) [62]. Co_β-methyl-imidazolylcobamide (**31**) (90% yield from Co_β-cyano-imidazolylcobamide) [68] and methyl-13-epicob(III)alamin (**47**) (88% yield from neovitamin B_{12}) [83] were synthesized by alkylation with methyl iodide. Also, alkyl bridged dimeric B_{12}-derivatives, such as tetramethylene-1,4-biscobalamin (**48**; see Fig. 9) and rotaxane, an organometallic B_{12}, were synthesized by similar methods [111, 112].

This methodology has been further expanded by the development of an electrochemical method for the preparation of Co_β-[(methoxycarbonyl)methyl]cob(III)alamin (**49**) via the alkylation of cob(II)alamin (**23**) [113]. The more easily reducible organocob(III)alamins are known to be reductively cleaved by direct electrochemical reduction or by reduction with cob(I)alamin (**40**$^-$) [90]. For **49**, in which an acceptor-substituted C-atom is directly bound to the Co-center, a peak potential of reduction of -0.90 V

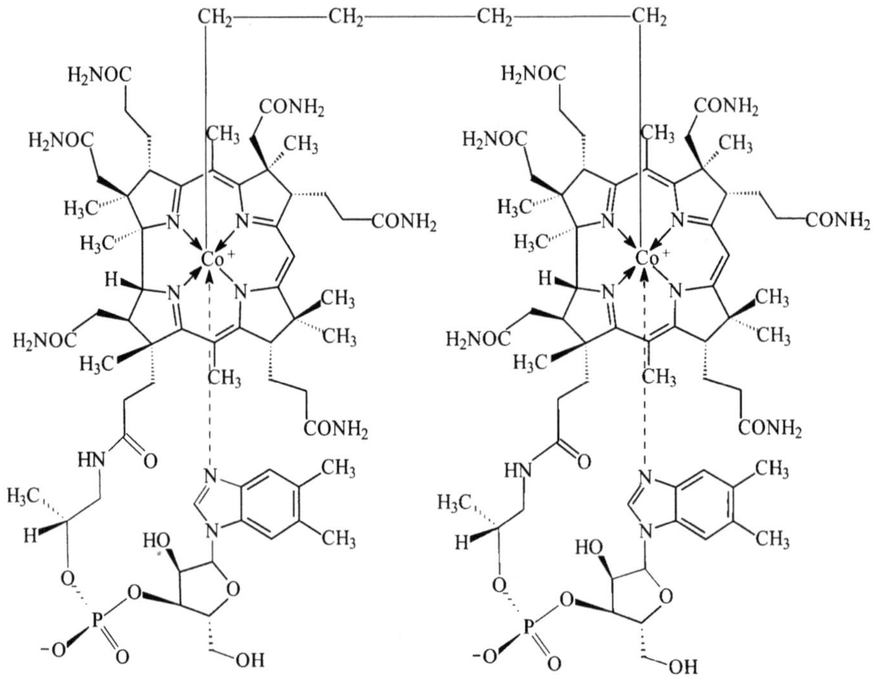

Fig. 9 Structural formula of the alkyl-bridged biscobalamin **48** [111]

vs. SCE (in DMF, room temperature) was determined [107]. This value is close to that of the redox couple (base-on) B_{12r} (**23**)/B_{12s} (**40**$^-$) (-0.85 V) and explains the difficulties encountered when preparing **49** via alkylation of **40**$^-$ [107]. However, under potentiostatic conditions aquocob(III)alamin chloride (**4**$^+$-Cl^-) was submitted to a controlled potential of -0.45 V (vs. 0.1 n CE) to give **23** and after the addition of methylbromoacetate the crystalline alkyl cob(III)alamin was isolated in 75% yield. The reaction was proposed to take place directly via **23** and radical intermediates [113].

The alkylation of complete Co(II)corrinoids is thus an efficient and alternative method to the more established synthetic procedures via Co(I)corrinoids for the synthesis of reduction-labile Co(III)organocorrinoids [113, 114].

4
Reactivity of B_{12}-Derivatives in Organometallic Reactions

The formation and cleavage of the Co – C bond in organometallic B_{12}-cofactors are crucial steps in the reactions catalyzed by B_{12}-derivatives and by B_{12}-dependent enzymes [1, 6, 25, 26, 115–118]. The reactivity of B_{12}-derivatives in organometallic reactions also holds the key to much of the understanding of the biological activity of the B_{12}-dependent enzymes. The preparation of such organometallic B_{12}-derivatives is usually based on the alkylation of Co(I)-corrins, one practical method is the electrochemical approaches as described in the previous section.

In solution cleavage and formation of the Co – C bond have been observed to occur in all of the basic oxidation levels for the cobalt center of the corrin core [115–118]. Two main paths for these organometallic reactions have been found:

1. The homolytic mode is formally a one-electron reduction/oxidation of the corrin-bound cobalt center and involves the cleavage or formation of a single axial bond, as is typical of the reactivity of coenzyme B_{12} [119–123]:

$$\text{5}'\text{-adenosyl-Co(III)-corrin} \rightleftharpoons \text{Co(II)-corrin} + \text{5}'\text{-adenosyl radical}.$$

2. The nucleophile induced, heterolytic mode, is formally a two-electron reduction/oxidation of the corrin-bound cobalt center and involves the cleavage or formation of two (*trans*-) axial bonds, as is typical of the reactivity of methylcobalamin [124–127]:

$$\text{methyl-Co(III)-corrin} + \text{nucleophile} \rightleftharpoons \text{Co(I)-corrin} + \text{methylating agent}.$$

As coenzyme B_{12} (AdoCbl, **2**) is considered to be a "reversible carrier of an alkyl radical" [119] (or a reversibly functioning "radical source"), the homolytic mode of the cleavage of the Co – C bond of **2** is of particular importance in its role as a cofactor. The strength of the Co – C bond of AdoCbl has been calculated to be about 30 kcal/mol by using detailed kinetic analyses of the thermal decomposition of **2** [119, 122, 123]. Considerable cage effects [123, 128], and with both base-on and base-off forms of **2** being present, caused complications in the quantitative treatment of the homolytic Co – C bond dissociation energy (BDE). In fact, the nucleotide coordinated base-on forms of several organocobalamins decomposed faster than their corresponding nucleotide-deficient organocobinamides or their protonated (base-off) forms of the organocobalamins [129–131]. The intramolecular coordination of the nucleotide was therefore considered to cause a "mechanochemical" means of labilizing the Co – C bond of organometallic B_{12}-derivatives [129–131]. The extension of this idea to the enzymatic reactions with **2** as cofactor was disputed [51] and, indeed, now seems less likely due to the crystallographic studies of several coenzyme B_{12}-dependent enzymes [18–20].

For the particular case of AdoCbl (**2**) it was found that the contribution of nucleotide coordination to the ease of homolytic cleavage was relatively small. On the basis of available thermodynamic data concerning the coordination of the nucleotide in **2** and of the homolysis product cob(II)alamin (**23**), the coordination of the nucleotide was estimated to weaken the Co – C bond by only 0.7 kcal/mol [75, 86]. In contrast to that, the intramolecular coordina-

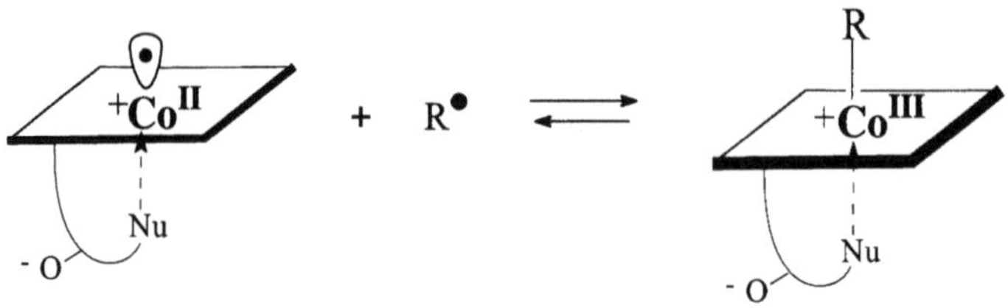

23

Fig. 10 The "radical trap" cob(II)alamin (**23**) rapidly combines with radicals on the "upper" β-face

tion of the nucleotide in methylcobalamin (**3**) was determined to even slightly increase the homolytic Co – CH_3 BDE of **3**, after studies of the methyl group transfer equilibrium between methylcobalamin (**3**)/cob(II)inamide (**42**) and methylcobinamide (**45**)/cob(II)alamin (**23**) [86].

For the homolytic mode of formation of the Co – C bond in coenzyme B_{12} (**2**) the structure [51] and reactivity of cob(II)alamin (**23**) gave crucial information. The radicaloid **23** has a pentacoordinated Co(II) center and is considered to fulfill all the structural criteria of a highly efficient radical trap (see Fig. 10), since its reactions with alkyl radicals occur with negligible restructuring of the DMB-nucleotide coordinated cobalt-corrin moiety [51]. From this it is understandable that the remarkably high reaction rate of **23** with alkyl radicals (such as the 5′-deoxy-5′-adenosyl radical) and the diastereospecificity for the reaction to occur at the β-face, are both consistent and explainable due to the structure of cob(II)alamin. The coordination of the DMB-nucleotide in **23** controls the (α/β)-diastereoface selectivity (in both a kinetic and thermodynamic sense) in alkylation reactions at the Co(II) center.

The stereochemical situation however, is appreciably more complex in incomplete corrins, such as cob(II)ester (**24**) and base-off forms of complete corrins. The axial ligand at the corrin-bound Co(II) center is expected to direct the formation of the Co – C bond. In this way kinetic control can lead with high efficiency to the "rare" α-alkyl-Co(III)-corrins [84, 128]. In such radical recombination reactions the axial ligand at the α- or β-side of the metal center will not only steer the diastereoselectivity of the alkylation but also may contribute to significant altering of the cage effects [122, 123].

The second important type of organometallic reactivity of B_{12}-derivatives concerns the highly nucleophilic/nucleofugal Co(I)-corrins [75, 91, 132]. These provide the basis of the heterolytic mode of formation/cleavage of the Co – C bond, important in methyl-corrinoids in enzyme-catalyzed methyl-transfer reactions [125–127]. This mode is represented by the reaction of Co(I)-corrins with alkylating agents in the formation of the Co – CH_3 bond and the nucleophile-induced demethylation of methyl Co(III)-corrins for the

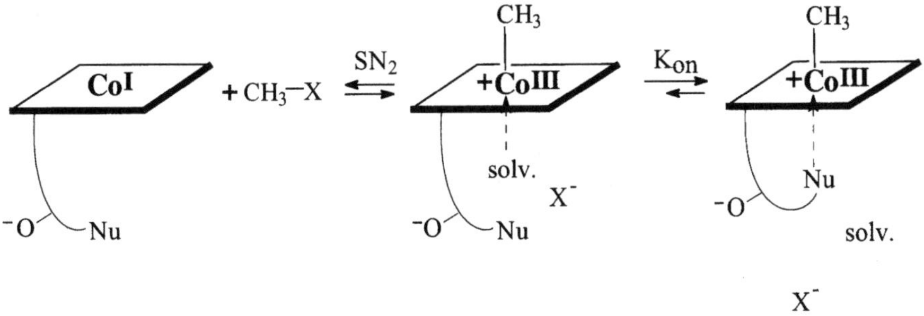

Fig. 11 Methylation of the DMB-containing cob(I)alamin B_{12s} (**40**$^-$) SN_2-mode is directed to the "upper" β-face (by both, kinetic and thermodynamic reasons)

cleavage of the Co – CH_3 bond. Overall an oxidative *trans* addition/reductive *trans* elimination occurs at the corrin-bound cobalt center [133, 134].

Alkylation at the corrin-bound Co(I) center normally proceeds via the "classical" bimolecular nucleophilic substitution (SN_2) mechanism, where the Co(I)-corrin acts as a "supernucleophile" [91, 132]. However, in certain cases alkylation occurs via a two-step one-electron transfer path, where Co(I)-corrins act as strong one-electron reducing agents and the process goes via Co(II)-corrin intermediates [84]. In complete corrins, such as B_{12s} (**40**$^-$), either pathway results in methylation at the β-face (see Fig. 11), which allows the nucleotide to coordinate at the α-face of methylcobalamin (**3**) [86, 134]. When the nucleotide base has been changed from a DMB-base to an imidazole little effect on the thermodynamics of the methyl transfer reaction occurs [68].

The studies of Co(I)-corrins, like B_{12s} (**40**$^-$), have shown the following reactivity patterns relevant for the SN_2-alkylation pathway:

1. The nucleophilicity of Co(I)-corrins is virtually independent of the presence of the DMB-nucleotide, both complete and incomplete Co(I)-corrins react preferentially at their β-face, which is essentially more nucleophilic [132, 134]. The immediate product of the β-alkylation may be a pentacoordinate (or already solvated and effectively hexacoordinate) Co_β-alkyl-Co(III)-corrin.
2. In aqueous solution and at room temperature the base-on (hexacoordinate) methylcob(III)alamin is more stable by about 4 kcal/mol than the base-off Co_α-aquo-Co_β-methylcob(III)alamin [135]. From NMR studies, the latter has been estimated to still be more stable in water, by around 7 kcal/mol, than the corresponding base-off and dehydrated form of Co_β-methylcob(III)alamin, which has a pentacoordinate Co_β-methyl-Co(III)-center [136].

With incomplete cobalt-corrins the situation is again more complex, with two diastereoisomeric alkylation products often formed [74, 84, 134]. In specific cases, under suitable kinetic control, one of the alkyl-Co(III)-corrin diastereoisomers can form with high selectivity. For example with the lipophilic Co(I)-heptamethylcobyrinate (**34**) [84], the SN_2-pathway can provide β-methylation with high diastereoselectivity (> 96%), while the one-electron transfer mechanism permits the formation of the Co_α-methylated product with high diastereoselectivity (> 98%) [75, 84]. Though, overall, methyl group transfer reactions (involving Co(I)-, Co(II)- and unalkylated Co(III)-corrins as methyl group acceptors) are often complicated due to rapid equilibration [134].

The reverse process, the nucleophile-induced dealkylations of methyl-Co(III)-corrins, has been less studied due to the impediment of the intramolecular coordination of the nucleotide base [134, 137]. Indeed, thiolates demethylate methylcobinamide (**45**) to cob(I)inamide (**43**) approximately 1000 times faster than MeCbl (**3**) to B_{12s} (**40**$^-$) [137], reflecting the strong stabilizing effect of the coordinated nucleotide in **3** [86, 134]. This is of relevance for enzymatic methyl group transfer reactions involving protein bound Co(I)- and methyl-Co(III)-corrins, where considerable axial base effects are expected [125, 138].

The two most relevant modes of formation and cleavage of the Co – C bond of the cobalt center differ significantly in the structural requirements (see Fig. 12):

1. In the heterolytic mode of cleavage and formation of the Co – C bond significant reorganization at both faces of the corrin-bound cobalt center. Cleavage of the Co – CH_3 bond is brought about by attack of a nucleophile at the readily accessible carbon of the cobalt-bound methyl group.
2. In the homolytic mode of cleavage the cobalt-corrin portion of complete cob(III)amides (such as **2** and **23**) hardly changes structure.

Other basic modes of formation/cleavage of Co – C bonds in alkyl Co(III)-corrins involve nucleophilic alkylating agents and the electrophilic properties of the Co(III)-corrins [115–118]. Coordination of the DMB-nucleotide modifies the reactivity of the metal center by enhancing the ease of abstraction of the cobalt-bound alkyl group, in both a kinetic and thermodynamic sense [75, 138]. The Co – C bond of alkyl-Co(III)-corrins is rather inert against proteolytic cleavage under physiological conditions. The acid-induced heterolytic cleavage of the Co – C bond of coenzyme B_{12} occurred less readily when compared to 2′-deoxycoenzyme B_{12} (**50**) and 2′,3′-dideoxycoenzyme B_{12} (**51**) [139, 140]. This significant reactivity difference can be traced back largely to the effect of the ease of protonation of the cobalt-bound organic group [139].

A little recognized mode of cleavage of the Co – C bond of organometallic B_{12}-derivatives, may be represented by the thermodynamically favorable radical-induced substitution at the cobalt-bound carbon center (see Fig. 13) [111, 141]. This type of reactivity holds interest due to some unusual

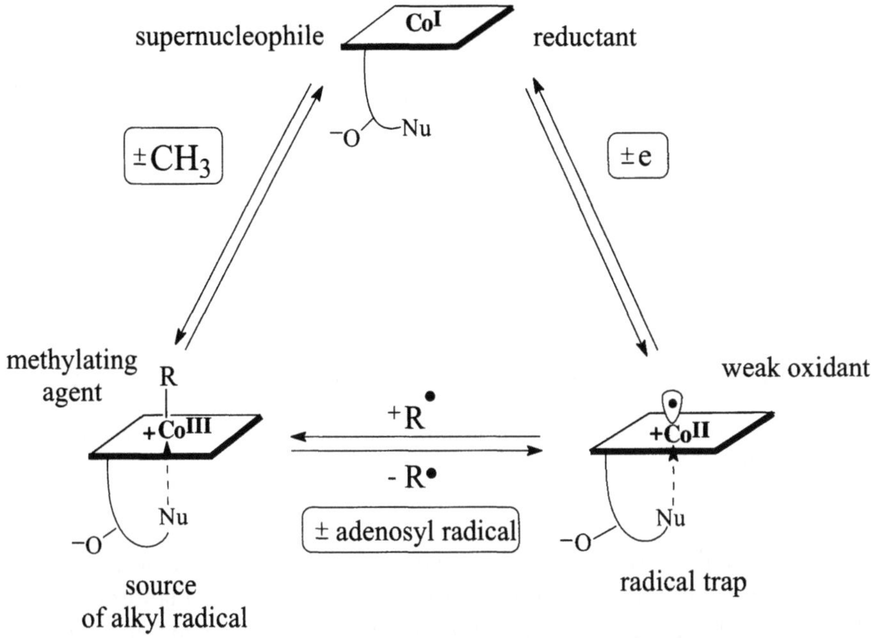

Fig. 12 Formal analysis of elementary reaction steps of "complete" corrinoids in organometallic and redox transformations, characterizing their patterns of reactivity relevant for their cofactor function in B_{12}-dependent enzymes

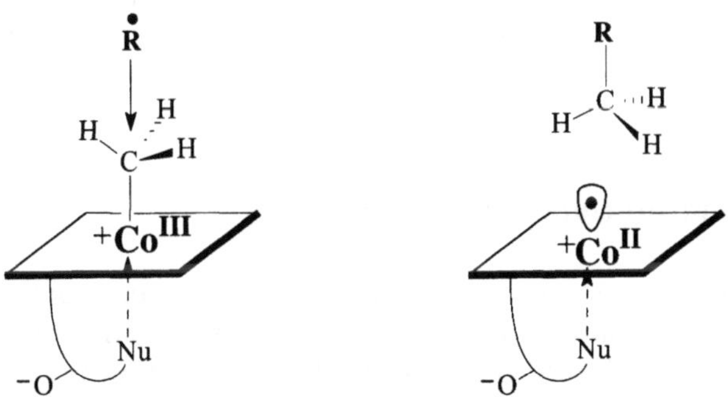

Fig. 13 Methylcobalamin (3) as a methylating agent for an organic radical [141]

C–C bond forming methylation reactions at seemingly inactivated carbon centers [142].

In a formally related radical abstraction reaction, the cobalt-bound methyl group of methylcobalamin (**3**) and other methylcorrinoids is rapidly transformed to co(II)corrinoids, such as cob(II)inamide (**42**$^+$), (giving methylcob(III)inamide, **45**$^+$) and cob(II)alamin (B_{12r}, **23**) (see Fig. 14) [86, 134]. Under appropriate conditions (aprotic solvents), this type of reaction is not sensitive to the presence of molecular oxygen and does not involve free methyl radicals [134].

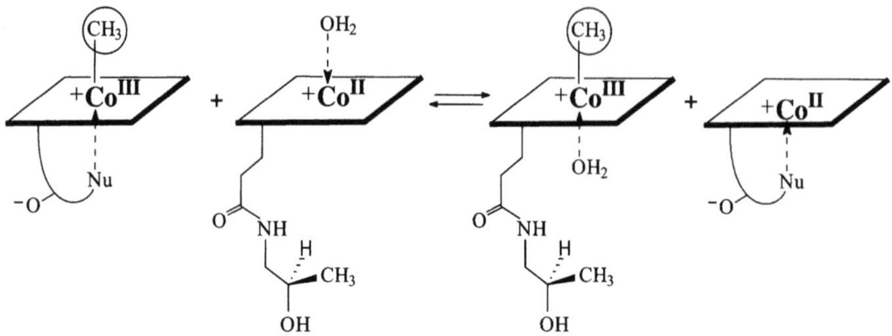

MeCbl (**3**) Cbi(II) (**42**$^+$) MeCbi (**45**$^+$) B_{12r} (**23**)

Fig. 14 Methyl transfer reaction involving MeCbl (**3**) and MeCbi (**45**$^+$) as methyl group donors and B_{12r} (**23**) and Cbi(II) (**42**$^+$) as methyl group acceptors

Organocobalamins have long been know to be sensitive to visible light [135], which induces homolytic cleavage of the Co – C bond with a quantum yield of about 0.3 [143]. Organocob(III)amides are also labile to strong one-electron reducing agents, as it has been found that after one electron reduction of organyl-Co(III)-corrins the Co – C bond is considerably weakened. As noted above, this aspect may render it difficult to prepare organocob(III)amides, via alkylation of the strongly reducing cob(I)amides, that have electron-withdrawing substituents [145].

5 Occurrence and Structure of Natural Corrinoids

Natural B_{12}-derivatives can occur either as complete corrinoids where a nucleotide function is attached to the propionic acid substituent at C17, or as incomplete corrinoids, which lack the nucleotide function and generally represent biosynthetic intermediates on the way to the complete corrinoids [23, 29].

In complete B_{12}-analogues the constitution of the nucleotide base can vary and the known classes of nucleotide functionalities found are benzimidazoles, such as the 5,6-dimethylbenzimidazole (DMB) of the cobalamins, purines, such as adenine and 2-methyladenine found in pseudovitamin B_{12} (**10**) and factor A (**11**), respectively, and phenols, such as *p*-cresol found in dicyano-*p*-cresolylcobamide (**52**) [145]. Sewage sludge is a particular rich, classical source of corrinoids [23]. Recent studies on the corrinoids from anaerobic microorganisms have shown a wide range of purine bases and benzimidazoles [1, 6, 37].

The functional B_{12}-cofactors are thus unique due to their unusual α-nucleotide function; all known complete corrinoids obey this common stereochemical feature including the B_{12}-derivatives with noncoordinating phenolic (pseudo)nucleotides [22, 23, 145, 146]. This unusual α-configuration, first of all, allows the heterocyclic base to coordinate to the lower α-axial coordination site of the corrin-bound cobalt center in an intramolecular fashion [7], but may also be of significance for the discriminative recognition and binding by the B_{12}-binding apoenzymes. The selective and tight binding of complete corrinoids by human (and other) B_{12}-binding proteins, points to the importance of the structure of the nucleotide function for B_{12}-uptake and transport (but the mode of binding is not yet experimentally characterized) [147, 148].

In solution the intramolecular coordination of the nucleotide function to the lower α-axial coordination site of the corrin-bound cobalt center of complete corrins occurs with little build-up of strain [149]. This allows the (coordinating) nucleotide to steer the reactivity, as well as the face-selectivity, of certain organometallic reactions involving the corrin-bound cobalt center [75]. Experiments by Eschenmoser demonstrated that cobalamins can self-constitute in solution from the B_{12}-nucleotide portion and incomplete cobyrinic acid derivatives to show a remarkable kinetic and thermodynamic preference for the specific formation of the B_{12}-structure, and to a pre-enzymatic origin of the basic structural elements of the complete corrins [149].

The variety in the structure of the (pseudo)nucleotide unit of complete corrinoids appears to be largely the consequence of the particular biosynthetic availability in the various microorganisms. The known purine bases of complete corrinoids are mostly adenine derivatives or related heterocycles [76, 150] as found also in RNA [151]. For an alternative "functional rationalization", the differing properties of the complete corrinoids can be considered, as reflected by the tendency of their biologically relevant organometallic forms to be base-off or base-on in aqueous solution [150]. The phenolyl-cobamides are of course complete corrinoids with base-off structure, helpful for binding in a base-off/His-on form to B_{12}-binding apoenzymes [152]. As shown in the x-ray structure analysis of the B_{12}-binding domain of methionine synthase [17], the base-off form of the B_{12}-cofactor MeCbl (**3**) presents a molecular surface to the protein which is significantly larger than that of the base-on form of **3**.

6
B_{12}-Dependent Methyl Transferases

The B_{12}-dependent methyltransferases play an important role in amino acid metabolism in many organisms (including humans) as well as in one-carbon metabolism and CO_2 fixation in anaerobic microbes [153]. The reactivity of the "supernucleophilic" Co(I)-corrins and of methyl-Co(III)-corrins make B_{12}-derivatives ideal as cofactors in such enzymatic methyl group transfer reactions [1]. B_{12}-dependent methionine synthase has been particularly well studied (see [125, 153–155]) as have methyl transferases in aerobic acetogenesis (see [156, 157]), methanogenesis (see [126]) and anaerobic catabolism of acetic acid to methane and CO_2 (see [158]). Various substrates act as sources of methyl groups, such as, methanol, methyl amines, aromatic methyl esters, methylated heavy metals or N^5-methyltetrahydropterins (such as N^5-methyltetrahydromethanopterin or N^5-methyltetrahydrofolate) [145, 153]. For N^5-methyltetrahydrofolate as a source of the methyl group it has been suggested the methyl group donor is more likely to be the protonated form of N^5-methyltetrahydrofolate [159]. Thiols are the methyl group acceptors in methionine synthesis (homocysteine) [125, 154] and methanogenesis (coenzyme M) [126]. In the anaerobic biosynthesis of acetyl-coenzyme

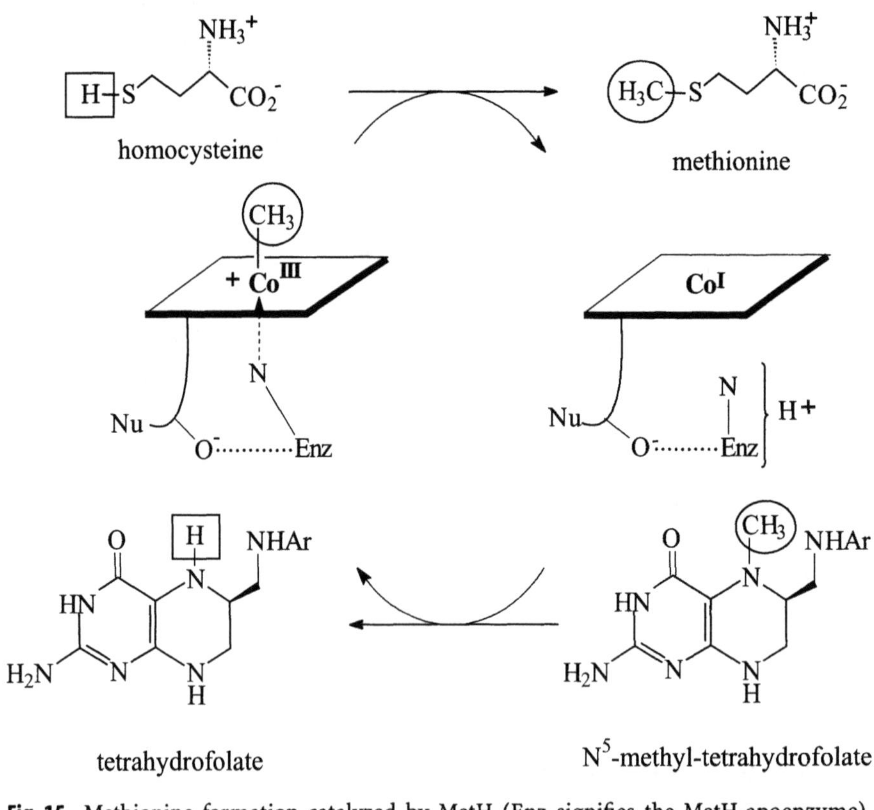

Fig. 15 Methionine formation catalyzed by MetH (Enz signifies the MetH-apoenzyme), where the bound corrinoid shuttles between MeCbl (3), in a "base-off/His-on" form, and cob(I)alamin (B_{12s}, 40^-) [125]

A from one-carbon precursors the methyl group acceptor is suggested to be the nickel center attached to an Fe/S-cluster [157].

The methyl group transfers catalyzed by methionine synthase from *E. coli* [154, 160], in cell-free extracts of the methanogen *Methanosarcina barkeri* [161] and in the assembly of the two-carbon unit of acetyl coenzyme A by the acetogens *Sporomusa ovata* [145] and *Moorella thermoaceticum* [162] are all indicated to proceed with an overall retention of configuration (i.e., consistent with two nucleophilic displacement steps, each with inversion of configuration). These stereochemical findings exclude free methyl cations or radicals as intermediates, even though in a formal sense the methyl transfer reactions catalyzed by B_{12}-enzymes involve (nucleophilic-bound) methyl "cations" and heterolytic cleavage/formation of the $Co-CH_3$ bond. The methyl group transfer, in fact, relies on the catalytic properties of enzyme-bound Co(I)corrins and methyl-Co(III)corrins [154] and is amenable to considerable control from the protein environment [163], due to the great structural changes expected to accompany the transitions from (tetracoordinate) Co(I)corrins to (hexacoordinate) methyl-Co(III)corrins [1, 75].

6.1 Methionine Synthase

Methionine synthase (MetH) of *E. coli* represents the most thoroughly studied B_{12}-dependent methyl transferase and is one of the essential roles of B_{12} in mammalian metabolism [125, 153, 154]. It is a modular enzyme containing separate binding domains for homocysteine, N^5-methyltetrahydrofolate, S-adenosyl-methionine (SAM) and the B_{12}-cofactor [125, 153–155]. The B_{12}-binding domain in its different oxidation states must interact punctually and specifically with each of the other three domains: The Co(I) form with the N^5-methyltetrahydrofolate binding domain, the Co(II) form with the SAM binding domain, and the $CH_3-Co(III)$ form with the homocysteine binding domain [153, 155].

MetH catalyzes the methylation of the bound and reduced cob(I)alamin cofactor by (N^5-protonated) N^5-methyltetrahydrofolate to give enzyme-bound methylcobalamin (**3**) in a base-off/His-on form (see later) [125, 153–155]. The methyl-Co(III)corrinoid is demethylated by homocysteine, whose sulfur is activated and deprotonated due to the coordination to a zinc ion (held by three *cysteine* residues) of the homocysteine binding domain [164] (see Fig. 15). The two methyl-transfer reactions occur in a sequential mechanism [124, 125, 153, 154]. Intermittently, the bound Cob(I)alamin (**40**$^-$) is oxidized to enzymatically inactive cob(II)alamin (**23**) and requires reactivation by reductive methylation with SAM and a flavodoxin as a reducing agent [125, 153–155, 165].

The x-ray crystal analysis of the B_{12}-binding domain of MetH provided the first insight into the three-dimensional structure of a B_{12}-binding protein [17,

138, 163, 166]. The astounding revelation of this work was that the cobalt-coordinating DMB-nucleotide tail of the protein-bound cofactor MeCbl (**3**) was displaced by a histidine imidazole and extended into the core of the "Rossmann fold" [17, 138, 163]. Consequently, in methionine synthase the corrinoid cofactor is bound by histidine ligation to the metal center and in a base-off constitution, i.e., bound in a base-off/His-on mode. The crucial cobamide-ligating histidine residue [17] is part of a Gly–X–X–His–X–Asp sequence, which was noticed earlier as a common sequence in some B_{12}-binding proteins [167]. The B_{12}-binding domain of MetH therefore, provides both an anchoring site for the nucleotide tail and cobalt-ligation via the residues of the His–Asp–Ser triad (the "regulatory" unit) [17, 138, 163], holding the corrinoid cofactor with its "catalytic" β-side exposed at an interdomain interface.

The crystallographic studies also helped to confirm the suspected domain alternation used as a means of control for the two ways of methylating the bound corrinoid [165]. More recently, the crystal structure of the N-terminal substrate-binding modules of MetH has been described [168]. These are the two domains, which bind and activate homocysteine and N^5-methyltetrahydrofolate. The substrates are bound in orientations that position them for reaction with the bound corrinoid, but the two active sites are separated by ≈ 50 Å. To complete the catalytic cycle, the B_{12}-binding domain thus must shuttle back and forth between these distant active sites [168]. In the active site of the homocysteine binding domain the substrate forms a metal-ligand cluster with approximately tetrahedral geometry. This result agrees with the measurements showing four sulfur ligands to zinc in homocysteine complexes of *E. coli* MetH [164] (as mentioned earlier in the section).

The crystallographic results on the structure of MetH and the finding of the base-off/His-on binding of the cofactor in a B_{12}-dependent methyl transferase were consistent with earlier ESR spectroscopic evidence for histidine binding to the cobalt center of *p*-cresolyl-cobamide (**52**) in the acetogen *Sporomusa ovata* [145, 169]. Various other B_{12}-dependent methyltransferases are indicated to have either a base-off/His-on bound methyl-Co(III)-corrinoid, or even a methyl-corrinoid cofactor in base-off form (where His-coordination is absent) [156].

6.2
B_{12}-Cofactors in Enzymatic Methyl-Group Transfer

In a catalytic cycle of B_{12}-dependent methyl transferases the corrinoid is indicated to cycle between a methyl-Co(III)-corrin and a Co(I)-corrin [125, 126, 153, 155]. The changing between the hexacoordinate methyl-Co(III)-form and (presumably) tetracoordinate Co(I)-form is accompanied by constitutional/conformational changes, which are highly likely to provide a means for controlling the organometallic reactivity of the bound cofactor [170], subject to H^+-uptake or H^+-release (see Fig. 15). In response, a H^+-mediated switch mechanism may result, mediated via the "regulatory" His–Asp–Ser triad, which provides the crucial conformational alterations associated with the enzyme [125, 138, 153, 154, 171]. The nucleophile-induced methyl group transfers, involving heterolytic cleavage and formation of the organometallic $Co-CH_3$ bond at the corrin-bound cobalt center, are expected to be in-line attacks (incoming nucleophile/CH_3 group/leaving group) and to be subject to strict geometric control: a main role of the His–Asp–Ser triad appears to be participating in maintaining conformational control of the mutual placement of the corrinoid cofactors and the enzyme-bound substrates [125, 153, 154].

An important second role of the His–Asp–Ser triad in heterolytic organometallic reactions is associated with the thermodynamic effect of the α-axial base-coordination on the strength of the $Co_\beta-CH_3$ bond. Solution studies showed a significant thermodynamic *trans*-effect of the DMB-coordination in methylcobalamin (**3**) [1, 75, 86] and of the imidazole-coordination in $Co_\beta-CH_3$-imidazoyl-cobamide (**31**) [68] on heterolytic methyl group transfer reactions. The result showed the stronger coordinating (nitrogen-)ligand stabilizes the methyl-Co(III)-form against nucleophilic abstraction of the methyl group by about 4 kcal/mol in **3** [86]. This may be seen mainly as an "electronic" effect [1, 75], consistent with the observation of anomalous structural *trans*-effects in other methyl-Co(III)-complexes [72]. More recent studies with **31** suggested the imidazole base exerts similar electronic effects as the DMB-base in **3** but **31** is more basic and, therefore, imidazolylcobamides (or the base-off/His-on form of **3**) are more readily protonated near neutral pH [68]. The His–Asp–Ser triad may then represent a "relay" for H^+-uptake/release, assumed to function in the enzymatic methylation/demethylation cycles [172].

In conclusion, the axial $Co-N_\alpha$-bond in the methyl-Co(III)-form of the protein-bound cofactor of MetH (and other B_{12}-dependent methyl transferases) seems to have three important consequences. The weakening of this bond activates both (1) the methyl group for heterolytic abstraction by a nucleophile and (2) the Co(II)-form for reduction to the Co(I)-form and (3) helps to position the methyl-cob(III)amide cofactor for methyl group transfer [125, 138, 171].

7
Coenzyme B_{12}-Dependent Enzymes

About ten coenzyme B_{12}-dependent enzymes are now known (see Table 1) [6, 25, 153, 173]. These enzymes are four carbon skeleton mutases (methylmalonyl-CoA mutase [174], glutamate mutase [175, 176], methylene glutarate mutase [175] and isobutyryl-CoA mutase [177]), diol dehydratase [178],

Table 1 Enzymatic reactions involving natural substrates and adenosyl-cobamide-dependent enzymes

(*R*)-methylmalonyl-CoA/succinyl CoA (see Fig. 17) [120, 174]

(*S*)-glutamic acid/(2*S*,3*S*)-methylaspartic acid (see Fig. 18) [173, 175, 176]

2-Methyleneglutarate/(*R*)-3-methylitaconate [175]

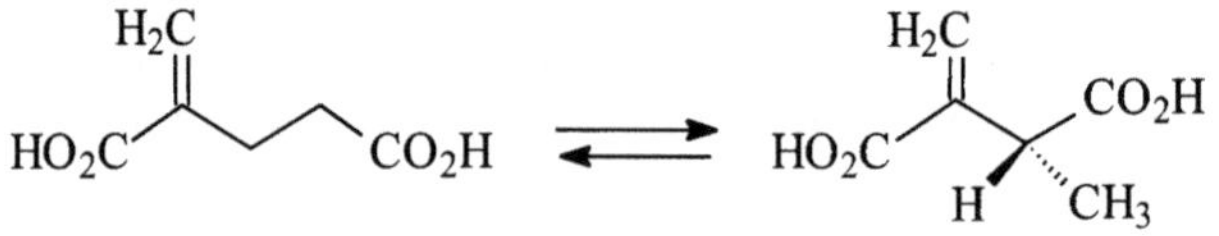

Isobutyryl-CoA/*n*-butyryl-CoA [177]

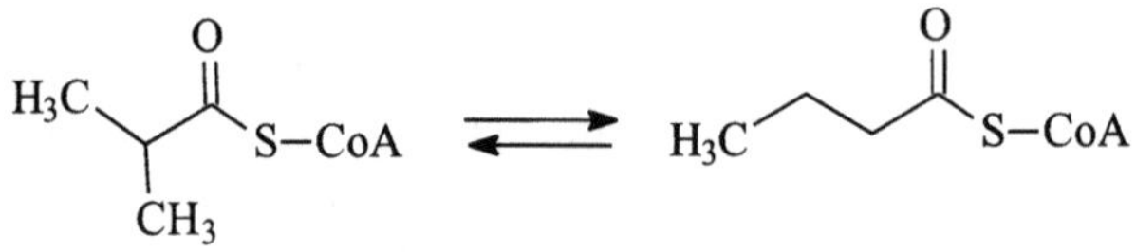

D-ornithine/(2*R*,4*S*)-2,4-diaminovaleric acid [180]

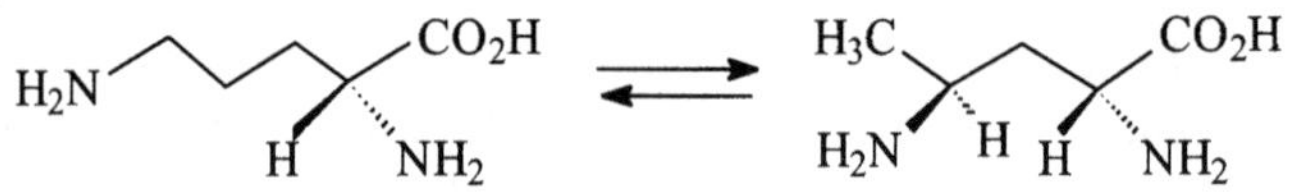

D-(*α*)-lysine/2,5-diaminohexanoic acid [180]

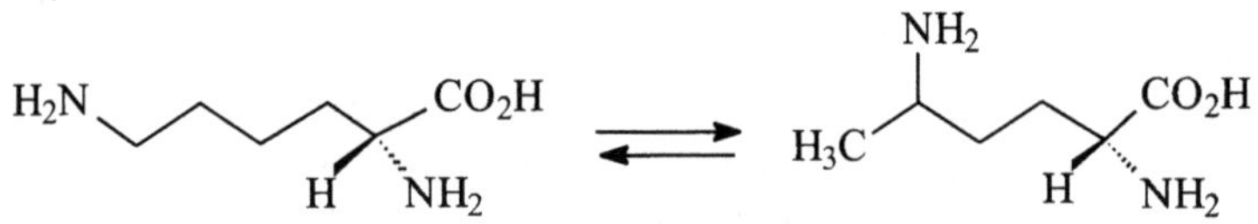

L-(*β*)-lysine/(3*R*,5*S*)-3,5-diaminohexanoic acid [180]

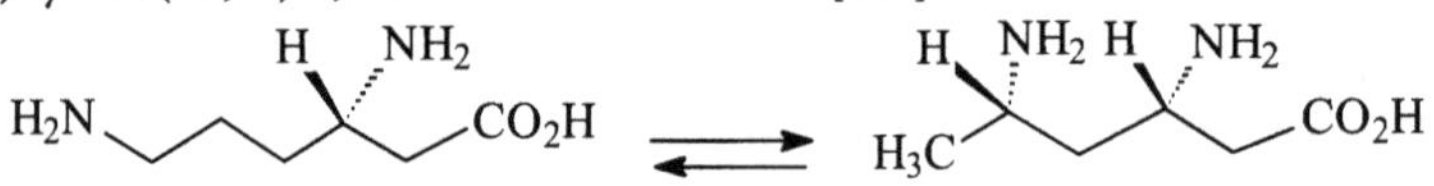

1,2-Propanediol/propanal (see Fig. 19) [210]

Glycerol/3-hydroxypropanal (see Fig. 19) [210]

2-Hydroxyethylamine (ethylamine)/acetaldehyde [179]

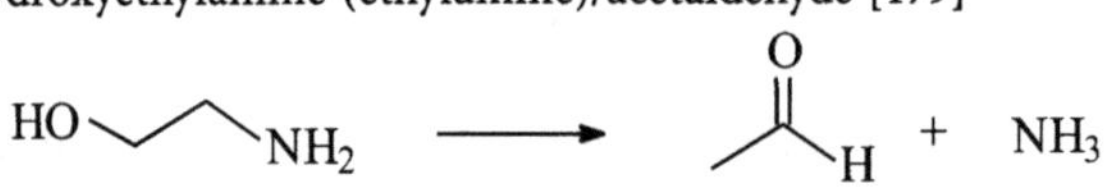

Ribonucleotide/2′-deoxyribonucleotide (see Fig. 20) [182]

glycerol dehydratase [178], ethanolamine ammonia lyase [178, 179], two amino mutases [180, 181] and B_{12}-dependent ribonucleotide reductase [182]. The coenzyme B_{12}-dependent enzymes are disproportionately distributed in the living world. Only methylmalonyl-CoA mutase is indispensable in human metabolism. In methanogens a functional role of coenzyme B_{12}-dependent enzymes is suspected but has not (yet) been clearly revealed [126].

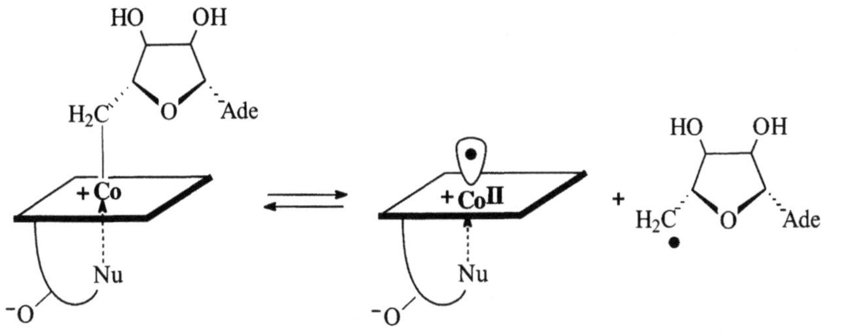

coenzyme B_{12} (**2**) cob(II)alamin (**23**) 5'-deoxy-5'-adenosyl radical

Fig. 16 Coenzyme B_{12} (AdoCbl, **2**) a reversible source of the 5′-deoxy-5′-adenosyl radical and of cob(II)alamin (B_{12r}, **23**)

The coenzyme B_{12}-dependent enzymes perform chemical transformations that are difficult to achieve by typical organic reactions. With the exception of the enzymatic ribonucleotide reduction [182], the results of coenzyme B_{12}-catalyzed enzymatic reactions correspond to isomerizations with vicinal exchange of a hydrogen atom and of a group with heavy atom centers. Homolytic cleavage of the Co – C bond of the protein-bound AdoCbl (**2**) to a 5′-deoxy-5′-adenosyl radical and cob(II)alamin (**23**) was indicated to be the entry to H-abstraction reactions induced by the 5′-deoxy-5′-adenosyl radical [183]. Therefore, homolysis of the Co – C bond of **2**, which is the thermally most easily achieved reaction of **2** in solution (homolytic Co – C BDE of about 30 kcal/mol [119, 123]) appears to be its biologically most significant reactivity: coenzyme B_{12} (**2**) is quoted to be a "reversible free radical carrier" [119]) (see Figs. 10, 12, 16). However, the homolysis of the Co – C bond of the protein-bound coenzyme needs to be accelerated by a factor of about 10^{12} to agree with the observed rates of reaction of catalysis by the coenzyme B_{12}-dependent enzymes [119, 123]. Consequently, the major functions of the enzyme concern not only the catalysis of its proper reactions but also the reversible generation of the radical intermediates and the protection of its proteinic environment from nonspecific radical chemistry, dubbed "negative catalysis" [183, 184].

To conclude the coenzyme B_{12}-dependent enzymes all appear to based on the reactivity of bound organic radicals, which are formed (directly or indirectly) by a H-atom abstraction by the 5′-deoxy-5′-adenosyl radical, that originates form the homolysis of the Co – C bound of AdoCbl (**2**). In these enzymatic reactions, the 5′-deoxy-5′-adenosyl radical is the established reactive partner in the actual enzymatic reaction, so that **2** should be looked at as a "pre-catalyst" (or catalyst precursor) [75]. Coenzyme B_{12} (**2**) might then be considered to be a structurally highly sophisticated, reversible source for an alkyl radical, whose Co – C bond is labilized in the protein bound state [119].

7.1
Carbon Skeleton Mutases

In the four known carbon skeleton rearrangement reactions, catalyzed by coenzyme B_{12}-dependent mutases, two vicinal groups (a hydrogen atom and an organic substituent) exchange their positions in a (pseudo)intramolecular fashion [173]. The B_{12}-cofactor is bound base-off/His-on at an interface between two modules, the B_{12}-binding and substrate activating domains (or subunits) as proven by the analysis of the crystal structures of methylmalonyl-CoA mutase (MMCM) [18] and glutamate mutase (GM) [19]. The B_{12}-binding motif (Gly–X–X–His–X–Asp) [167] occurs in MMCM and GM, as well as in 2-methyleneglutarate mutase (MGM) [185, 186] and isobutyryl-CoA mutase (ICM) [177]. The B_{12}-binding domains of (in MMCM and MGM) and the B_{12}-binding subunits (in GM and ICM) in fact exhibit considerable sequence homology, which includes even the B_{12}-binding domain of MetH. Such homology does not extend to the other coenzyme B_{12}-depending enzymes or even to the substrate binding domains (subunits) of the carbon skeleton mutases [177, 187].

7.1.1
Methylmalonyl-CoA Mutase

Methylmalonyl-CoA mutase (MMCM) interconverts *R*-methylmalonyl-CoA and succinyl-CoA [120, 174] (see Fig. 17). Binding of the substrate triggers the homolysis of the Co – C bond of the bound adenosyl-corrinoid. The 5′-deoxy-5′-adenosyl radical prepares the substrate for the rearrangement reaction by abstracting an H-atom from the methyl group of enzyme-bound methylmalonyl-CoA. A large deuterium isotope effect of the (substrate) H-atoms to be abstracted on the rate of homolysis of the Co – C bond was observed [188]. Labilization of the Co – C bond towards homolysis is largely due to a decrease of the enthalpy of activation by about 16 kcal/mol [189]. Recently, spectroscopic and computational studies [190] on the activation of Co – C bond for homolysis indicated that, although the enzyme may serve to activate the cofactor on its Co(III)-ground state, the dominant contribution presumably comes through stabilization of the 5′-deoxy-5′-adenosyl radical and cob(II)alamin (**23**, B_{12r}), consistent with the earlier suggestion based on the crystal structure of **23** [51].

H-atom abstraction gives the 2-methylmalon-2′-yl-CoA radical, which rearranges rapidly to the succin-3-yl-CoA radical [174, 191]. Both, fragmentation/recombination and intramolecular addition/elimination, via a cyclo-

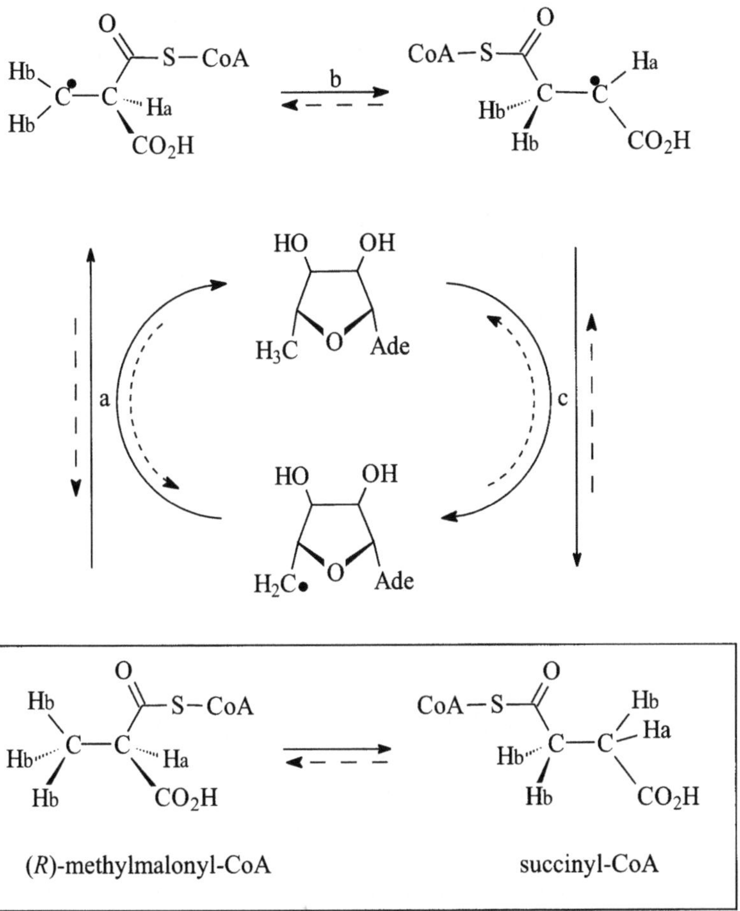

Fig. 17 Methylmalonyl-CoA mutase (MMCM) interconverts (*R*)-methylmalonyl-CoA and succinyl-CoA. Proposed reaction mechanism of the carbon skeleton rearrangement, catalyzed by MMCM involving H-atom abstraction (step a), radical rearrangement (step b) and back transfer of H-atom (step c). (The experimentally supported substrate triggered formation of the 5′-deoxy-5′-adenosyl radical and of cob(II)alamin (**23**, B_{12r}) by homolysis of protein bound AdoCbl (**2**) is omitted here, see Fig. 16 [120, 173, 191])

propyloxyl radical, pathways have been considered for this rearrangement. However, computational studies indicate that the energetic barrier for the addition/elimination pathway is lower than the dissociative pathway [192–194]. Experimental evidence has also been used to support this theory, in which (1) the succin-3-yl-CoA radical arises from an intramolecular radical rearrangement and (2) occurs without noticeable participation of the bound cob(II)alamin (**23**) [191]. The succin-3-yl-CoA radical (resulting from the rearrangement) then re-abstracts an H-atom from 5′-deoxyadenosine to give succinyl-CoA and the 5′-deoxyadenosyl radical, that recombines with **23** to give coenzyme-bound AdoCbl (**2**).

The x-ray analysis of MMCM was the first crystal structure of a coenzyme B_{12}-dependent enzyme [18, 163, 194, 195]. The study concerned the 150 kDa heterodimeric MMCM from *P. shermanii* and showed the B_{12}-cofactor to be bound base-off/His-on. The α-side of the corrin-bound cobalt center was coordinated to the histidine of the regulatory triad His–Asp–Lys. As in MetH the nucleotide tail of the bound corrinoid was tightly inserted into the protein and the corrinoid was bound at an interface between two domains. A rather "flat" corrin ligand with a ligand-folding comparable to that in imidazolylcobamides was revealed [31, 68].

In other crystallographic work MMCM was investigated in a substrate free form as well as with bound pseudosubstrate: both structures showed the adenosyl group of the bound cofactor to be again repositioned (presumably with the help of the bound substrate), indicating the organometallic group to be particularly strained or detached from the cobalt center [195–197]. The data suggest substrate binding assists the homolysis of the Co – C bond by squeezing the adenosyl group off from the cobalt corrin [163, 197].

7.1.2
Glutamate Mutase

Barker, in the early 1960s, identified glutamate mutase (GM) from *Clostridium tetanomorphum*, which was the first known coenzyme B_{12}-dependent enzyme [198], and this eventually led to the discovery of the organometallic nature of AdoCbl (**2**). GM from *Cl. tetanomorphum* and also from *Cl. cochlearium* have been comprehensively studied, but other microorganisms are predicted to have enzymes with GM-activity [175]. GM catalyzes the reversible rearrangement between (*S*)-glutamate and (2*S*,3*R*)-3-methylaspartate, where the equilibrium favors glutamate (by about 10) [175, 187, 199, 200], and is the only known carbon skeleton isomerase in which a saturated carbon moiety undergoes migration [201–203] (see Fig. 18).

GM proteins from *Cl. tetanomorphum* and *Cl. cochlearium* consist of a pair of B_{12}-binding subunits (σ) and a larger substrate binding unit (ε). Two molecules of adenosyl-cobamide are bound in the heterotetramers ($\varepsilon_2\tau_2$). The cofactors from the two *Clostridia* were identified as pseudocoenzyme B_{12} (**37**) [151, 201] and adenosyl-factor A (**38**) [151], but coenzyme B_{12} (**2**) also functions as a cofactor [175, 200]. Binding of the substrate triggers homolysis of the Co – C bond of the adenosyl cofactor and in a similar situation to MMCM, homolysis and H-atom abstraction of the bound substrate are kinetically coupled. A pre-steady state D-isotope effect of about 30 is observed [173, 202] and more recently the secondary tritium isotope effect was measured giving a surprisingly high value of ≈ 0.72 [203]. After abstraction of an H-atom from the methyl group of enzyme bound methyl aspartate, the rearrangement of the 3-methylaspart-3′-yl radical to the glutam-4-yl radical [175, 176] was postulated to take place via a fragmentation recombination

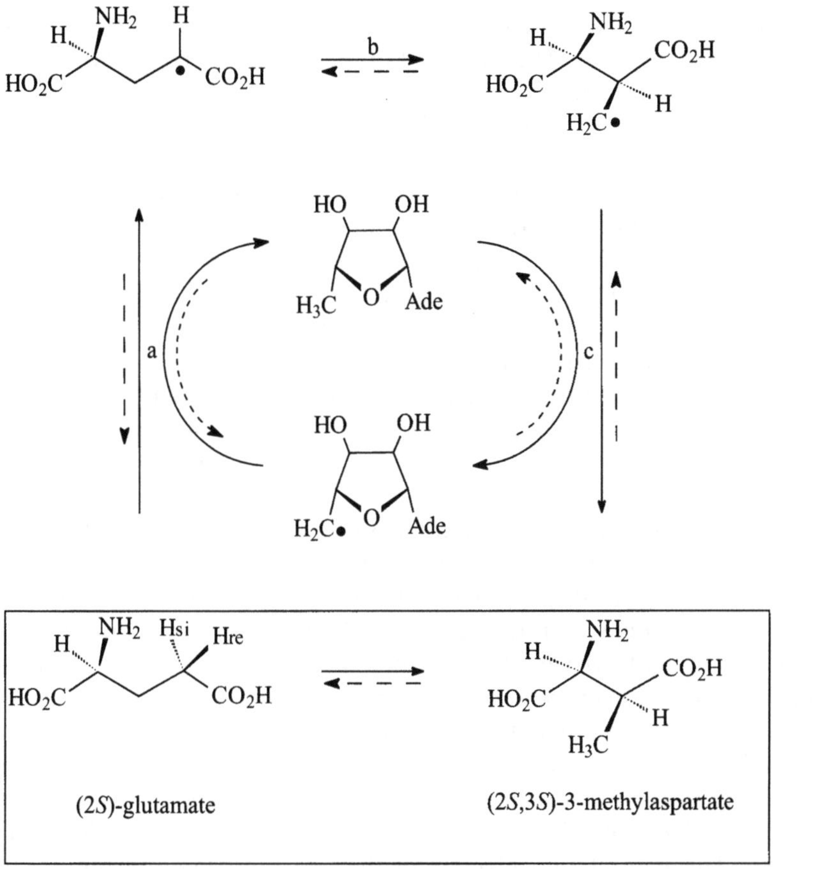

Fig. 18 Glutamate mutase (GM) interconverts (*S*)-glutamate and (2*S*,3*S*)-3-methylaspartate. Proposed reaction mechanism of the carbon skeleton rearrangement, catalyzed by GM involving H-atom abstraction (step a), radical rearrangement (step b) and back transfer of H-atom (step c). (The experimentally supported substrate triggered formation of the 5′-deoxy-5′-adenosyl radical and of cob(II)alamin (**23**, B_{12r}) by homolysis of protein bound AdoCbl (**2**) is omitted here, see Fig. 16 [173, 175, 176])

mechanism (involving acrylate and a glycyl radical) by Buckel and Golding [199]. A cyclopropyl intermediate, as proposed in the related reaction of MMCM, cannot be formed in GM as the migrating carbon is sp^3 hybridized [176]. Further experimental [204] and theoretical studies [173, 176, 205] indeed support the fragmentation recombination pathway [202]. The glutam-4-yl radical re-abstracts an H-atom from 5′-deoxyadenosyl to give the rearrangement product, (*S*)-glutamate, and the 5′-deoxyadenosyl radical, that recombines with cob(II)alamin (**23**) to give enzyme bound **2**.

The crystal structure of GM from *Cl. cochlearium* has provided a detailed structural view of the enzyme, in which the corrinoid cofactor is bound base-off/His-on and at the interface between the σ and ε subunits [19]. The cobalt-coordinating histidine (His16) is part of the H-bonded regulatory His–Asp–Ser triad. Detailed analysis of GM with bound coenzyme B_{12} revealed the position of the ribose part of the 5′-deoxyadenosol moiety to be disordered and present in two conformations, related to each other by a pseudorotation of the furanose ring ($C_{2'}$-endo *vs* $C_{3'}$-endo) [63]. In one conformation the 5′-methylene group of adenosine is positioned above the cobalt center, but at a distance of 3.1 Å, and thus appears to have the expected features for the direct product of the homolysis of the Co – C bond of the bound cofactor AdoCbl (2). In the other species the 5′-methylene carbon is 4.5 Å from the metal center and is displaced towards the substrate-binding site where it is in van der Waals contact to the bound substrate and would be ideally positioned for hydrogen atom exchange. In this way, GM can achieve a controlled and energetically facile transportation of the 5′-radical center from cobalt to the substrate [63, 176].

The solution structures of the cofactor-free B_{12}-binding σ-subunits of GM from *Cl. tetanomorphum* [206, 207] and *Cl. cochlearium* [208] were analyzed by heteronuclear NMR spectroscopy. The studies revealed a close overall similarity to the x-ray structure and the σ-subunits to be largely pre-organized for B_{12}-binding. However, the apoprotein was seen to include a flexible loop and a "nascent" helix, which were both suggested to structure only upon binding of the cofactor [209]. From these analyses a model for the events in binding of the base-off/His-on corrinoid by the σ-subunit was derived. According to the model, the nucleotide tail of the base-off cofactor is first trapped and the attached nucleotide stabilizes the "nascent" helix of the protein. The natural B_{12}-cofactors of the two *Clostridia* are the adeninyl-cobamides, pseudocoenzyme B_{12} (**37**) and adenosyl-factor A (**38**), for which the base-off form predominates in aqueous solution [76]. In aqueous solution AdoCbl (2) prefers to be in the base-on form (with only $\approx$ 1% in the base-off form) [34].

7.1.3
Other B_{12}-Dependent Carbon Skeleton Mutases

Besides MMCM and GM, two other coenzyme B_{12}-dependent carbon skeleton mutases are known. These are (1) methylene glutarate mutase (MGM) from the anaerobe *Eubacterium* (*Clostridium*) *barkeri*, which catalyzes the equilibration of 2-methylene-glutarate with (*R*)-3-methylitaconate as part of a degradative path of nicotinic acid [175, 199] and (2) isobutyryl-CoA mutase (ICM), which is observed in species of gram-positive bacteria *Streptomyces* and catalyzes the reversible rearrangement of iso-butyryl-CoA and n-butyryl-CoA [177]. The isomerization of iso-butyryl-CoA and n-butyryl-CoA in ICM is relevant in the biosynthesis of polyketide antibiotics [177].

7.2
Diol Dehydratases and Ethanolamine Ammonia Lyase

Diol dehydratase (DD) and glycerol dehydrates (GD) are isofunctional enzymes that catalyze the dehydration of glycerol, ethane-1,2-diol and propane-1,2-diol to 3-hydroxypropanal, acetaldehyde and propanal, respectively (other glycols can be dehydrated in an analogous fashion) [210] (see Fig. 19). DD has about a twofold preference for propane-1,2-diol to glycerol, whereas GD prefers glycerol to propane-1,2-diol as a substrate [178]. Conversion of glyc-

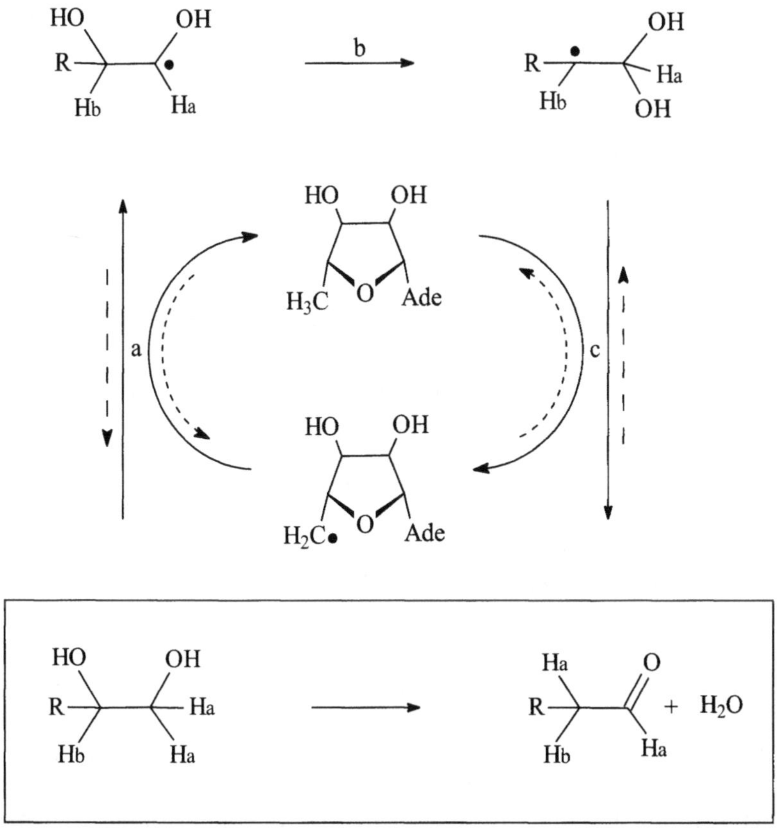

Fig. 19 Diol dehydratase and glycerol dehydratase isomerize vicinal diols – via 1,2-migration of a hydroxyl group – to geminal diols (such as 1,2-propane-diol to 1,1-propane-diol, which looses water to give propanal). Possible reaction mechanism of the hydroxyl group migration, catalyzed by DD (or GD) and involving H-atom abstraction (step a), radical rearrangement (step b) and back transfer of H-atom (step c); substrates for diol dehydratase: propane-1,2-diol ($R = CH_3$) or ethane-1,2-diol ($R = H$); for glycerol dehydratase: glycerol ($R = CH_2 - OH$). (The experimentally supported substrate triggered formation of the 5′-deoxy-5′-adenosyl radical and of cob(II)alamin (**23**, B_{12r}) by homolysis of protein bound AdoCbl (**2**) is omitted here, see Fig. 16 [178, 210])

erol (or even diols) causes an irreversible deactivation of the enzymes but this can be repaired with the assistance of a reactivating enzyme [178, 210].

The diol dehydratases use adenosyl-cobamides (such as AdoCbl, **2**) as cofactors but in contrast to the B_{12}-dependent carbon skeleton mutases, the cofactor is bound in a (conventional) base-on form. This was made likely by analysis of the protein sequences of DD [211] and GD [212], which lack the diagnostic "B_{12}-binding" (Gly–X–X–His–Asp) sequence, ESR-investigations [213–215] and finally conformation by x-ray analysis [20, 215]. In addition, in the active site of the enzyme a potassium ion is bound, which has been ascribed the role of a Lewis acid in the radical rearrangement [20, 216]. The AdoCbl-catalyzed reaction is again based on a substrate-assisted homolysis of the Co – C bond of the cofactor and formation of the 5′-deoxy-5′-adenosyl radical and cob(II)alamin (**23**). In diol dehydratase the adenosyl radical undergoes a counter clockwise rotation of approximately 94° where it can then abstract the *pro-R* and *pro-S* hydrogens of the (*R*)- and (*S*)-isomers of propane-1,2-diol, respectively [178, 217]. Thus, forming the substrate derived 1,2-diol-1-yl radical and 5′-deoxyadenosine. The substrate-derived radicals undergo OH group migration from C2 to C1 by a concerted pathway through a cyclic transition, resulting in rearrangement to the product-derived 1,1-diol-2-yl radicals. From the results of site directed mutagenesis and theoretical calculations, His143 and Glu170 play important roles as a general base and a general acid, respectively in the stabilization of the transition state [178, 218]. The K^+-ion seems to play an essential role in orientation of the substrates and increasing the substrate-binding energy [178]. C2 of the 1,1-diol-2-yl radicals then comes close to the CH_3 group of 5′-deoxyadenosine and back-abstracts a hydrogen atom, producing 1,1-*gem*-diols and the adenosyl radical. The stereospecific dehydration of the 1,1-*gem*-diols to the final product propionaldehyde is catalyzed by Asp335 and Glu170 [178].

Three crystal structures of DD [20, 217, 219] and two of GD [215, 220] have been studied, confirming the base-on nature of the bound corrinoid. Analysis of the crystal structure of DD reconstituted with the coenzyme B_{12} analogue 5-adeninyl-pentyl-cobalamin (**53**) [217] helped identify an adenine-binding pocket in the "substrate-binding" α-subunit. The observed mode of adenine binding allowed the inference of a build-up of strain in the bound adenosyl-cobamide cofactor and activation towards homolysis of the Co – C bond.

Ethanolamine ammonia lyase (EAL) converts ethanolamine (2-hydroxy-ethylamine) to acetaldehyde, with the loss of ammonia. The cobamide bound to the enzyme of *Clostridium* sp. was identified as a pseudocoenzyme B_{12} (**37**) [221] but a range of other adenosyl-cobamides are accepted as cofactors. The active enzyme is multimeric, has an apparent molecular mass of about 560–600 kDa, but further studies are much less advanced with EAL than those of the AdoCbl-dependent enzymes described above. From ESR studies [222], and the fact that the cobalamin binding motif (Gly–X–X–His–X–Asp) is absent [223], EAL was concluded to be a base-on B_{12}-dependent enzyme. A radical mechanism, similar to that of diol dehydratase, is proposed for the isomerization of the vicinal aminoalcohol substrates (ethanolamine, (*R*)- and (*S*)-aminopropanol) [1, 224–227].

7.3
B_{12}-Dependent Amino Mutases

Two coenzyme B_{12}-dependent amino mutases are currently known and have been characterized, D-ornithine aminomutase and lysine 5,6-aminomutase. Both enzymes catalyze the migration of ω-amino groups to the (ω-1)-position in diamino-acids, using coenzyme B_{12} and pyridoxal-phosphate (PLP) as co-factors [180, 181]. D-ornithine aminomutase catalyzes the migration of the terminal amino group of D-ornithine to the 4-position, to give (2*R*,4*S*)-diaminovaleric acid. The migration of the amino group is initiated by the 5′-deoxy-5′-adenosyl radical, which abstracts a H-atom from the pyridoxal-conjugate of the substrate. The required pyridoxal cofactor assists the migration by forming a Schiff's base with the migrating amino function by way of an azacyclopropylcarbinyl radical.

Lysine 5,6-aminomutase (5,6-LAM) catalyzes the isomerization of D-lysine to 2,5-diaminohexanoic acid [180, 181]. The mechanism proposed is analogous to that of D-ornithine amino mutase [228, 229]. 5,6-LAM was predicted to be a base-off/His-on B_{12}-dependent enzyme [230], as was recently conformed by the crystal structure [231].

7.4
B_{12}-Dependent Ribonucleotide Reductase

In all organisms, ribonucleotide reductases (RNRs) play an essential role in the biosynthesis of DNA by catalyzing the reduction of all four nucleoside di- or triphosphates to the corresponding 2′-deoxynucleotides [182, 232]. Despite their central role in primary metabolism, RNRs have evolved to use a diverse array of cofactors to initiate the radical reaction, which eventually leads to nucleotide reduction (see Fig. 20) [233, 234].

The reductase from *Lactobacillus leichmanii* (RNR-*Ll*) belongs to the class II RNRs, which make use of adenosyl-cobamides to initiate the radical reac-

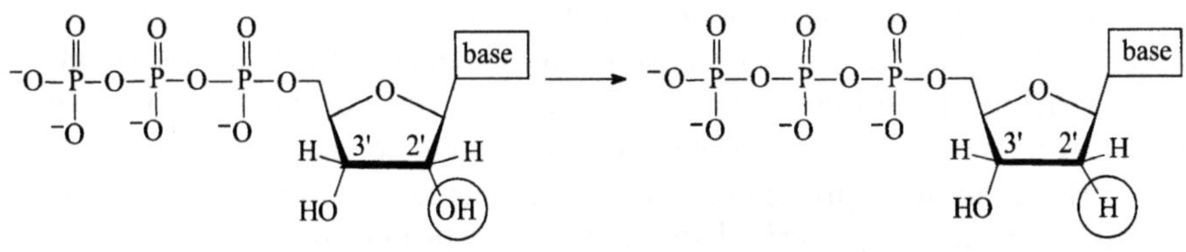

Fig. 20 Schematic illustration of the reduction of ribonucleoside triphosphates to 2′-desoxyribonucleoside triphosphates, as catalyzed by ribonucleotide reductases [182]

tion, and is the most studied of the group that depend on B_{12}-cofactors. RNR-*Ll* uses nucleoside triphosphates (NTPs) as substrates and 2′-deoxynucleotide triphosphates (dNTPs) as allosteric effectors [182]. The B_{12}-binding motif, (Gly–X–X–His–X–Asp) sequence, is not found and indeed, by ESR spectroscopy [235] and later by the crystal structure [236], base-on binding of the corrinoid cofactor was confirmed. Uniquely in RNR-*Ll*, with relation to other B_{12}-dependent enzymes, a protein-centered thiyl-radical is generated (at cysteine-408) from the homolysis of the Co – C bond of the bound coenzyme, and this induces the radical reaction that formally leads to the reductive substitution by hydrogen of the 2′-hydroxyl group of the ribonucleotide. Homolytic cleavage of the Co – C bond is known to be accelerated by about 10^{11}-fold in RNR [237].

The crystal structures of both dGTP-free AdoCbl-dependent RNR-*Ll* in the apoform, and complexed with Co_{β}-adeninylpentyl-cobalamin (**53**), have been solved [236]. As suspected earlier the bound corrinoid was found to be base-on.

7.5 B_{12}-Coenzymes in Enzymatic Radical Reactions

The homolytic cleavage of the Co – C bond of the protein-bound organometallic cofactor AdoCbl (**2**) is the initial step of the coenzyme B_{12}-catalyzed enzymatic reactions. Halpern quoted that adenosyl cobamides can be considered as "reversibly functioning sources for organic radicals" [119]. A neutral aqueous solution of **2** is remarkably stable with a half-life of 10^{10} s (in the dark at room temperature), but decomposes, mainly with the homolysis of the Co – C bond, at higher temperatures [119, 123]. The coenzyme B_{12}-catalyzed enzyme reactions occur with maximal rates of approximately 100 s^{-1} [173, 239]. Rapid formation of Co(II)corrins occurs only with addition of substrate to a solution of holoenzyme (or of apoenzymes and **2**), as demonstrated in most of the known coenzyme B_{12}-dependent enzymes, e.g., in methyl-malonyl-CoA mutase [121], glutamate mutase [202] and ribonucleotide reductase [239].

Thus, an intriguing feature of the coenzyme B_{12}-dependent enzymes is the dramatic ($> 10^{12}$-fold) labilization of the bound organometallic cofactor towards homolysis of the Co – C bond [119, 123, 173]. The mechanism of the enzyme (and substrate-induced) labilization of this Co – C bond is still a key problem, and much discussed, in B_{12}-chemistry. Evidence for covalent restructuring of the bound cofactor (except for the formation of the base-off/His-on form in the carbon skeleton mutases) is not available [75, 119, 123, 173, 194]. In addition protein and solvent molecules can only weakly stabilize a radical center [240]. Steric distortions of the protein-bound cofactor were discussed as means for the enhanced rate of Co – C bond homolysis [51, 119, 163, 217]. Halpern's theory of an "upwards conformational distortion" of the corrin ligand of AdoCbl (**2**) by the proteinic environment [119] and the related idea of a sterically induced distortion of the corrin ring by movement of the bulky dimethylbenzimidazole base ([241]) were particularly attractive. However, in view of available crystal structures of cob(II)alamin (**23**) [51] and of several coenzyme B_{12}-dependent enzymes, an upwards conformational distortion of the cobalt-corrin part of **2** is not considered of relevance any longer. It appears now that labilization comes about largely from a protein and substrate-induced strain on the organometallic group, separation of the largely nonstrained homolysis fragments and strong binding by the protein of the separated pair, 5′-deoxy-5′-adenosyl radical and **23** in base-off/His-on or base-on form [62, 163, 195]. Fixed placement of the corrin moiety at the interfaces of the B_{12}-binding and substrate-binding/activating domains appears to be of high significance and movements of the corrin moiety are not required. The regulatory triads logically appears not to be involved in proton-transfer steps and may conserve its structure largely during enzymatic turnover [1]. Electronic effects of the axial *trans* ligand on the homolytic Co – C bond dissociation energy in **2** and MeCbl (**3**) are seen to be of less importance [81, 86].

The rearrangement steps of B_{12}-dependent enzymatic rearrangements are now assumed to be accomplished by tightly protein-bound radicals that are controlled in their reaction space [178, 184, 199] but (practically) unassisted by the Co(II)-corrin fragment of the coenzyme (which has a "spectator" role) [184, 199]. In the coenzyme B_{12}-dependent enzymes, the main role of the bound cofactor thus is "only" the production and controlled presentation of the 5′-deoxy-5′-adenosyl radical from homolysis of the Co – C bond of AdoCbl (**2**), with little structural reorganization occurring in the cobalt-corrin part.

8
B_{12}-Dependent Reductive Dehalogenases

The ability of methanogens and acetogens containing B_{12}-dependent reductive dehalogenases play an important role in the detoxification of aromatic and aliphatic chlorinated compounds [153, 242], which include "priority organic pollutants" (the so-called "dirty dozen") such as toxaphene [243]. Several B_{12}-dependent dehalogenases have been purified (see [244, 245] for further details) with nearly all containing one or more iron-sulfur clusters. In the anaerobic bacterium *Sulfurospirillum* (formerly *Dehalospirillum*) *multivorans*, which catalyzes the reductive dehalogenation of tetrachloroethene (PCE) and trichloroethene (TCE) to *cis*-1,2-dichloroethene (DCE) [245] (see Fig. 21), a novel corrinoid cofactor was found that had different catalytic properties to all known cobamides [246, 247]. This cofactor has since been identified as norpseudovitamin B_{12} (**9**, Co_β-cyano-7″-adeninyl-176-norcobinamide or 176-norpseudovitamin B_{12}) [37]. The corrinoid is a homo-

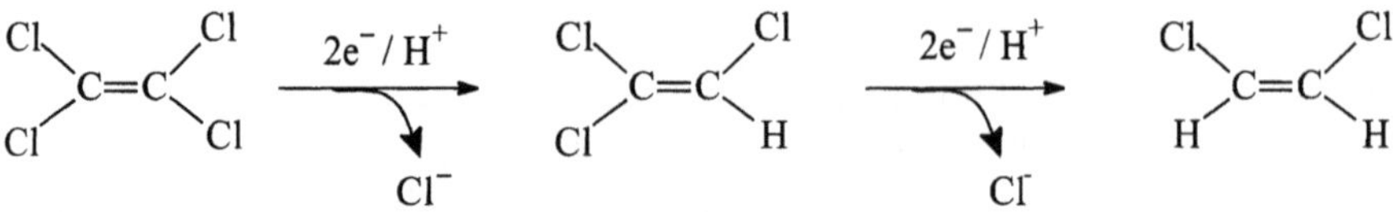

Fig. 21 Schematic illustration of the reductive dechlorination of tetrachloroethene (to trichloroethene and to *cis*-dichloroethene) by tetrachloroethene reductive dehalogenase

logue of pseudovitamin B_{12} (**10**) with the notable difference being the lack of the methyl group attached to carbon 176 and therefore is the first example of a naturally occurring "complete" B_{12} cofactor lacking a characteristic peripheral methyl group of the cobamide ligand.

The role of the B_{12}-cofactor in reductive dehalogenases appears to be significantly different from that of coenzyme B_{12}-dependent enzymes and B_{12}-dependent methyltransferases, but mechanistic studies are less developed. The B_{12}-catalyzed reductive dechlorination of PCE and TCE is probably the most studied. It is generally believed that the first step involves a dissociative electron transfer from cob(I)alamin (**40**$^-$) to PCE, leading to cob(II)alamin (**23**), chloride elimination and the formation of a trichlorovinyl radical [248, 249]. The further transformation of TCE is not as clear, although spectroscopic studies have suggested the formation of organometallic B_{12}-derivatives [250]. In the work of McCauley and coworkers [70] chlorovinylcobalamin (**32**) and vinylcobalamin (**33**) were synthesized and the possible roles of organocobamides in reductive dehalogenation reactions investigated. Their findings showed that chlorinated organometallic derivatives could be possible intermediates, as **40**$^-$ promoted reactions can reduce such compounds back to the active form of the catalyst [70].

9 B_{12} in Toxicology and Medicine

9.1 Toxicology

The high nucleophilicity of cob(I)alamin (**40**$^-$) makes it an ideal tool for the trapping and detection of electrophilic reagents. Such analytical methods are required for facilitating in vitro and in vivo studies of genotoxic compounds in cancer risk assessment. Most genotoxic compounds are directly electrophilically reactive or form reactive metabolites. Therefore it is of importance to have methods to study the formation and disappearance of reactive compounds/metabolites. The main advantage of cob(I)alamin to other standard nucleophiles, e.g., thiosulfate and nicotinamide [251, 252] is that it is, by orders of magnitude, more reactive.

The use of cob(I)alamin (**40**$^-$) as an analytical tool has been investigated in the trapping of oxiranes, metabolites of alkenes, to form alkyl-Cbl complexes (Fig. 22) [252–255]. It is presumed that the reaction proceeds according to an SN_2 reaction following attack at the least hindered carbon according to a base-catalyzed reaction [256, 257].

In the work of Fred et al. [255], the 1,2-epoxide metabolites (oxiranes) of 1,3-butadiene were studied. For each metabolite a specific alkyl-Cbl complex was formed and it was possible to discriminate between the products by HPLC-UV and by LC-MS [255]. The cob(I)alamin (**40**$^-$), used in this study, had the advantage of reacting about 400 000 times faster than, e.g., nicotinamide, and therefore gives a better on-the-spot account [252, 255].

In vitro studies have been described by Watson et al., who aimed to mimic the chemical reactions that could deplete vitamin B_{12} as a result of human exposure to electrophilic xenobiotics (styrene, chloroprene and 1,3-butadiene) [258]. It was shown that NADPH in liver microsomes converts hydroxycobalamin not only to cob(II)alamin (**23**) but also to cob(I)alamin (**40**$^-$), both of which react with the genotoxic epoxides of chloroprene and 1,3-butadiene to form their respective organometallic derivatives. With styrene, the metabolically formed styrene oxide reacted with **40**$^-$ alone. The findings also showed though, that when glutathione was added, these reactions are blocked due to the formation of glutathionylcobalamin and it was postulated that this acts as a protective reservoir for vitamin B_{12} (**1**) by inhibiting alkylation by epoxides and alkyl halides [258].

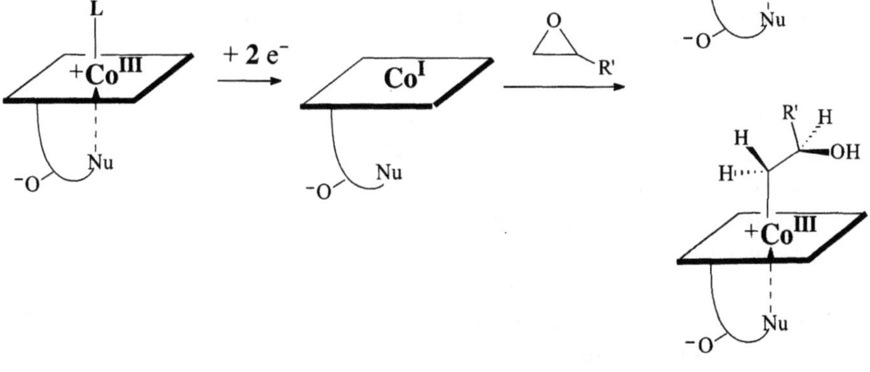

Fig. 22 Illustration of the reduction of a cobalamin to cob(I)alamin (**40**$^-$) and further reaction with an oxirane [255]

Another use of cob(I)alamin (40^-) as a tool in toxicology is for the analysis of DNA-phosphate adducts. Utilizing the nucleophilicity of 40^-, alkyl groups from the phosphotriester configuration in DNA were transferred with the formation of alkyl-cobalamin complexes [251].

9.2 Medical Aspects

An average supply of food with about 3–4 μg of vitamin B_{12} (**1**) is considered necessary for sustaining physical well being [148]. The proteins known to be involved in uptake and transport of cobalamin in humans are intrinsic factor (IF), transcobalamin (TC) and haptocorrin (HC) [148, 259]. These three soluble proteins ensure that the needed amount of cobalamin reaches the two intracellular enzymes methionine synthase (in cytosol) and methylmalonyl-CoA mutase (in mitochondria) [148, 260].

Tumors often require more vitamin B_{12} (**1**) than normal tissue, as rapidly dividing cells require abnormally high quantities of cobalamins for the synthesis of thymidine to support DNA replication prior to cell division [253]. It has therefore been hypothesized that cobalamin bound to transcobalamin (TC) could be used as a carrier molecule to target cancerous cells. The B_{12}-pharmacophore in covalently modified cobalamin derivatives is still favorably recognized by TC [261].

This observation has been used to target the transfer of chemotherapeutic agents [261–265] and radionuclides [39, 261, 266] to cancer cells and also cobalamin conjugates suitable for the imaging of cobalamin receptors [267, 268].

The strategies used require the derivatization of vitamin B_{12} (**1**), usually by one of the following methods: conjugation to peripheral acid groups (prepared by controlled amide hydrolysis), coordination to the 5′-OH group in the ribose ring of the nucleotide loop, or where the molecule is attached directly to the cobalt(III) center, normally by the reduction to Co(I) followed by oxidative alkylation resulting in the formation of an organometallic B_{12}-derivative.

Two alkylcob(III)alamin bioconjugates have been synthesized as potential chemotherapeutic agents. Firstly the bioconjugate of cob(III)alamin and the alkylating agent chlorambucil has been produced to give 2-[4-[4′-[bis(2-chloroethyl)amino]phenyl]butyroxy]-ethylcob(III)alamin [262]. The active agent was shown to be released by sonolysis, providing a possible new method for the selective release of anticancer drugs and thus potentially reduce systemic toxicity [262]. The second was where a colchicine derivative was conjugated to vitamin B_{12} through an acid-labile hydrazone linker [265]. The cobalamin moiety leads to preferential uptake of the cobalamin-colchicine prodrug by cancer cells, whereupon the hydrazone linker undergoes hydrolysis in the lysosome to unmask colchicine, which acts as a potent cytotoxin. The bioconjugate is stable in cell culture media and at neutral pH but undergoes hydrolysis with a half-life of 138 min at pH 4.5. The colchicine-cobalamin bioconjugate exhibits nanomolar LC_{50} values against breast, brain, and melanoma cancer cell lines in culture [265].

Fluorescent derivatives of cobalamin (CobalaFluors) have been prepared by linking fluorophores to cobalamin through a propylamide spacer. Fluorescein, naphthofluorescein, and Oregon Green derivatives have been prepared by the reaction of the fluorophore NHS-esters with organometallic β-(3-aminopropyl)cobalamin (**54**). The CobalaFluors have the potential to be suitable for the in vitro and in vivo imaging of transcobalamin receptors on cancer cells [267].

Kunze and coworkers have shown that the Co(III)-coordinated cyanide group of vitamin B_{12} bridges to rhenium(I) and technetium(I) centers, to yield robust complexes with the central structural feature $\{Co - CN - Re(Tc)\}$ (**12**) [39]. This concept allows the direct labelling of vitamin B_{12} (**1**) with $[^{99m}Tc(OH_2)(L^2)(CO)_3]$ complexes for radiodiagnosis or with rhenium as a mediator between **1** and additional biomolecules.

References

1. Kräutler B, Ostermann S (2003) In: Guilard R (ed) The Porphyrin Handbook, vol 11. Academic Press, San Diego, p 227
2. Rickes EL, Brink NG, Koniuszy FR, Wood TR, Folkers K (1948) Science 107:396
3. Smith EL, Parker LFJ (1948) Biochem J 43:VIII
4. Ellenbogen L, Cooper BA (1984) In: Machlin LJ (ed) Handbook of Vitamins, Nutritional & Clinical Aspects; Food Science and Technology. Marcel Dekker, New York, p 491
5. Zagalak B, Friedrich B (1979) Vitamin B_{12}. In: Proceedings of the 3rd European Symposium on Vitamin B_{12} and Intrinsic Factor). Walter de Gruyter, Berlin
6. Kräutler B, Arigoni D, Golding BT (eds) (1998) Vitamin B_{12} and B_{12}-Proteins. Wiley, Weinheim
7. Hodgkin DC, Pickworth J, Robertson JH, Trueblood KN, Prosen RJ, White JG (1955) Nature (London) 176:325
8. Hodgkin DC, Kamper J, Mackay M, Pickworth J, Trueblood KN, White JG (1956) Nature (London) 178:64
9. Lenhert PG, Hodgkin DC (1961) Nature 192:937
10. Eschenmoser A, Wintner CE (1977) Science 196:1410
11. Eschenmoser A (1982) Nova Acta Leopoldina 55:5
12. Woodward RB (1979) In: Friedrich W (ed) Vitamin B_{12}, Proceedings of the 3rd European Symposium on Vitamin B_{12} and Intrinsic Factor. Walter de Gruyter, Berlin, p 37
13. Battersby AR (1998) In: Kräutler B, Arigoni D, Golding BT (eds) Vitamin B_{12} and B_{12}-Proteins. Wiley, Weinheim, p 47
14. Blanche F, Cameron B, Crouzet J, Debussche L, Thibaut D, Vuilhorgne M, Leeper FJ, Battersby AR (1995) Angew Chem Int Ed Engl 34:383
15. Scott AI (1998) In: Kräutler B, Arigoni D, Golding BT (eds) Vitamin B_{12} and B_{12}-Proteins. Wiley, Weinheim, p 81

16. Scott AI, Roessner CA, Santander PJ (2003) In: Guilard R (ed) The Porphyrin Handbook, vol 12. Academic Press, San Diego, p 211
17. Drennan CL, Huang S, Drummond JT, Matthews RG, Ludwig ML (1994) Science 266:1669
18. Mancia F, Keep NH, Nakagawa A, Leadlay PF, McSweeney S, Rasmussen B, Diat O, Evans PR (1996) Structure 4:339
19. Reitzer R, Gruber K, Jogl G, Wagner UG, Bothe H, Buckel W, Kratky C (1999) Structure 7:891
20. Shibata N, Masuda J, Tobimatsu T, Toraya T, Suto K, Morimoto Y, Yasuoka N (1999) Structure 7:997
21. Sintchak MD, Arjara G, Kellogg BA, Stubbe J, Drennan CD (2002) Nature Struct Biol 9:293
22. Pratt JM (1972) Inorganic Chemistry of Vitamin B_{12}. Academic Press, New York
23. Friedrich W (1975) In: Dirscherl W (ed) Fermente, Hormone und Vitamine, vol III/2. Georg Thieme Verlag, Stuttgart, p 173
24. Dolphin D (ed) (1982) B_{12}. Wiley, New York
25. Banerjee R (ed) (1999) Chemistry and Biochemistry of B_{12}. Wiley, New York
26. Brown KL (2005) Chem Rev 105:2075
27. Glusker JP (1982) In: Dolphin D (ed) B_{12}, vol I. Wiley, New York, p 23
28. Kratky C, Kräutler B (1999) In: Banerjee R (ed) Chemistry and Biochemistry of B_{12}. Wiley, New York, p 9
29. Kräutler B (1998) In: Kräutler B, Arigoni D, Golding BT (eds) Vitamin B_{12} and B_{12}-Proteins. Wiley, Weinheim, p 517
30. Moss GP (1987) Pure Appl Chem 6:779
31. Kräutler B, Konrat R, Stupperich E, Färber G, Gruber K, Kratky C (1994) Inorg Chem 33:4128
32. Randaccio L, Furlan M, Geremia S, Šlouf M, Srnova I, Toffoli D (2000) Inorg Chem 39:3403
33. Stoeckli-Evans H, Edmond E, Hodgkin DC (1972) J Chem Soc Perkin Trans 2:605
34. Brown KL, Evans DR, Zubkowski JD, Valente EJ (1996) Inorg Chem 35:415
35. Brown KL, Zou X, Wu G-Z, Zubkowski JD, Valente EJ (1995) Polyhedron 14:1621
36. Gruber K, Jogl G, Klintschar G, Kratky C (1998) In: Kräutler B, Arigoni D, Golding BT (eds) Vitamin B_{12} and B_{12}-Proteins. Wiley, Weinheim, p 335
37. Kräutler B, Fieber W, Ostermann S, Fasching M, Ongania K-H, Gruber K, Kratky C, Mikl C, Siebert A, Diekert G (2003) Helv Chim Acta 86:3698
38. Kopf J, von Deuten K, Bieganowski R, Friedrich W (1981) Z Naturforsch 36c:506
39. Kunze S, Zobi F, Kurz P, Spingler B, Alberto R (2004) Angew Chem Int Ed Engl 43:5025
40. Kratky C, Färber G, Gruber K, Wilson K, Dauter Z, Noltin H-F, Konrat R, Kräutler B (1995) J Am Chem Soc 117:4654
41. Randaccio L, Geremia S, Stener M, Toffoli D, Zangrando E (2002) Eur J Inorg Chem 2002(1):93
42. Garau G, Geremia S, Marzilli LG, Nardin G, Randaccio L, Tauzher G (2003) Acta Cryst B59:51
43. Perry CB, Fernandes MA, Brown KL, Zou X, Valente EJ, Marques HM (2003) Eur J Inorg Chem 2003(11):2095
44. Venkatesan K, Dale D, Hodgkin DC, Nockols CE, Moor FH, O'Connor BH (1971) Proc Roy Soc London Ser A 323:455
45. Fischli A, Daly JJ (1980) Helv Chim Acta 63:1628
46. Kamiya K, Kennard O (1982) J Chem Soc Perkin Trans 1:2279

47. Markwell AJ, Pratt JM, Shaikjee MS, Toerien JG (1987) J Chem Soc Dalton Trans 1987(6):1349
48. Kräutler B, Caderas C, Konrat R, Puchberger M, Kratky C (1995) Helv Chim Acta 78:81
49. Lewis NJ, Nussberger R, Kräutler B, Eschenmoser A (1983) Angew Chem Int Ed Engl 22:736
50. Nussbaumer C, Arigoni D (1983) Angew Chem Int Ed Engl 22:737
51. Kräutler B, Keller W, Kratky C (1989) J Am Chem Soc 111:8936
52. Langen P, Lehmann M, Wilkinson C, Jogl G, Kratky C (1999) Acta Cryst D55:51
53. Kräutler B, Keller W, Hughes M, Caderas C, Kratky C (1987) J Chem Soc Chem Comm 1987(21):1678
54. Lenhert PG (1968) Proc Roy Soc London A303:45
55. Savage HFJ, Lindley PF, Finney JL, Timms PA (1987) Acta Cryst B43:296
56. Bouquiere JP, Finney JL, Lehmann MS, Lindley PF, Savage HFJ (1993) Acta Cryst B49:79
57. Ouyang L, Rulis P, Ching WY, Nardin G, Randaccio L (2004) Inorg Chem 43:1235
58. Brown KL, Cheng S, Zou X, Li J, Chen G, Valente EJ, Zubkowski JD, Marques HM (1998) Biochem 37:9704
59. Sando GN, Blakley RL, Hogenkamp HP, Hoffmann B (1975) J Biol Chem 250:8774
60. Krouwer JS, Holmquist B, Kipnes RS, Babior BM (1980) Bioch Biophys Acta 612:153
61. Pagano TG, Marzilli GL, Flocco MM, Tsai L, Carrell HL, Glusker JP (1991) J Am Chem Soc 113:531
62. Gschösser S, Hannak RB, Konrat R, Gruber K, Mikl C, Kratky C, Kräutler B (2005) Chem Eur J 11:81
63. Gruber K, Reitzer R, Kratky C (2001) Angew Chem Int Ed Engl 40:3377
64. Alcock NW, Dixon RM, Golding BT (1985) J Chem Soc Chem Commun 1985(9):603
65. Rossi M, Glusker JP, Randaccio L, Summers MF, Toscano PJ, Marzilli LG (1985) J Am Chem Soc 107:1729
66. Brown KL, Cheng S, Zou X, Zubkowski JD, Valente EJ, Knapton L, Marques HM (1997) Inorg Chem 36:3666
67. Wagner T, Afshar CE, Carrell HL, Glusker JP (1999) Inorg Chem 38:1785
68. Fasching M, Schmidt W, Kräutler B, Stupperich E, Schmidt A, Kratky C (2000) Helv Chim Acta 83:2295
69. McCauley KM, Pratt DA, Wilson SR, Shey J, Burkey TJ, van der Donk WA (2005) J Am Chem Soc 127:1126
70. McCauley KM, Wilson SR, van der Donk WA (2003) J Am Chem Soc 125:4410
71. Purcell KF, Kotz JC (1977) Inorganic Chemistry. Holt-Saunders, Philadelphia
72. De Ridder DJA, Zangrando E, Bürgi H-B (1996) J Mol Struct 374:63
73. Pett VB, Liebman MN, Murray-Rust P, Prasad K, Glusker JP (1987) J Am Chem Soc 109:3207
74. Brown KL (1999) In: Banerjee R (ed) Chemistry and Biochemistry of B_{12}. Wiley, New York, p 197
75. Kräutler B (1998) In: Kräutler B, Arigoni D, Golding BT (eds) Vitamin B_{12} and B_{12}-Proteins. Wiley, Weinheim, p 3
76. Fieber W, Hoffmann B, Schmidt W, Stupperich E, Konrat R, Kräutler B (2002) Helv Chim Acta 85:927
77. Konrat R, Tollinger M, Kräutler B (1998) In: Kräutler B, Arigoni D, Golding BT (eds) Vitamin B_{12} and B_{12}-Proteins. Wiley, Weinheim, p 349
78. Summers MF, Marzilli LG, Bax A (1986) J Am Chem Soc 108:2093
79. Bax A, Marzilli LG, Summers MF (1987) J Am Chem Soc 109:566

80. Benn R, Mynott R (1985) Angew Chem Int Ed Engl 24:333
81. Pagano TG, Yohannes PG, Hay BP, Scott JR, Finke RG, Marzilli LG (1989) J Am Chem Soc 111:1484
82. Tollinger M, Konrat R, Kräutler B (1999) Helv Chim Acta 82:1596
83. Kontaxis G, Riether D, Hannak RB, Tollinger M, Kräutler B (1999) Helv Chim Acta 82:848
84. Kräutler B, Caderas C (1984) Helv Chim Acta 67:1891
85. Randaccio L, Furlan M, Geremia S, Slouf M (1998) Inorg Chem 37:5390
86. Kräutler B (1987) Helv Chim Acta 70:1268
87. Kräutler B (1999) In: Banerjee R (ed) Chemistry and Biochemistry of B_{12}. Wiley, New York, p 315
88. Ragsdale SW (1991) Crit Rev Biochem Mol Biol 26:261
89. Steiger B, Ruhe A, Walder L (1990) Anal Chem 62:759
90. Lexa D, Savéant J-M (1983) Acc Chem Res 16:2235
91. Schrauzer GN, Deutsch E, Windgassen RJ (1968) J Am Chem Soc 90:2441
92. Chemaly SM, Pratt JM (1984) J Chem Soc Dalton Trans 4:595
93. Endicott JF, Netzel TL (1979) J Am Chem Soc 101:4000
94. Lexa D, Savéant JM, Zickler J (1977) J Am Chem Soc 99:2786
95. Lexa D, Savéant JM (1976) J Am Chem Soc 98:2652
96. de Tacconi NR, Lexa D, Savéant JM (1979) J Am Chem Soc 101:467
97. Faure D, Lexa D, Savéant JM (1982) J Electroanal Chem 140:269
98. Faure D, Lexa D, Savéant JM (1982) J Electroanal Chem 140:285
99. Faure D, Lexa D, Savéant JM (1982) J Electroanal Chem 140:297
100. Bard AJ, Faulkner LR (1980) Electrochemical Methods. Wiley, New York
101. Rubinson KA, Itabashi E, Mark HB (1982) Inorg Chem 21:3571
102. Rubinson KA, Parekh HV, Itabashi E, Mark HB (1983) Inorg Chem 22:458
103. Lexa D, Savéant JM (1975) J Chem Soc Chem Commun 1975(21):872
104. Lexa D, Savéant JM, Zickler J (1980) J Am Chem Soc 102:2654
105. Lexa D, Saveant J-M, Zickler J (1980) J Am Chem Soc 102:4851
106. Lexa D, Savéant JM (1978) J Am Chem Soc 100:3220
107. Zhiu D-L, Tinembart O, Scheffold R, Walder L (1990) Helv Chim Acta 73:2225
108. Martin BD, Finke RG (1992) J Am Chem Soc 114:585
109. Waddington MD, Finke RG (1993) J Am Chem Soc 115:4629
110. Hannak RB (1996) Dissertation, University of Innsbruck
111. Kräutler B, Dérer T, Liu P, Mühlecker W, Puchberger M, Gruber K, Kratky C (1995) Angew Chem Int Ed Engl 34:84
112. Hannak RB, Färber G, Konrat R, Kräutler B (1997) J Am Chem Soc 119:2313
113. Puchberger M, Konrat R, Kräutler B (2003) Helv Chim Acta 86:1453
114. Tollinger M, Dérer T, Konrat R, Kräutler B (1997) J Mol Catal A116:147
115. Brown KL (1982) In: Dolphin D (ed) B_{12}, vol I. Wiley, New York, p 245
116. Hogenkamp HPC (1982) In: Dolphin D (ed) B_{12}, vol I. Wiley, New York, p 295
117. Halpern J (1982) In: Dolphin D (ed) B_{12}, vol I. Wiley, New York, p 501
118. Golding BT (1982) In: Dolphin D (ed) B_{12}, vol I. Wiley, New York, p 543
119. Halpern J (1985) Science 227:869
120. Rétey J (1982) In: Dolphin D (ed) B_{12}, vol II. Wiley, New York, p 357
121. Rétey J (1998) In: Kräutler B, Arigoni D, Golding BT (eds) Vitamin B_{12} and B_{12}-Proteins. Wiley, Weinheim, p 273
122. Hay BP, Finke RG (1986) J Am Chem Soc 108:4820
123. Finke RG (1998) In: Kräutler B, Arigoni D, Golding BT (eds) Vitamin B_{12} and B_{12}-Proteins. Wiley, Weinheim, p 383

124. Matthews RG, Banerjee RV, Ragsdale SW (1990) BioFactors 2:147
125. Matthews RG (1999) In: Banerjee R (ed) Chemistry and Biochemistry of B_{12}. Wiley, New York, p 681
126. Sauer K, Thauer RK (1999) In: Banerjee R (ed) Chemistry and Biochemistry of B_{12}. Wiley, New York, p 655
127. Harms U, Thauer RK (1998) In: Kräutler B, Arigoni D, Golding BT (eds) Vitamin B_{12} and B_{12}-Proteins. Wiley, Weinheim, p 157
128. Brown KL, Zhou L, Zhao D, Cheng S, Zou X (1998) In: Kräutler B, Arigoni D, Golding BT (eds) Vitamin B_{12} and B_{12}-Proteins. Wiley, Weinheim, p 417
129. Pratt JM (1982) In: Dolphin D (ed) B_{12}, vol I. Wiley, New York, p 325
130. Grate JG, Schrauzer GN (1979) J Am Chem Soc 101:4601
131. Chemaly SM, Pratt JM (1980) J Chem Soc Dalton Trans 1980(11):2259
132. Schrauzer GN, Deutsch E (1969) J Am Chem Soc 91:3341
133. Bonnett R (1982) In: Dolphin D (ed) B_{12}, vol I. Wiley, New York, p 201
134. Kräutler B (1988) In: Glockling F (ed) The Biological Alkylation of Heavy Elements. Royal Society of Chemistry, London, p 168
135. Pailes WH, Hogenkamp HPC (1968) Biochem 7:4160
136. Milton P, Brown TL (1977) J Am Chem Soc 99:1390
137. Hogenkamp HPC, Bratt GT, Sun S (1985) Biochem 24:6428
138. Drennan CL, Matthews RG, Ludwig ML (1994) Curr Opin Struct Biol 4:919
139. Kräutler B (1995) In: Maskill H (ed) Organic Reactivity: Physical and Biological Aspects. Royal Society of Chemistry, London, p 209
140. Jensen MP, Halpern J (1999) J Am Chem Soc 121:2181
141. Mosimann H, Kräutler B (2000) Angew Chem Int Ed Engl 39:393
142. Galliker PK, Gräther O, Rümmler M, Fitz W, Arigoni D (1998) In: Kräutler B, Arigoni D, Golding BT (eds) Vitamin B_{12} and B_{12}-Proteins. Wiley, Weinheim, p 447
143. Cole AG, Yoder LM, Shiang JJ, Anderson NA, Walker LAI, Banaszak Holl MM, Sension RJ (2002) J Am Chem Soc 124:434
144. Tinembart O, Walder L, Scheffold R (1988) Ber Bunsenges Phys Chem 92:1225
145. Stupperich E, Konle R, Lehle M (1998) In: Kräutler B, Arigoni D, Golding BT (eds) Vitamin B_{12} and B_{12}-Proteins. Wiley, Weinheim, p 179
146. Renz P (1999) In: Banerjee R (ed) Chemistry and Biochemistry of B_{12}. Wiley, New York, p 557
147. Nexø E (1998) In: Kräutler B, Arigoni D, Golding BT (eds) Vitamin B_{12} and B_{12}-Proteins. Wiley, Weinheim, p 461
148. Stupperich E, Nexø E (1991) Eur J Biochem 199:299
149. Eschenmoser A (1988) Angew Chem Int Ed Engl 27:5
150. Hoffmann B, Oberhuber M, Stupperich E, Bothe H, Buckel W, Konrat R, Kräutler B (2000) J Bacteriol 182:4773
151. Limbach PA, Crain PF, McCloskey JA (1994) Nucleic Acids Res 22:2183
152. Poppe L, Hull WE, Nitsche R, Graf T, Stupperich E, Rétey J (1999) Helv Chim Acta 82:1250
153. Banerjee R, Ragsdale SW (2003) Annu Rev Biochem 72:209
154. Matthews RG (2001) Acc Chem Res 34:681
155. Bandarian V, Ludwig ML, Matthews RG (2003) Proc Nat Acad Sci USA 100:8156
156. Ragsdale SW, Kumar M, Zhao S, Menon S, Seravalli J, Doukov T (1998) In: Kräutler B, Arigoni D, Golding BT (eds) Vitamin B_{12} and B_{12}-Proteins. Wiley, Weinheim, p 167
157. Drennan et al. (2004) J Biol Inorg Chem 9:511
158. Ferry JG (1995) Annu Rev Microbiol 49:305

159. Doukov T, Seravalli J, Stezowski JJ, Ragsdale SW (2000) Structure 8:817
160. Zydowsky TM, Courtney LF, Frasca V, Kobayashi K, Shimizu H, Yuen LD, Matthews RG, Benkovic SJ, Floss HG (1986) J Am Chem Soc 108:3152
161. Zydowsky LD, Zydowsky TM, Haas ES, Brown JW, Reeve JN, Floss HG (1987) J Am Chem Soc 109:7922
162. Lebertz H, Simon H, Courtney LF, Benkovic SJ, Zydowsky LD, Lee K, Floss HG (1987) J Am Chem Soc 103:3173
163. Ludwig ML, Evans PR (1999) In: Banerjee R (ed) Chemistry and Biochemistry of B_{12}. Wiley, New York, p 595
164. Peariso K, Goulding CW, Huang S, Matthews RG, Penner-Hahn JE (1998) J Am Chem Soc 120:8410
165. Bandarian V, Pattridge KA, Lennon BW, Huddler DR, Matthews RG, Ludwig ML (2002) Nature Struct Biol 9:53
166. Drennan CL, Dixon MM, Hoover DM, Jarrett JT, Goulding CW, Matthews RG, Ludwig ML (1998) In: Kräutler B, Arigoni D, Golding BT (eds) Vitamin B_{12} and B_{12}-Proteins. Wiley, Weinheim, p 133
167. Marsh ENG, Holloway DE (1992) FEBS Lett 310:167
168. Evans JC, Huddler DP, Hilgers MT, Romanchuk G, Matthews RG, Ludwig ML (2004) Proc Nat Acad Sci USA 101:3729
169. Stupperich E, Eisinger HJ, Albrecht PJ (1990) Eur J Biochem 193:105
170. Kräutler B (1994) In: Schneider G (ed) Carbon Dioxide Fixation and Reduction in Biological and Model Systems: Proceedings of the Royal Swedish Academy of Sciences Nobel Symposium 1991. Oxford University Press, Oxford, p 83
171. Kräutler B, Kratky C (1996) Angew Chem Int Ed Engl 35:168
172. Jarrett JT, Amaratunga M, Drennan CL, Scholten JD, Sands RH, Ludwig ML, Matthews RG (1996) Biochem 35:2464
173. Marsh ENG, Drennan CL (2001) Curr Opin Struct Biol 5:499
174. Banerjee R, Chowdhury S (1999) In: Banerjee R (ed) Chemistry and Biochemistry of B_{12}. Wiley, New York, p 707
175. Buckel W, Bröker G, Bothe H, Pierik AJ (1999) In: Banerjee R (ed) Chemistry and Biochemistry of B12. Wiley, New York, p 757
176. Gruber K, Kratky C (2002) Curr Opin Struct Biol 6:598
177. Zerbe-Burckhardt K, Ratnatilleke A, Vrijbloed JW, Robinson JA (1999) In: Banerjee R (ed) Chemistry and Biochemistry of B_{12}. Wiley, New York, p 859
178. Toraya T (2003) Chem Rev 103:2095
179. Bandarian V, Reed GH (1999) In: Banerjee R (ed) Chemistry and Biochemistry of B_{12}. Wiley, New York, p 811
180. Frey PA, Chang CH (1999) In: Banerjee R (ed) Chemistry and Biochemistry of B_{12}. Wiley, New York, p 835
181. Baker JJ, Stadtman TC (1982) In: Dolphin D (ed) B_{12}, vol II. Wiley, New York, p 203
182. Stubbe J, Licht S, Gerfen G, Silva D, Booker S (1998) In: Kräutler B, Arigoni D, Golding BT (eds) Vitamin B_{12} and B_{12}-Proteins. Wiley, Weinheim, p 321
183. Rétey J (1999) In: Banerjee R (ed) Chemistry and Biochemistry of B_{12}. Wiley, New York, p 271
184. Rétey J (1990) Angew Chem Int Ed Engl 29:355
185. Bothe H, Bröker G, Müller U, Schall I, Textor S, Golding BT, Buckel W (1998) In: Kräutler B, Arigoni D, Golding BT (eds) Vitamin B_{12} and B_{12}-Proteins. Wiley, Weinheim, p 253
186. Beatrix B, Zelder O, Linder D, Buckel W (1994) Eur J Biochem 221:101
187. Marsh ENG (2000) Bioorg Chem 28:176

188. Chowdhury S, Banerjee R (2000) J Am Chem Soc 122:5417
189. Chowdhury S, Banerjee R (2000) Biochem 39:7998
190. Brooks AJ, Vlasie M, Banerjee R, Brunold TC (2004) J Am Chem Soc 126:8167
191. Golding BT, Anderson RJ, Ashwell S, Edwards CH, Garnett I, Kroll F, Buckel W (1998) In: Kräutler B, Arigoni D, Golding BT (eds) Vitamin B_{12} and B_{12}-Proteins. Wiley, Weinheim, p 201
192. Smith DM, Golding BT, Radom L (1999) J Am Chem Soc 121:1383
193. Loferer MJ, Webb BM, Grant GH, Liedl KR (2003) J Am Chem Soc 125:1072
194. Evans PR, Mancia F (1998) In: Kräutler B, Arigoni D, Golding BT (eds) Vitamin B_{12} and B_{12}-Proteins. Wiley, Weinheim, p 217
195. Gruber K, Kratky C (2001) In: Wieghart K (ed) Handbook of Metalloproteins. Wiley, Chichester, p 995
196. Mancia F, Evans PR (1998) Structure 6:711
197. Mancia F, Smith GA, Evans PR (1999) Biochem 38:7999
198. Barker HA, Weissbach H, Smyth RD (1964) J Biol Chem 239:3260
199. Buckel W, Golding BT (1996) Chem Rev Soc 25:329
200. Holloway DE, Marsh ENG (1994) J Biol Chem 269:20425
201. Barker HA, Weissbach H, Smith RD (1958) Proc Nat Acad Sci USA 44:1093
202. Marsh ENG, Ballou DP (1998) Biochem 37:11864
203. Cheng M-C, Marsh ENG (2004) Biochem 43:2155
204. Chih H-W, Marsh ENG (2000) J Am Chem Soc 122:10732
205. Wetmore SD, Smith DM, Golding BT, Radom L (2001) J Am Chem Soc 123:7963
206. Tollinger M, Konrat R, Hilbert BH, Marsh EN, Kräutler B (1998) Structure 6:1021
207. Hoffmann B, Tollinger M, Konrat R, Huhta M, Marsh ENG, Kräutler B (2001) ChemBioChem 2:643
208. Hoffmann B, Konrat R, Bothe H, Buckel W, Kräutler B (1999) Eur J Biochem 263:178
209. Tollinger M, Eichmüller C, Konrat R, Huhta M, Marsh ENG, Kräutler B (2001) J Mol Biol 309:777
210. Toraya T (1999) In: Banerjee R (ed) Chemistry and Biochemistry of B_{12}. Wiley, New York, p 783
211. Tobimatsu T, Hara T, Sakaguchi M, Kishimoto Y, Wada Y, Isoda M, Sakai T, Toraya T (1995) J Biol Chem 270:7142
212. Tobimatsu T, Azuma M, Matsubara H, Takatori H, Niida T, Nishimoto K, Satoh H, Hayashi R, Toraya T (1996) J Biol Chem 271:22352
213. Yamanishi M, Yamada S, Muguruma H, Murakani Y, Tobimatsu T, Ishida A, Yamauchi J, Toraya T (1998) Biochem 124:598
214. Abend A, Nitsche R, Bandarian V, Stupperich E, Rétey J (1998) Angew Chem Int Ed Engl 37:625
215. Yamanishi M, Yunoki M, Tobimatsu T, Sato H, Matsui J, Dokiya A, Iuchi Y, Oe K, Suto K, Shibata N, Morimoto Y, Yasuoka N, Toraya T (2002) Eur J Biochem 269:4484
216. Toraya T, Yoshizawa K, Eda M, Yamabe T (1999) J Biochem 126:650
217. Masuda J, Shibata N, Morimoto Y, Toraya T, Yasuoka N (2000) Structure 8:775
218. Speranza G, Buckel W, Golding BT (2004) J Porphyrins Phthalocyanines 8:290
219. Shibata N, Masuda J, Morimoto Y, Yasuoka N, Toraya T (2002) Biochem 41:12607
220. Liao D-I, Dotson G, Turner I, Reiss L, Emptage M (2003) J Inorg Biochem 93:84
221. Kaplan BH, Stadtman ER (1968) J Biol Chem 243:1794
222. Abend A, Bandarian V, Nitsche R, Stupperich E, Rétey J, Reed GH (1999) Arch Biochem Biophys 370:138
223. Faust LRP, Connor JA, Roof DM, Hoch JA, Babior BM (1990) J Biol Chem 265:12462
224. Semialjac M, Schwarz H (2002) J Am Chem Soc 124:8974

225. Wetmore SD, Smith DM, Bennett JT, Radom L (2002) J Am Chem Soc 124:14054
226. Semialjac M, Schwarz H (2003) J Org Chem 68:6967
227. Semialjac M, Schwarz H (2004) Chem Eur J 10:2781
228. Tang K-H, Harms A, Frey PA (2002) Biochem 41:8767
229. Tang K-H, Casarez AD, Wu W, Frey PA (2003) Arch Biochem Biophys 418:49
230. Chang CH, Frey PA (2000) J Biol Chem 275:106
231. Berkovitch F, Behshad E, Tang K-H, Enns EA, Frey PA, Drennan CL (2004) Proc Nat Acad Sci USA 101:15870
232. Blakley RL (1982) In: Dolphin D (ed) B_{12}, vol II. Wiley, New York, p 381
233. Stubbe JA, van der Donk WA (1998) Chem Rev 98:705
234. Stubbe J (2000) Curr Opin Struct Biol 10:731
235. Lawrence CC, Gerfen GJ, Samano V, Nitsche R, Robins MJ, Rétey J, Stubbe J (1999) J Biol Chem 274:7039
236. Sintchak MD, Arjara G, Kellogg BA, Stubbe J, Drennan CD (2002) Nature Struct Biol 9:293
237. Licht SS, Lawrence CC, Stubbe J (1999) Biochem 38:1234
238. Uhlin U, Eklund H (1994) Nature 370:533
239. Licht SS, Booker S, Stubbe J (1999) Biochem 38:1221
240. Griller D, Wayer DDM (1989) Pure & Appl Chem 61:717
241. Bresciani-Pahor N, Forcolin M, Marzilli LG, Randaccio L, Summers MF, Toscano PJ (1985) Coord Chem Rev 63:1
242. Holliger C, Wohlfahrt G, Diekert G (1999) FEMS Microbiol Rev 22:383
243. Ruppe S, Neumann A, Diekert G, Vetter W (2004) Environ Sci Technol 38:3063
244. Maillard J, Schumacher W, Vazquez F, Regeard C, Hagen WR, Holliger C (2003) Appl Environ Microbiol 69:4628
245. Wohlfahrt G, Diekert G (1999) In: Banerjee R (ed) Chemistry and Biochemistry of B_{12}. Wiley, New York, p 871
246. Neumann A, Siebert A, Trescher T, Reinhardt S, Wohlfahrt G, Diekert G (2002) Arch Microbiol 177:420
247. Siebert A, Neumann A, Schubert T, Diekert G (2002) Arch Microbiol 178:443
248. Glod G, Angst W, Holliger C, Schwarzenbach RP (1997) Environ Sci Technol 31:253
249. Shey J, van der Donk WA (2000) J Am Chem Soc 122:12403
250. Glod G, Brodmann U, Angst W, Holliger C (1997) Environ Sci Technol 31:3154
251. Haglund J, Rafiq A, Ehrenberg L, Golding BT, Tørnqvist M (2000) Chem Res Toxicol 13:253
252. Haglund J, Magnusson A-L, Ehrenberg L, Tørnqvist M (2003) Toxicol Environ Chem 85:81
253. Schneider Z, Stroinski A (1987) Comprehensive B_{12}: Chemistry, Biochemistry, Nutrition, Ecology, Medicine. Walter de Gruyter, Berlin
254. Alsberg T, Minten J, Haglund J, Tørnqvist M (2001) Rapid Commun Mass Spectrom 15:2438
255. Fred C, Haglund J, Alsberg T, Rydberg P, Minten J, Tørnqvist M (2004) J Sep Sci 27:607
256. Zhou DL, Walder R, Scheffold R (1992) Helv Chim Acta 75:995
257. Fessenden RJ, Fessenden JS (1994) Organic Chemistry. Brooks/Cole Publishing Company, Pacific Grove CA, p 320
258. Watson WP, Munter T, Golding BT (2004) Chem Res Toxicol 17:1562
259. Alpers DH, Russel-Jones GJ (1999) In: Banerjee R (ed) Chemistry and Biochemistry of B_{12}. Wiley, New York, p 411

260. Rosenblatt DS, Fenton WA (1999) In: Banerjee R (ed) Chemistry and Biochemistry of B_{12}. Wiley, New York, p 367
261. Hogenkamp HPC, Collins DA, Grissom CB, West FG (1999) In: Banerjee R (ed) Chemistry and Biochemistry of B_{12}. Wiley, New York, p 385
262. Howard WA Jr, Bayomi A, Natarajan E, Aziza MA, El-Ahmady O, Grissom CB, West FG (1997) Bioconjugate Chem 8:498
263. Hogenkamp HPC, Collins DA, Live D, Benson LM, Naylor S (2000) Nucl Med Biol 27:89
264. Bauer JA, Morrison BH, Grane RW, Jacobs BS, Dabney S, Gamero AM, Carnevale KA, Smith DJ, Drazba J, Seetharam B, Lindner DJ (2002) J Natl Cancer Inst 94:1010
265. Bagnato JD, Eilers AL, Horton RA, Grissom CB (2004) J Org Chem 69:8987
266. Hogenkamp HPC, Collins DA (1998) In: Kräutler B, Arigoni D, Golding BT (eds) Vitamin B_{12} and B_{12}-Proteins. Wiley, Weinheim, p 505
267. Smeltzer CC, Cannon MJ, Pinson PR, Munger JD Jr, West FG, Grissom CB (2001) Org Lett 3:799
268. Adamczyk M, Johnson DD, Mattingly PG, Moore JA, Pan Y (2004) Bioorg Med Chem Lett 14:3917

Top Organomet Chem (2006) 17: 57–82
DOI 10.1007/3418_003

Published online: 12 April 2006

Catalytic Nickel–Iron–Sulfur Clusters: From Minerals to Enzymes

Anne Volbeda · Juan C. Fontecilla-Camps (✉)

Laboratoire de Cristallographie et de Cristallogenèse des Protéines,
Institut de Biologie Structurale J.P. Ebel (CEA-CNRS-UJF), 41 rue Jules Horowitz,
38027 Grenoble Cédex 1, France
juan.fontecilla@ibs.fr

Abstract The geochemical theory of the origin of life proposes that primordial, pre-biotic reactions were carried out in a metal-sulfide-rich environment similar to that found near hot springs at the ocean floor. Many contemporary experiments have shown that reactions reminiscent of those carried by extant anaerobic microorganisms involving gases such as CO, CO_2 and H_2, can indeed take place abiotically in the presence of iron and nickel sulfides. Here we discuss some of these reactions and compare them to those catalyzed by NiFeS-containing enzymes. In addition, we compare three NiFeS active sites and their protein environment and show that they share a significant number of structural features. We also comment on possible catalytic mechanisms.

Keywords Acetyl coenzyme A synthase · Carbon monoxide dehydrogenase · Geochemical theory of the origin of life · Hydrogenase · NiFeS-enzymes

1 Introduction

Many anaerobic microorganisms can use CO or CO_2 as a sole source of carbon and CO and/or H_2 for the generation of energy [1]. Thus, acetogens generate acetyl coenzyme A, an activated acetic acid that serves as a "universal" precursor for the generation of biomass (see below), and acetic acid from two CO_2 molecules [2]. Carboxydotrophic bacteria oxidize CO to CO_2 generating two reducing equivalents [3] but CO_2 is also used by methanogens as the final electron acceptor [4]. The above reactions are coupled to the generation of either a Na^+ or H^+ gradient across the cytoplasmic membrane [5–7]. The ion gradient may be then either converted to chemical energy as ATP or be used to drive endergonic reactions such as the synthesis of acetyl-coenzyme A. The latter is generated directly in the supposedly very ancient Wood/Ljungdahl pathway of carbon fixation [1, 8] that is used by acetogens:

$$2CO_2 + HSCoA + 4H_2 \rightleftharpoons CH_3COSCoA + 3H_2O\,. \tag{1}$$

The central enzyme of this pathway is the bifunctional carbon monoxide dehydrogenase/acetyl coenzyme A synthase (CODH/ACS). The CODH active site catalyzes the reduction of CO_2 to CO ($E^{0'} = -512$ mV):

$$CO_2 + 2H^+ + 2e^- \rightleftharpoons CO + H_2O\,. \tag{2}$$

When the CO concentration is kept very low, as observed under normal turnover conditions of CODH/ACS [9, 10], H_2 may be used as the electron donor, after its oxidation by a hydrogenase ($E^{0'} = -414$ mV):

$$H_2 \rightleftharpoons 2H^+ + 2e^-\,. \tag{3}$$

ACS takes up the CO produced by CODH to catalyze the non-redox reaction:

$$CH_3{}^+ + CO + SCoA^- \rightleftharpoons CH_3COSCoA\,. \tag{4}$$

The methyl cation is derived from a second CO_2 molecule through a series of two-electron reduction reactions mediated by tetrahydrofolate-containing enzymes.

The active sites of CODH and ACS contain NiFeS clusters that are somewhat similar to the active site of [NiFe]-hydrogenase [11]. It has been noted recently that these biological NiFeS clusters resemble the Ni-containing form of the mineral greigite, which has been assigned an important role in the geochemical theory of the origin of life [12]. Because thiolate-ligated [4Fe – 4S] clusters form spontaneously from a solution of $FeCl_2$, HS^- and $HOCH_2CH_2SH$ [13], it is tempting to postulate that such mineral clusters were incorporated into oligopeptides that predated CODH, ACS and [NiFe]-hydrogenase. Here we will (i) discuss the origin of mineral FeS and NiFeS clusters and their catalytic properties within the context of carbon fixation as activated acetic acid; and (ii) compare these centers to those found in the highly complex extant enzyme active sites.

2 Mineral FeS and NiFeS Clusters

The conditions under which life on Earth first developed are currently assumed to have been very harsh, so that the only suitable place for the origin of life may have been the ocean floor [14]. Very hot (> 350 °C) acidic springs were a source of Ni(II) and Fe(II) sulfides, most of which, because of the almost total absence of O_2, would have remained in solution in the HCO_3^--containing acidic ocean. Wächtershäuser has proposed that the exergonic formation of pyrite from FeS provided a source of reducing power for life emergence [15]:

$$FeS + H_2S \rightleftharpoons FeS_2 + H_2(g) . \quad (5)$$

The reported standard free energy (ΔG^0, at 25 °C) of this reaction is – 38.4 kJ/mol [16]. Pyrite-pulled metabolism would have required acidic pH and a high H_2S concentration. In experiments carried out under such conditions, in the presence of pyrrhotite ($Fe_{1-x}S$) at 50–100 °C, CH_3SH was generated from CO_2 and H_2S [17]. After H_2 production through pyrite formation as in Eq. 5 the reaction may be written as:

$$CO_2 + H_2S + 3H_2 \rightleftharpoons CH_3SH + 2H_2O . \quad (6)$$

Subsequently, it was observed that an equimolar amount of precipitated NiS and FeS catalyzed the formation of acetate from CO and CH_3SH at 100 °C with an optimal pH of 6.5 [18], which is slightly basic at this temperature [19]:

$$CH_3SH + CO + H_2O \rightleftharpoons CH_3COO^- + H^+ + H_2S . \quad (7)$$

In addition, a small amount of the thioester CH_3COSCH_3 was also formed, which was proposed to act as an activated acetic acid intermediate in the mechanism of Eq. 7, in analogy with the formation of acetyl-CoA catalyzed by ACS (Eq. 1). In the absence of Ni no reaction was observed, indicating that Fe alone does not afford carbon fixation under the reported conditions. Table 1 shows the free energies of these and other selected reactions.

More recently, iron sulfide has been shown to play a catalytic role in the synthesis of pyruvate from alkyl thiols and carbon monoxide at 250 °C and pressures between 500 and 2000 bar (conditions that could be found at the bottom of the ocean or in a shallow oceanic crust) [20]. The overall reaction may be written as:

$$CH_3SH + 2CO + H_2O \rightleftharpoons CH_3COCOO^- + H^+ + H_2S + H_2 . \quad (8)$$

A number of carbonylated organometallic intermediates were also detected. The solubility of CO is greatly increased at high pressure and this may favor reaction Eq. 8. Although the overall reaction is exergonic (Table 1), the yield of pyruvate, which plays a central role in many biosynthetic pathways, was

Table 1 Free energies of reactions involving simple carbon compounds

Reaction	$\Delta G^{0'}$ (kJ/mol) [a]	ΔG (kJ/mol) [b]
$2CO_2 + HSCoA + 4H_2 \rightleftharpoons CH_3COSCoA + 3H_2O$	- 146.7	- 45.7
$CH_3COSCoA + H_2O \rightleftharpoons HSCoA + CH_3COO^- + H^+$	- 35.7	- 52.8
$CO_2 + H_2S + 3H_2 \rightleftharpoons CH_3SH + 2H_2O$	- 121.3	- 52.0
$CH_3SH + CO + H_2O \rightleftharpoons CH_3COO^- + H^+ + H_2S$	- 72.2	- 70.5
$CH_3SH + 2CO + H_2O \rightleftharpoons CH_3COCOO^- + H^+ + H_2S + H_2$	- 39.5	- 37.8
$CO + H_2O \rightleftharpoons CO_2 + H_2$	- 11.1	- 23.9
$2CO_2 + 4H_2 \rightleftharpoons CH_3COO^- + H^+ + 2H_2O$	- 182.4	- 98.5
$CO_2 + 4H_2 \rightleftharpoons CH_4 + 2H_2O$	- 193.7	- 126.6
$CO_2 + H_2 \rightleftharpoons HCOO^- + H^+$	- 22.5	- 7.9
$CH_3COO^- + H^+ \rightleftharpoons CO_2 + CH_4$	- 11.3	- 28.1
$3CO_2 + 5H_2 \rightleftharpoons CH_3COCOO^- + 3H_2O + H^+$	- 156.6	- 41.0
$CH_3COO^- + CO_2 + H_2 \rightleftharpoons CH_3COCOO^- + H_2O$	25.8	57.5

[a] Standard free energy (25 °C, 1 bar, pH 7, 1 molar activities and unit activity for pure water) calculated from tabulated free energies of formation of the reactants [19, 146, 147], corrected for pH 7 (where appropriate): $RT\ln 10^{-7} = -39.9$ kJ/mol. Note that a 1 M concentration for the dissolved gases is only possible at pressures much higher than 1 bar.
[b] Calculated free energy ($\Delta G = \Delta G^0 + RT\ln K$) at pH 7 and 25 °C at physiologically more relevant activities (taking gas solubility into account) at 1 atm: $[CO_2] = 5.6$ mM, $[H_2] = [CO] = 0.5$ mM, $[CH_4] = 0.2$ mM, $[CH_3COO^-] = [H_2S] = [HCOO^-] = [CH_3SH] = [CH_3COCOO^-] = [CH_3COSCoA] = [HSCoA] = 1$ mM.

very low. The potential of various transition metal sulfides to fix carbon was tested under similar conditions [21], using the following reaction:

$$CH_3(CH_2)_8SH + CO + H_2O \rightleftharpoons CH_3(CH_2)_8COOH + H_2S\,. \quad (9)$$

The highest yields of the decanoate product were observed with nickel and cobalt sulfides, but significant activity was also observed with pyrite. In addition, the formation of methyl nonyl sulfide was observed with the methyl group originating from the reduction of CO. In this case, copper and iron sulfides were the most effective catalysts. It should be noted, however, that many important bio-molecules, such as thioesters and nucleic acids, are unstable in hot water, making some aspects of the evolution of life as we know it highly unlikely at high temperatures.

Russell and co-workers [14] have argued that the metal sulfides providing putative catalytic surfaces for abiotic carbon fixation would not have precipitated from the solutions that emanated from hot acidic springs into the equally acidic and anoxic ocean. Instead, a moderate temperature alkaline spring would have provided much more likely substrates for the emergence of life [22]. Alkaline springs originate from the convection of ocean water within a hot ocean crust containing minerals such as $MgFeSiO_3$ (orthopyroxene) and $CaMg_6FeSi_{12}O_{24}$ (diopside). Their reaction with water forms $Mg_3Si_2O_5(OH)_4$ (serpentine), SiO_2 (silica), Fe_3O_4 (magnetite), OH^- and H_2. Reaction of another mineral, olivine ($MgFeSiO_4$), with ocean water yields similar products:

$$3MgFeSiO_4 + 4H_2O \rightleftharpoons 3SiO_2 + Fe_3O_4 + 3Mg^{2+} + 6OH^- + H_2 \,. \tag{10}$$

The resulting alkaline and reducing fluid dissolves sulfide minerals present in the crust and, consequently, these hydrothermal systems carry high concentrations of HS^-. Precipitation of HS^- with Fe(II) present in the ocean would have led to the formation of bubbles of hydrothermal fluid enclosed by semi-permeable membranes consisting mainly of FeS. Simple compounds such as CO, CN^-, NH_3, CH_3SH and HCHO would have been significant additional constituents of the hydrothermal fluid, which upon mixing with CO_2-containing ocean water provided the building blocks for pre-biotic reactions. The FeS membranes would have separated a mildly oxidizing acidic ocean solution from a reducing alkaline spring solution, thus creating a gradient in the form of both a proton motive force and an electrostatic potential. This is postulated to have provided favorable conditions for endergonic reactions to occur, as in living cells. For example, if hydrogen oxidation (Eq. 3) took place in a compartment at pH 10, the potential of the resulting electrons would be low enough to reduce CO_2 (Eq. 2) in another compartment at pH 6, assuming the existence of a conducting FeS membrane between the two compartments. Because all these reactions would have taken place within the iron-sulfur vesicles, a rapid escape of products would have been prevented, thereby allowing the gradual formation of more complex molecules [23].

Although the exact structure of the proposed iron-sulfur bubbles remains undefined, FeS should have initially precipitated as disordered mackinawite [24]. This is a highly reactive phase, which gradually converts to more stable species such as greigite (Fe_3S_4) and pyrite (FeS_2). The latter mineral plays a prominent role in Wächtershäuser's theory of pyrite-pulled surface metabolism [25] but it is not expected to be a good catalyst, because in its crystal structure each Fe(II) is bound by six sulfur pairs [26] and the resulting octahedral coordination does not leave room for an additional ligand to iron. However, surface defects in the crystal lattice could liberate metal-binding coordination sites with catalytic potential. Because aldehydes seem both to inhibit pyrite formation and enhance greigite formation [27], and simple aldehydes were plausible components of the hydrothermal fluid, it has been postulated that the iron-sulfur membranes contained at least a fraction of greigite. In this respect it is also interesting to note that both mackinawite and greigite, but not pyrite, are found in present-day magnetotactic bacteria [28]. Other metals such as nickel and cobalt may also have been incorporated into the pre-biotic FeS membrane structures. For example, in the presence of nickel, greigite may be converted from the Fe_3S_4 to the $NiFe_5S_8$

form [29]. In the laboratory, synthesis of peptide bonds in the presence of CO has been observed under alkaline conditions in the presence of a FeS/NiS mixture of, however, ill-defined structure [30].

In conclusion, many of the conditions required to support reactions resembling those taking place in extant metabolism may have existed near the ocean floor of the early Earth. Obviously, the emergence of life would have involved many other reactions and, more importantly, their regulation. The fact is that, although there are many theories about the origin of life (e.g. [31–33]), very little is actually understood about its initial conditions [34]. At any rate, there are surprising similarities between some of the molecules arising from the abiotic chemistry discussed above and the NiFeS active sites of enzymes that function in anaerobic autotrophic carbon fixation. These sites will be the subject of the following reviews.

3 NiFeS Clusters in Enzymes

Three enzymes are known to use NiFeS active sites to catalyze reactions that involve simple, primordial, gases: carbon monoxide dehydrogenase (CODH) reduces CO_2 to CO (Eq. 2, see introduction), acetyl coenzyme A synthase (ACS) combines CO with a methyl group (Eq. 4) and NiFe hydrogenase oxidizes H_2 (Eq. 3). Depending on either the organism or the physiological requirements, these enzymes can catalyze the reverse reactions as well. In acetogens like *Moorella (M.) thermoacetica*, ACS and CODH constitute the α and β-subunits of an $\alpha_2\beta_2$ bi-functional enzyme complex (see [35] and references therein) that catalyzes the sum of Eqs. 2 and 4:

$$CO_2 + CH_3^+ + SCoA^- + 2H^+ + 2e^- \rightleftharpoons CH_3COSCoA + H_2O\,. \tag{11}$$

In methanogens like *Methanosarcina thermophila,* they are part of the so-called acetyl-CoA decarbonylase/synthase (ACDS) $(\alpha\beta\gamma\delta\varepsilon)_8$ multi-enzyme complex that allows these organisms to grow on acetate, after it is first converted to acetyl-CoA:

$$\begin{aligned} &CH_3COSCoA + H_4SPt + H_2O \\ &\rightleftharpoons SCoA^- + CO_2 + CH_3 - H_4SPt + 2H^+ + 2e^-\,. \end{aligned} \tag{12}$$

H_4SPt is the tetrahydrosarcinapterin cofactor. The $\alpha_2\varepsilon_2$ units correspond to CODH, β is ACS, $\gamma\delta$ is a corrinoid cobalt-containing iron–sulfur protein (CoFeSP) that transfers a methyl group to H_4SPt. The electron acceptor is a ferredoxin (see [36] and references therein). A similar complex is used by obligate chemo-autotrophic methanogens, such as *Methanococcus jannaschii,* to catalyze the formation of acetyl-CoA from CO_2 and H_2 in the reverse reaction [35].

Methanogens from the genus *Methanosarcina* have one soluble and two membrane-bound [NiFe] hydrogenases [6,37]. The latter are probably involved in energy conservation through the generation of a proton gradient. The acetogen *M. thermoacetica* may also contain several hydrogenases, but they have not been well characterized [2]. The presence of several hydrogenases is typical of a large number of microorganisms [38] and in many cases they are directly associated with other enzymes. An example of this is the coupling of CO oxidation and H_2 production by a [NiFe] hydrogenase and CODH in *Carboxydothermus (C.) hydrogenoformans* [39] and *Rubrivivax (R.) gelatinosus* [7]:

$$CO + H_2O \rightleftharpoons CO_2 + H_2 \,. \tag{13}$$

Here, we will focus on those NiFe-containing enzymes for which crystal structures have been reported.

3.1 [NiFe] Hydrogenases

Heterodimeric [NiFe]-hydrogenase crystal structures have been reported for four closely related sulfate-reducing bacteria from *Desulfovibrio* sp.: *D. gigas* [40,41], *D. vulgaris* (Miyazaki) [42–44], *D. fructosovorans* [45,46] and *D. desulfuricans* [47]. Overall, the structures are very similar being roughly

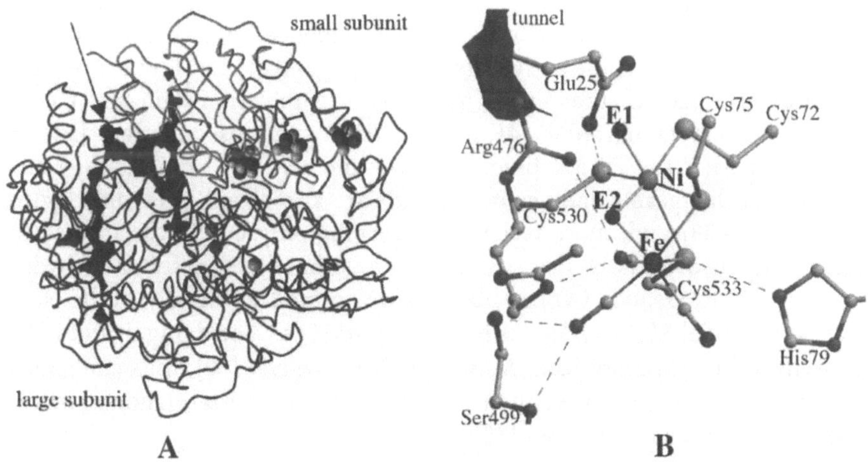

Fig. 1 Structure of [NiFe]-hydrogenase. **A** Polypeptide fold. The *arrow* indicates a hydrophobic tunnel network, shown in *dark grey*. Spheres highlight metal and inorganic sulfur sites: three FeS clusters in the small subunit and a Mg-site as well as the Ni – Fe active site in the large subunit. **B** Zoomed depiction of the active site, shown as a ball-and-stick model. *Dashed lines* indicate putative H-bonds. Exogenous ligand binding sites are labeled E1 and E2

spherical with a radius of about 35 Å (Fig. 1A). The NiFe active site is located close to the molecular center in the large subunit, which also contains a cation site close to the surface (either Mg or Fe, see below). A water ligand to the cation may be involved in a putative proton transfer pathway after molecular hydrogen cleavage from the active site, according to Eq. 3 [48]. The small subunit transfers electrons between the active site and the molecular surface through a proximal [4Fe – 4S], a median [3Fe – 4S] and a distal [4Fe – 4S] cluster. The structure of the related [NiFeSe]-hydrogenase of *Desulfomicrobium (Dm.) baculatum* shows several differences [49]: (i) a [4Fe – 4S] cluster replaces the median [3Fe – 4S] cluster of the other enzymes; (ii) a selenocysteine replaces a terminal cysteine ligand to the active site Ni; and (iii) Fe replaces Mg in the large subunit. A remarkable feature, common to the five enzymes of known structure, is the presence of a conserved, largely hydrophobic, tunnel network connecting the active site to several sites at the molecular surface. Crystal xenon-binding experiments, and subsequent molecular dynamics simulations, have shown that these tunnels are likely to provide access (and exit) pathways for H_2 gas [45].

The active site Ni and Fe ions are bridged by two cysteine thiolates. The Ni also has two additional terminal cysteine thiolate ligands, leaving two sites, here called E1 and E2, potentially available for exogenous molecules (Fig. 1B). The terminal E1 site is *trans* to the apical thiolate ligand and perpendicular to the plane defined by the three other S ligands, whereas the E2 site lies in this plane and bridges Ni and Fe. An exogenous Ni – Fe bridging ligand has only been observed in unready and ready states of oxidized inactive enzyme where it has been assigned to either an oxygen species (OH^- or OOH^-) [50, 51] or a sulfur atom [42, 47]. There is also recent evidence for a chemical modification of up to two of the active site cysteine ligands [50, 51], but it remains to be determined whether this could be due to radiation damage. A combination of crystallographic [41, 46] and FTIR spectroscopic data [52–55] has shown that the Fe ion binds one CO and two CN^- molecules. These two non-exchangeable endogenous ligands are also found in Fe-only hydrogenases [56–58] and, more recently concerning only CO, in the iron-sulfur cluster-free hydrogenase [59]. CO is also a competitive inhibitor of hydrogenases and exogenous CO has been shown to bind terminally to the Ni E1 site [44]. Special EPR techniques have detected H_2 as a bridging ligand in the H_2-sensing hydrogenase from *Ralstonia (R.) eutropha* [60], thus confirming our initial speculation based on the first hydrogenase crystal structure that E2 could bind a hydrogen species [40].

The presence of CN^- and CO ligands of the active site Fe raises the question as to how these potentially toxic molecules are incorporated in the enzyme. Significant progress has been made in the elucidation of the biosynthetic pathway of cyanide from a carbamoyl phosphate precursor [61]. On the other hand, the origin of the CO ligand remains unclear [62]. Many proteins are involved in the biosynthesis and the maturation of the active site of [NiFe]-

hydrogenases, but their specific roles have not yet been completely elucidated. For more details on these interesting aspects, the reader is referred to a recent review by Vignais and Colbeau [63].

A wealth of spectroscopic (EPR, FTIR, Mössbauer, XAS), kinetic and mutated enzyme data has been published concerning the Ni – Fe active site and many review articles are available (e.g. [11, 48, 64–72]). In addition, many density functional theory (DFT) studies of the Ni – Fe active site have been reported, but consensus on the nature of the various structures and on the catalytic mechanism is lacking (e.g. [73–79]). These theoretical studies will not be further discussed here.

A summary of most of the known states of standard [NiFe] hydrogenases and their proposed relationships is depicted in Fig. 2. Four $S = 1/2$ paramagnetic states have been detected by EPR; they are called Ni – A, Ni – B, Ni – C and Ni – L because the spin density is mainly localized on the Ni. Five diamagnetic states, called Ni – SU, Ni – "S", Ni – SI, SI – CO and Ni – R, have been characterized by FTIR, thanks to the vibration bands of the triple bonds in the CO and CN^- ligands found in the 1900–2100 cm^{-1} range. The active Ni – C and Ni – R states are directly involved in the catalytic cycle whereas the unready Ni – A and Ni – SU states are inactive and require a long reductive activation. The ready Ni – B and Ni – SI states are also inactive but can be immediately activated by H_2 in the absence of O_2. An additional active

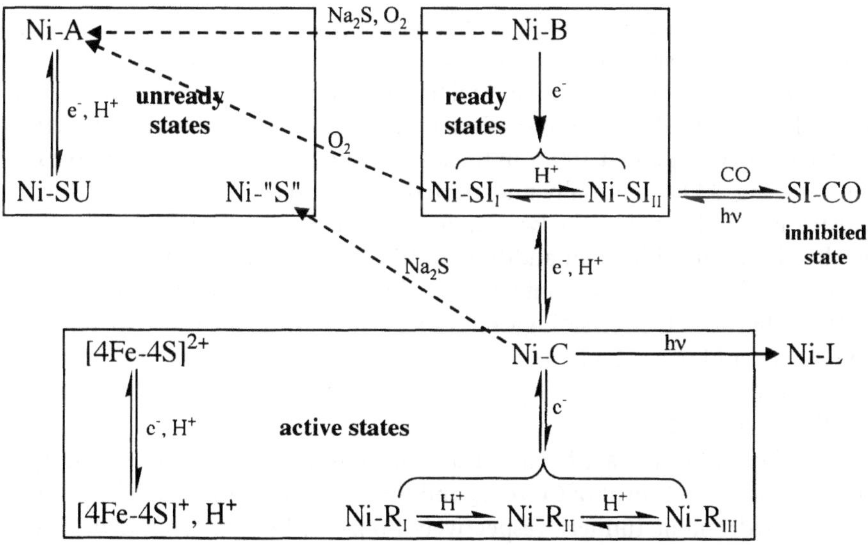

Fig. 2 Overview of the stable intermediates of [NiFe]-hydrogenase. The EPR-silent Ni – "S" state is obtained by treating active enzyme anaerobically with Na_2S [80]. Aerobic treatment of the ready Ni – B state with Na_2S leads to its conversion to the unready Ni – A state [51]. The unready state can also be obtained by addition of O_2 to enzyme in a Ni – SI state and with a reduced [3Fe – 4S] cluster (e.g. [148])

Ni – SI state with an FTIR spectrum identical to that of ready enzyme has also been characterized (not shown) [80]. The paramagnetic Ni – L state is obtained upon photolytic cleavage of the bound putative hydrogen species of the Ni – C species. Finally, SI – CO is a CO-inhibited state (e.g. [81, 82]). The picture is further complicated by the presence of different protonation states in some of these stable intermediates [54, 80] and by more than one Ni – L species [83].

Synthetic chemists have tried for many years to obtain simple structural and functional models of hydrogenases, but with limited success, especially in the case of the [NiFe] enzymes (see [84–86] for recent reviews). One possible reason for this situation is the difficulty in obtaining stable complexes with two orthogonal (*cis*) Ni empty coordination sites, a characteristic of the active site of all the [NiFe] hydrogenase crystal structures reported so far. Another problem for biomimetic modeling is the tendency of thiolates to bridge metal ions, leading to the formation of polynuclear clusters. On the other hand, good progress has been made in the understanding of the redox properties of Ni and the modeling of the iron center and its ligands. Furthermore, many of the properties of [NiFe] hydrogenase biomimetic models may also be relevant for the NiFeS active sites of CODH and ACS discussed below.

3.2
Nickel-Containing Carbon Monoxide Dehydrogenases

Four crystal structures of Ni CODHs have been determined from the following organisms: *C. hydrogenoformans* [87, 88], *Rhodospirillum (R.) rubrum* [89] and the bifunctional CODH/ACS from *M. thermoacetica* [90, 91]. In each case CODH has a very similar homodimeric quaternary structure with a diameter of about 100 Å in the largest dimension and a total of five FeS clusters (Fig. 3A). An initially unexpected [4Fe – 4S] center, now called the D-cluster, is coordinated by the two subunits very close to the molecular surface. Each subunit also binds an additional [4Fe – 4S] center, called the B-cluster, as well as the catalytic Ni-containing C-cluster. There is an electron transfer pathway between the physiological redox partner, the exposed D-cluster, then the B-cluster of one subunit and finally the C-cluster of the other subunit. The electron flow direction will depend on whether the enzyme reduces CO_2 or oxidizes CO (Eq. 2).

Putative pathways have been characterized in *C. hydrogenoformans* CODH for the respective transit of CO/CO_2 and the H_2O product through hydrophobic and hydrophilic tunnels, respectively [87]. The bi-functional CODH/ACS from *M. thermoacetica* contains several hydrophobic tunnels that connect the two CODH C-cluster active sites to each other and to the ACS active site named the A-cluster [90]. High-pressure, xenon-binding experiments carried out in a CODH/ACS crystal have shown that these tunnels can trap many xenon atoms [91]. In addition, putative proton transfer pathways connecting

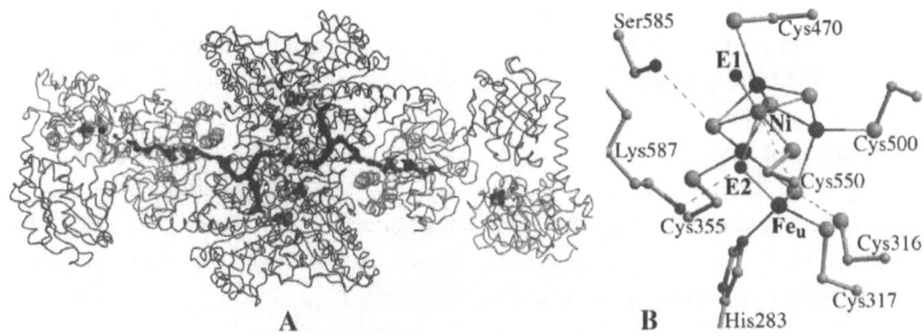

Fig. 3 Structure of the carbon monoxide dehydrogenase/acetyl coenzyme A synthase (CODH/ACS) hetero-tetramer. **A** Polypeptide fold of the CODH dimer (*center*) and of ACS in the closed (*left*) and open subunit conformation (*right*). Metal sites and inorganic sulfurs are shown as *spheres*; an extensive hydrophobic tunnel network is highlighted in *dark grey*. **B** Zoomed depiction of the CODH active site. *Dashed lines* indicate putative H-bonds

the C-cluster to the molecular surface have been proposed using site-directed mutagenesis [92].

The catalytic C-cluster consists of a [3Fe – 4S] sub-site linked through three labile S^{2-} ions to a Ni – Fe site (Fig. 3B): two of these sulfides are ligands to the Ni ion, whereas the third one binds the unique Fe (Fe_u), which is further coordinated by a histidine imidazole and a cysteine thiolate ligand. Given its unusual configuration, Fe_u most likely corresponds to the Ferrous Component II (FCII) defined by Mössbauer spectroscopy [93]. In addition, the Ni and the Fe ions of the [3Fe – 4S] sub-site are bound to the protein through cysteine thiolates. We have recently analyzed the available C-cluster structures and have found some differences that are most likely due to dissimilar conditions in the growth, purification and subsequent treatment of the enzymes [94]. In the structure of *R. rubrum* CODH, the Ni cysteine ligand also binds to Fe_u [89] at a site here called E2 by analogy with [NiFe]-hydrogenases (see above). In the *C. hydrogenoformans* enzyme, E2 is occupied by an exogenous bridging atom, modeled as a fifth inorganic S atom, which disappeared when the enzyme was treated with excess CO [88]. Because this treatment also resulted in loss of activity, it was argued that structures that lack the fifth labile sulfide are non-functional. However, Lindahl and co-workers [95] have shown that the addition of exogenous sulfide as Na_2S leads to the reversible inhibition of *R. rubrum* and *M. thermoacetica* CODHs. In fact, the bridging sulfide of *C. hydrogenoformans* CODH could be a SH^-, resulting from the reduction of COS, which is an alternative substrate of CODH [96]:

$$COS + H^+ + 2e^- \rightleftharpoons CO + SH^- . \qquad (14)$$

Like the Ni ion in hydrogenase, the C-cluster Ni has two *cis* binding sites available for exogenous ligands: a Ni – Fe bridging E2 site and a terminal E1 site. Also as in hydrogenases, E1 points to the end of a hydrophobic tunnel. In a CO-treated crystal of *M. thermoacetica* CODH/ACS E1 is occupied by a ligand that was tentatively assigned to a partially occupied CO molecule [91].

Several different stable active site intermediates of the C-cluster have been reported (see [97] and references therein). Gradual reduction of the oxidized enzyme results in the following species, as characterized by EPR spectroscopy: an initial diamagnetic inactive C_{ox} state, a paramagnetic C_{red1}, a putative diamagnetic C_{int} and a paramagnetic C_{red2} state. It is generally assumed that the two paramagnetic states are involved in catalysis, with C_{int} being a hypothetical intermediate between C_{red1} and C_{red2}. In addition, a large number of substrate and inhibitor complexes have been characterized. Many CO-bound states and putative formate or CO_2 complexes were detected in a recent FTIR study of CO binding to CODH/ACS of *M. thermoacetica* [98]. No high frequency bands were detected in the untreated enzyme, thus excluding the presence of intrinsic CO, as was proposed for the *R. rubrum* enzyme [99].

Several small molecules bind to the C_{red1} but not to the C_{red2} state, providing EPR-detectable paramagnetic species. These include CN^- [100, 101], SCN^-, OCN^-, N_3^- [102], SH^- [95] and CS_2 [103]. The electronic structure and binding mode of the tri-atomic anions SCN^-, OCN^- and N_3^- (EPR g_{av} between 2.15 and 2.17) differ from those of the C_{red1} and C_{red2} states and the other inhibitor complexes (EPR g_{av} between 1.66 and 1.86). In the latter, the unpaired electron appears to be localized mainly in the [3Fe – 4S] sub-site (see [97] and references therein). The crystal structures of most of the spectroscopically characterized C-cluster intermediates and complexes remain to be determined. The high temperature factors of many of the C-cluster atoms and ligands in the known structures suggest a substantial degree of disorder that may be due to the presence of a mixture of states in the crystals. Unfortunately, it has been generally difficult to obtain well-diffracting crystals of homogeneous states.

3.3 Acetyl Coenzyme A Synthases

Three crystal structures of ACS have been reported, two of CODH/ACS from *M. thermoacetica* [90, 91] and one of the monomeric ACS from *C. hydrogenoformans* [104]. ACS consists of three globular domains with the catalytic A-cluster bound to domain 3 at its interface with domain 1. As mentioned above, CODH/ACS has a hydrophobic tunnel network that allows CO, the product of Eq. 2 catalyzed by the C-cluster, to diffuse to the A-cluster where it combines with the methyl group donated by CFeSP to form an acetyl group that, in turn, binds to coenzyme A (Eq. 4). The presence of a tunnel connecting the two active sites was predicted before the structure was determined because, under normal turnover conditions, no CO was detected in the reaction medium [9, 10].

In the crystal structure determined by Doukov et al. the A-cluster was relatively buried and a change of ACS to a more open conformation was proposed to allow access of the methyl group and coenzyme A to the active site [90]. An open form was subsequently observed in one of the ACS subunits of the structure of a different crystal form of the same enzyme (the other sub-unit was in the already known closed conformation) [91] (Fig. 3A). A similar open form was later observed in the *C. hydrogenoformans* ACS crystal structure [104]. The rearrangement of the structure involves a rigid body rotation of domains 2 and 3 with respect to domain 1, which packs against the CODH dimer. In the open form, the movement of an α-helix of domain 1 blocks the tunnel to the A-cluster. We have proposed that these conformational changes provide a gating mechanism for access of substrates to the A-cluster when ACS is in its closed form that prevents CO leakage to the environment [105]. We also proposed that CO_2 could access the enzyme through the A-cluster and then travel to the C-cluster through the hydrophobic tunnels. In CODH/ACS mutants with a blocked tunnel, most of the ACS activity was lost in the presence of CO_2 and a reducing agent, confirming the important role of the tunnel in providing CO access to the A-cluster [106]. On the other hand, the CODH activity was not significantly affected, suggesting that other, so far undetected and maybe transient pathways allow diffusion of CO/CO_2 to and from the C-cluster.

The A-cluster consists of a standard [4Fe – 4S] cluster connected through one of its cysteine ligands to a proximal metal ion that is, in turn, bridged through two additional cysteine thiolates to a distal Ni ion (Ni_d). The crystal structures have shown that this metal ion can be Cu [90], Zn [91] or Ni [91, 104]. Cu and Zn are observed in the closed ACS conformations and Ni in the open ones (Fig. 4A). Following these observations, there was a short-lived debate as to which of these metals is part of the physiologically relevant active site (e.g. [107–111]). Currently, there is a consensus that the proximal active metal is Ni_p [91, 104, 112], as indicated by recent results obtained with the ACS subunit from the *Methanosarcina thermophila* ACDS complex [113, 114] and by an earlier finding that activity positively correlates with the presence of a labile Ni ion [115]. The labile nature of the catalytic Ni originates from its significant solvent exposure in the ACS open form. This has been confirmed by the recent observation that Zn can replace Ni_p at the A-cluster only during turnover, when ACS probably alternates between the closed and open forms [116].

Ni_d has square planar coordination involving two bridging cysteine thiolates and two main chain N atoms (Fig. 4A). A similar structure has been described recently for the active site of nickel superoxide dismutase (NiSOD) [117, 118] (Fig. 4B). In NiSOD, Ni(II) reacts with superoxide and is oxidized to Ni(III)-peroxide. On the other hand, ACS functions in reducing environments and Ni_d is thought to remain as Ni(II) throughout catalysis. Model chemistry supports this proposal indicating that Ni_p is much more

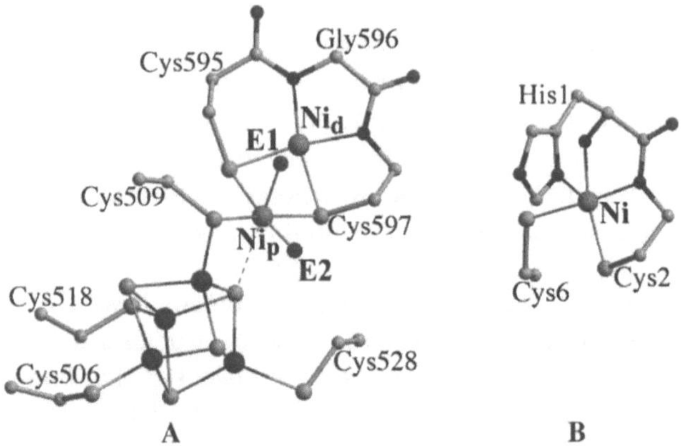

Fig. 4 **A** Active site of acetyl coenzyme A synthase (ACS). **B** Active site of Ni-superoxide dismutase (Ni – SOD)

likely to be reduced than Ni_d (for reviews see [119–122]). Reaction of the A-cluster with CO generates a paramagnetic species that has been assigned to a Ni(I) – CO complex. Isotopic substitution with ^{61}Ni, ^{13}CO or ^{57}Fe results in line broadening of the EPR spectrum and, consequently, the CO complex has been called the NiFeC species [123]. The isotopic effects can be best explained if the EPR signal arises from a proximal Ni(I) with bound CO.

In the open ACS conformation of *M. thermoacetica* CODH/ACS, Ni_p has square planar coordination involving three cysteine thiolates and an unidentified exogenous ligand at the E2 site that could correspond to either formate or SO_2 [105, 124]. Although these putative ligands are probably not functionally relevant, they underline the propensity of the E2 site to bind exogenous molecules directly from the solvent. An apical fifth potential coordination site, called E1, is *trans* to one of the labile S atoms of the [4Fe – 4S] cluster and points towards Phe512. A residual electron density peak close to this site, at 1.3 Å from the square planar Ni_p, suggests that a small fraction of the metal at the proximal site adopts a tetrahedral coordination, as already observed for both proximal Cu and Zn in the closed ACS conformation [90, 91]. This would require a rotation of Phe512, which is, as a matter of fact, partially disordered. We have not been able to identify the metal that occupies the minor tetrahedral site in the open ACS due to its low occupancy. In conclusion, Ni_p, like the Ni ions in [NiFe] hydrogenases and the C-cluster of CODH, has two *cis*-coordination sites available for substrate binding, although in this case both open sites are terminal as it may be required for the insertion/migration reaction of acetyl synthesis.

4 Catalytic Mechanisms

The structural similarities between the active sites of [NiFe]-hydrogenase, CODH and ACS suggest that the respective catalytic mechanisms may be similar as well. In each case, a hydrophobic tunnel points to the Ni apical E1 coordination site providing access and exit pathways for gaseous substrates and products. In addition, an orthogonal E2 site is also available for binding exogenous ligands. In CODH and hydrogenase, E2 is also a coordination site for the Fe ion. Although the structural differences between the three active sites are likely to be very important in determining the nature of the reaction, the protein environment may also play a major role in determining the catalytic properties of these enzymes. For example, the A-cluster of ACS cannot catalyze net two-electron redox reactions because there is no redox center at a suitable distance and, by the same token, hydrogenases cannot oxidize CO to CO_2 because there is no water channel leading to the buried NiFe active site.

A unique feature of the two major classes of hydrogenases is the presence of structural CN^- and CO ligands to Fe [41, 53–56, 125]. Recently, the so-called FeS-free hydrogenase, for a long time considered as a metal-free enzyme, has been shown to have a cofactor containing a low-spin iron, probably Fe(II), with bound CO [59, 126]. Thus, the association of CO to Fe appears to be central to the biological metabolism of molecular hydrogen. CO is a good π-acceptor that will favor transition metal binding to soft σ-donor ligands such as hydride *trans* to it. Indeed, in both NiFe and Fe-only enzymes, CO binds *trans* to a vacant coordination site to Fe. In the former, this is the bridging E2 site that binds either hydride or H_2 in the *R. eutropha* H_2-sensing hydrogenase [60]. Several catalytic mechanisms have been proposed for standard [NiFe]-hydrogenases but no consensus has been reached. It is often assumed that the Ni – Fe bridging site E2 that is occupied by O or S species in inactive states of the enzyme corresponds to the H_2 binding site during catalysis. This is an appealing proposition because in Fe-only hydrogenases the postulated H_2 binding site is a $FeS_2(CO)_2CN$ unit that can be closely superimposed to an equivalent $FeS_2CO(CN)_2$ portion in the [NiFe] enzyme active site [58]. However, the Ni – C state, which already has a hydrogen species bound to E2, is very active in H_2 uptake [127–129]. This would imply that hydrogen binds to the terminal Ni E1 site during turnover, consistent with the observation that the competitive inhibitor CO binds at this site [44]. In this case, the putative bridging hydride at E2 would function as the first base that is required for the heterolytic cleavage of H_2. These ideas are summarized in Scheme I shown inFig. 5.

In CODH, a crucial issue is the location of the two electrons in C_{red2}, which reduces CO_2 to CO according to Eq. 2. According to Mössbauer spectroscopic results, FCII is not the species reduced during the transition from

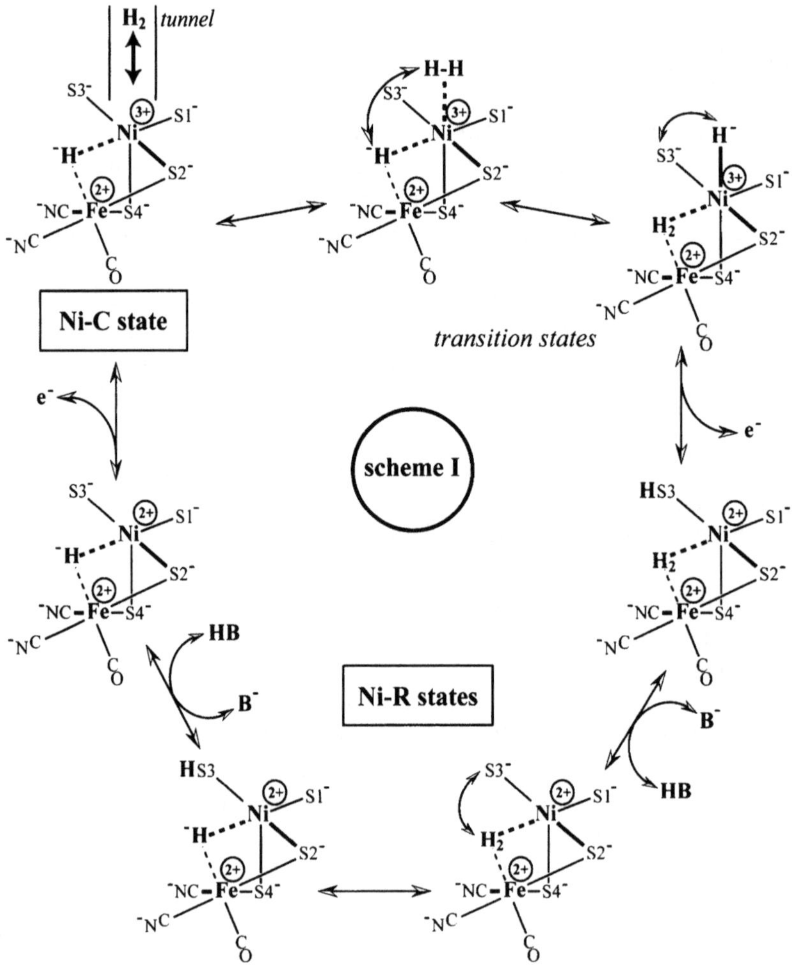

Fig. 5 Putative catalytic cycle of [NiFe]-hydrogenase. In this scheme, heterolytic H_2 cleavage at the terminal E1 site takes place simultaneously with H_2 bond formation at the Ni – Fe bridging E2 site, giving both exogenous *cis*-ligation sites of the Ni an indispensable role during catalysis. The resulting hydride at E1 immediately loses one electron to the Ni and one to the proximal cluster, whereas a nearby base that could be Cys530 accepts the remaining proton (Fig. 1B). The resulting Ni(II) – H_2 species may correspond to one of the three protonation states that have been detected for Ni – R, the most reduced stable state [80]. After proton transfer from Cys530 to Glu18, the next base in a likely pathway [130], Cys530 may assist in the heterolytic cleavage of the bridging H_2 by abstracting a proton. Finally, after electron transfer to the distal cluster, the proximal cluster may oxidize the active site to the Ni – C state, thus closing the cycle. An advantage of this scheme over others is that the bridging E2 site is always occupied, providing a stable coordination environment for the Ni ion in all the catalytic intermediates. In addition, it involves minimal conformational changes, thus allowing very rapid catalysis, in agreement with the observation that the turnover rate of the enzyme when coated on a spinning graphite electrode is limited by the rate of diffusion of H_2 to the active site [131]

the C_{red1} to the C_{red2} state [93]. In addition, neither the Mössbauer nor the UV/visible spectra are significantly different when these two states are compared, strongly suggesting that the electrons do not go to the [3Fe – 4S] subsite either [93, 132]. Another possibility is the reduction of the Ni center from Ni(II) to Ni(0). It is not clear how Ni^0 would be stabilized in the C-cluster but, if this proposition is taken only formally, the charge could be de-localized over the Ni ligands. This would agree with an XAS study of *C. hydrogenoformans* CODH that showed that the electron density at the Ni center does not change upon the two-electron reduction of the C-cluster [133]. A further scheme involves the reduction of a persulfide bond between Cys316 and an inorganic sulfur atom of the cluster, since this bond was observed as a minor component in the crystal structure of the *M. thermoacetica* enzyme [91]. Mutation of Cys316 to Ala resulted in inactive enzyme, but this turned out to be due to the loss of the C-cluster [92].

In the two known crystal structures of *M. thermoacetica* CODH/ACS, the $NiFe_u$ E2 bridging site does not display any detectable electron density [90, 91] (Fig. 3B), which apparently leaves only three protein ligands to coordinate Fe_u. Because such a paucity of ligands seems unlikely, it is possible that E2 is occupied by a hydride resulting from the two-electron reduction of a proton, which would thus result in a tetrahedrally coordinated Fe_u. This structure would then be analogous to the Ni – C state in [NiFe]-hydrogenases [60]. Indeed, a bridging hydride could explain the (low) H_2 evolution reported for CO-treated CODH in the absence of electron acceptors other than protons [98, 134]. Consequently, and mainly based on the crystal structures, we tend to favor a proton as the two-electron acceptor [94]. It should be mentioned, however, that there is also a potential problem with the hydride proposal: the absence of a strong ENDOR signal from a proton nucleus [101] is only compatible with a bridging hydride carrying no significant electron spin density. Assuming that the proton hyperfine coupling detected in the C_{red1} state is correctly assigned to a bridging hydroxide ligand [101], this would imply that C_{red2} and C_{red1} have significantly different electronic structures. Figure 6 depicts a plausible catalytic cycle involving a bridging hydride in C_{red2} (Scheme II). Similarities with the heterolytic H_2 cleavage reaction proposed in Fig. 5 include a bridging NiFe hydride and the participation of both E1 and E2 sites in the catalytic process.

Clearly many of the structural and electronic changes that occur during catalysis at the C-cluster are not yet completely understood and more studies are needed, including high-resolution structural characterizations of homogenous intermediate states of the catalytic cycle.

Although acetyl-CoA synthesis (Eq. 4), as catalyzed by the A-cluster, does not involve net electron consumption, the mechanism is thought to involve two redox steps. The first of these steps is the oxidative addition of the methyl cation that is transferred from CFeSP to Ni_p. The second redox step corresponds to the reductive elimination of the nascent acetyl group

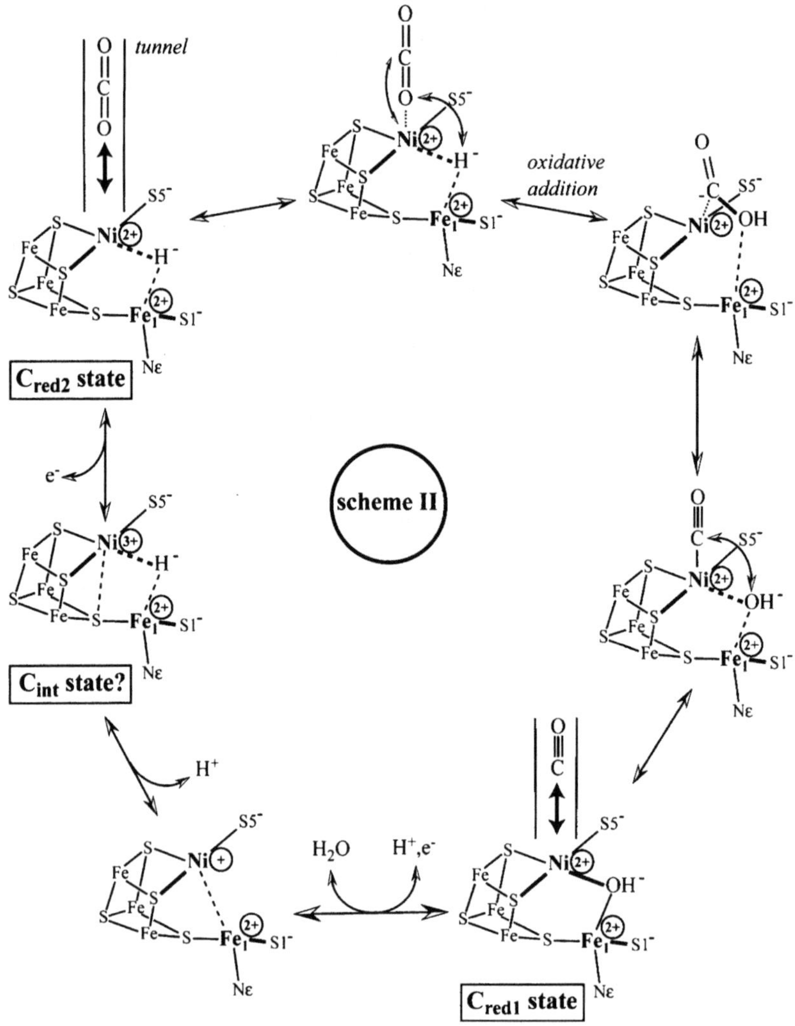

Fig. 6 Putative catalytic cycle of CODH involving a Ni – Fe bridging hydride. The bridging hydride is assumed to be stable enough to prevent its fast reaction with a proton. The reaction can be divided into several steps: 1) CO_2 binds end-on at the axial E1 site; 2) the hydride attacks the bound CO_2 forming a transient formate complex; 3) the C – OH bond is broken, generating an intermediate that has E1 occupied by CO and E2 by OH^-; 4) CO dissociates from Ni as a Ni(II)-bound axial CO is not very stable [135], generating the C_{red1} state that has a bridging hydroxide; 5) addition of two electrons via the D- and B-clusters and of two protons via a proton channel leads to dehydration and regeneration of the C_{red2} state. The hypothetical C_{int} form is depicted as having Ni(III). However, an alternative would be a species with Ni(II) and one of the cluster irons oxidized to Fe(III).

that is transferred to CoA. A crucial question concerns the identity of the two-electron donor/acceptor. Although non-redox, sulfur-based catalysis, in which the methyl cation binds to a thiolate ligand, has precedent in model compounds (e.g. [86]), and could represent an alternative mechanism, it is generally accepted that the transferred methyl group binds to a metal ion. A redox-dependent process is in agreement with the observation that the enzyme needs to be reductively activated [97]. In the following discussion we will therefore assume metal-based catalysis analogous to the industrial Monsanto process of acetate formation.

In the Monsanto process, which operates at 30 to 60 bars and 150 to 200 °C, the catalyst $[Rh^{I}(I)_2(CO)_2]^-$ is the initial electron donor: oxidative addition of CH_3I produces a transient $[Rh^{III}(I)_3(CO)_2CH_3]^-$ state that rapidly converts to a stable $[Rh^{III}(I)_3CO(COCH_3)]^-$ acetyl intermediate upon methyl migration. Addition of another CO followed by reductive elimination of CH_3COI and a final hydrolysis step leads to the production of CH_3COOH and HI and the regeneration of $[Rh^{I}(I)_2(CO)_2]^-$ [136]. ACS catalysis in *M. thermoacetica* takes place at 1 bar and 45–65 °C [2]. Possible electron donor/acceptor sites are Ni_d, Ni_p, the [4Fe – 4S] cluster and a putative pair of cysteines that could form a disulfide bond. Another unclear point is the sequence of events in vivo, i.e. which reaction occurs first, carbonylation or methylation, although both sequences are possible in vitro.

The absence of CO leakage to the medium during CO_2 reduction at the C-cluster and acetyl-CoA formation [9, 10] indicates that, in vivo, CO arrives to the A-cluster through the hydrophobic tunnel previously discussed. Therefore, the results of experiments where CO is added externally should be interpreted with caution. The situation may be different, however, for organisms like *C. hydrogenoformans* that are able to grow on CO as a carbon and reducing power source and which have mono-functional CODH and ACS [104]. As mentioned above, upon reaction with external CO, an apparently kinetically competent paramagnetic NiFeC species is observed in both *M. thermoacetica* CODH/ACS and *C. hydrogenoformans* ACS [137, 138]. However, no other paramagnetic intermediates have been observed upon subsequent methylation and CoA acetylation, and the involvement of the NiFeC species in catalysis has been questioned [114, 139, 140]. A summary of the NiFeC species-based "paramagnetic" mechanism is shown below:

$Fe_4S_4{}^{2+}Ni_p{}^{2+} + e^- \rightarrow Fe_4S_4{}^{+}Ni_p{}^{2+}$
(*reductive activation*)

$Fe_4S_4{}^{+}Ni_p{}^{2+} + CO \rightarrow Fe_4S_4{}^{2+}Ni_p{}^{+}CO$
(*carbonylation and rapid formation of the NiFeC state*)

$Fe_4S_4{}^{2+}Ni_p{}^{+}CO + CH_3{}^{+} \rightarrow Fe_4S_4{}^{2+}Ni_p{}^{3+}(CO)CH_3{}^-$
(*methylation by oxidative addition*)

$Fe_4S_4{}^{2+}Ni_p{}^{3+}(CO)CH_3{}^- + e^- \rightarrow Fe_4S_4{}^{2+}Ni_p{}^{2+}(CO)CH_3{}^-$
(*rapid one-electron reduction*)

$Fe_4S_4^{2+}Ni_p^{2+}(CO)CH_3^- \rightarrow Fe_4S_4^{2+}Ni_p^{2+}(COCH_3)^-$
(methyl migration or CO insertion)
$Fe_4S_4^{2+}Ni_p^{2+}(COCH_3)^- + CoA^- \rightarrow Fe_4S_4^{+}Ni_p^{2+} + \text{acetyl-CoA} + e^-$
(reductive elimination).

This mechanism requires at least two rapid electron transfer steps and, consequently, Ragsdale and co-workers initially proposed fast intramolecular electron transfer between CODH and ACS [137]. However, this is not compatible with the long distance that separates the ACS A-cluster and the CODH redox centers [90,91]. Under in vivo conditions another redox partner could be involved. In this scheme, it is tacitly assumed that CO is required for the first one-electron reaction as no paramagnetic species is detected upon reductive activation of the enzyme in the absence of CO [138].

Lindahl and co-workers have observed that, in vitro, acetyl-CoA synthesis may take place when methylation precedes carbonylation [140]. In this case, no stable paramagnetic state of the A-cluster, such as NiFeC, is detected by EPR. In addition, they also found that reduction of the [4Fe – 4S] cluster is slower than the methylation reaction and concluded that the former cannot be an electron donor for the latter reaction [141]. Therefore, the two electrons needed to form the CH_3 – Ni bond must come from somewhere else. One possibility involves the reduction of Ni_p to Ni^0 [91, 142]. A simplified summary of the corresponding "diamagnetic" mechanism is:

$Ni_p^{2+} + 2e^- \rightarrow Ni_p^0$
(reductive activation)
$Ni_p^0 + CH_3^+ \rightarrow Ni_p^{2+}CH_3^-$
(oxidative addition)
$Ni_p^{2+}CH_3^- + CO \rightarrow Ni_p^{2+}(COCH_3)^-$
(carbonylation + methyl migration or CO insertion)
$Ni_p^{2+}(COCH_3)^- + CoA^- \rightarrow Ni_p^0 + \text{acetyl-CoA}$
(reductive elimination).

This mechanism requires the coexistence of Ni^0 and a $[4Fe-4S]^{2+}$ center at the A-cluster. However, DFT calculations have shown that such an electronic configuration should convert to Ni^+ and $[4Fe-4S]^+$ [143]. In addition, Ni^0 would only be stable next to a reduced, EPR-active $[4Fe-4S]^+$ cluster [144], which is incompatible with the experimental results. A modification of the "Ni^0" mechanism involves two-electron transitions between a $[4Fe-4S]^{2+}Ni^{2+}/[4Fe-4S]^+Ni^+$ pair [144], against the evidence mentioned above that redox changes at the [4Fe – 4S] cluster are much slower than the turnover rate of the enzyme [141]. Involvement of an alternative $Ni_p^{2+}Ni_d^{2+}/Ni_p^{+}Ni_d^{+}$ pair is neither supported by DFT calculations nor by model chemistry [110].

Another proposed two-electron redox site involved the formation and cleavage of a disulfide bond between two cysteine residues located close to the A-cluster [140]. This proposal was initially discarded [90, 91] due to the absence of a pair of free cysteines near the A-cluster but, more recently, Svetlitchnyi and co-workers proposed that the two cysteines bridging Ni_p and Ni_d could function as a redox site [104]. Arguing against this possibility, we found that forming a disulfide bridge between these two residues is only possible after unreasonably large structural rearrangements (not shown). However, an alternative disulfide bond could be formed between Cys528, which binds one of the Fe ions of the cluster, and Cys597, through a simple rotation of the side chain of the former [105]. DFT calculations have indicated that this disulfide bond is stable in the methyl and acetyl complexes [144]. A simplified summary of a diamagnetic Cys528/Cys597 mechanism is:

$S^-S^-Ni^{2+} + CH_3^+ \rightarrow$ S-S $Ni^{2+}CH_3^-$

(*oxidative addition*)

S-S $Ni^{2+}CH_3^- + CO \rightarrow$ S-S $Ni^{2+}(COCH_3)^-$

(*carbonylation + methyl migration or CO insertion*)

S-S $Ni^{2+}(COCH_3)^- + CoA^- \rightarrow S^-S^-Ni^{2+}$ + acetyl-CoA

(*reductive elimination*).

One challenge with this mechanism is to determine the electronic state of the A-cluster before reductive activation. Another problem is the absence of a disulfide bond between Cys528 and Cys597 in all the structures solved so far. At any rate, this proposition makes sense because it is known from model chemistry that metal ions can oxidize coordinated thiolates, leading to the formation of a disulfide bond and a reduced metal center (e.g. [122]). A sulfur-based “D-site” could also be incorporated in a paramagnetic catalytic cycle:

$S^-S^-Ni^{2+} + CO + e^- \rightarrow S^-S^-Ni^+CO$

(*reductive activation and carbonylation*)

$S^-S^-Ni^+CO + CH_3^+ \rightarrow$ S-S $Ni^{2+}(COCH_3)^- + e^-$

(*oxidative addition and acetyl formation*)

S-S $Ni^{2+}(COCH_3)^- + CoA^- \rightarrow S^-S^-Ni^{2+}$ + acetyl-CoA

(*reductive elimination*).

What is the relevance of the NiFeC species? What can be deduced, based on the closed and open ACS conformations observed in the crystal structure, is that CO is more likely to bind in the closed form, prior to methyl group binding (in the open form), than the other way round [91, 105]. This is because the tunnel is blocked in the open form and apparently there is not enough space in the closed form for the binding of a methyl group at the E2 site.

A second question is, where does CO bind when added externally? It is known that acetyl-CoA synthesis is much faster in the presence of CO_2 plus reductant than with externally added CO [145]. This argues for different initial CO binding modes in the two cases and it is conceivable that the mechanism differs depending on whether CO is added externally or it arrives through the tunnel after CO_2 reduction at the C-cluster. More studies will be required before the ACS catalytic mechanism is fully elucidated.

5
Conclusions

NiFeS clusters and the origin of life. There are currently two opposite ways of considering the origin of life on Earth. One of these proposes the formation of pre-biotic structures in a "primordial soup" rich in organic molecules originally generated by meteoric activity. The other view postulates that pre-biotic metabolisms were iron–sulfur based. In this review we have analyzed this second proposition and have compared some inorganic reactions proposed to be ancestral to those found in extant, mostly anaerobic, microorganisms. Many of the active sites of enzymes catalyzing fundamental reactions such as hydrogen oxidation or carbon fixation have NiFeS cluster structures that are reminiscent of those of nickel-containing minerals, such as greigite. Although it is conceivable that the first stages in the evolution of catalysis took place in the absence of protein, the rather sophisticated control of diffusion of substrates and products by extant enzymes suggests that metal-polypeptide associations were early components in the evolution of life. Further characterization of NiFeS-based, pre-biotic catalysis will be needed before the plausibility of a mineral-based origin of life is confirmed.

Acknowledgements We thank Siem Albracht, Patricia Amara, Bart Faber, Christine Cavazza, Marie-Hélène Charon, Claudine Darnault, Victor Fernandez, Martin Field, Michel Frey, Elsa Garcin, Claude Hatchikian, Eun Jin Kim, Antonio de Lacey, Pierre Legrand, Paul Lindahl, Lydie Martin, Michael Matho, Yaël Montet, Yvain Nicolet, Marc Rousset, Winfried Roseboom and Xavier Vernède for their important contributions to our studies of NiFeS clusters in enzymes, and Michael Russell for the stimulating discussions on pre-life conditions and non-biological carbon fixation.

References

1. Wood HG (1991) FASEB J 5:156
2. Drake HL, Daniel SL (2004) Res Microbiol 155:869
3. Ragsdale SW (2004) Crit Rev Biochem Mol Biol 39:165
4. Thauer RK (1998) Microbiol 144:2377
5. Müller V (2003) Appl Environm Microbiol 69:6345

6. Deppenmeier U (2004) J Bioenerg Biomembr 36:55
7. Maness PC, Huang J, Smolinski S, Tek V, Vanzin G (2005) Appl Environ Microbiol 71:2870
8. Ljungdahl LG (1986) Annu Rev Microbiol 40:415
9. Maynard EL, Lindahl PA (1999) J Am Chem Soc 121:9221
10. Seravalli J, Ragsdale SW (2000) Biochemistry 39:1274
11. Volbeda A, Fontecilla-Camps JC (2005) Coord Chem Rev 249:1609
12. Russell MJ, Martin W (2004) Trends Biochem Sci 29:358
13. Bonomi F, Werth MT, Kurtz DM (1985) Inrg Chem 24:4331
14. Russell MJ, Hall AJ, Boyce AJ, Fallick AE (2005) Econ Geol 100:419
15. Wächtershäuser G (1988) Syst Appl Microbiol 10:207
16. Wächtershäuser G (1990) Proc Natl Acad Sci USA 87:200
17. Heinen W, Lauwers AM (1996) Orig Life Evol Biosph 26:131
18. Huber C, Wächtershäuser G (1997) Science 276:245
19. Amend JP, Shock EL (2001) FEMS Microbiol Rev 25:175
20. Cody GD, Boctor NZ, Filley TR, Hazen RM, Scott JH, Sharma A, Yoser HS Jr (2000) Science 289:1337
21. Cody GD, Boctor NZ, Brandes JA, Filley TR, Hazen RM, Yoser HS Jr (2004) Geochim Cosmochim Acta 68:2185
22. Russell MJ (2003) Science 302:580
23. Russell MJ, Hall AJ (1997) J Geol Soc (London) 154:377
24. Wolthers M, Van der Gaast SJ, Rickard D (2003) Am Mineral 88:2007
25. Wächtershäuser G (1988) Microbiol Rev 52:452
26. Finklea SL III, Cathey L, Amma EL (1976) Acta Cryst A32:529
27. Rickard D, Butler IB, Oldroyd A (2001) Earth Planet Sci Lett 189:85
28. Posfai M, Buseck PR, Bazylinski DA, Frankel RB (1998) Science 280:880
29. Vaughan DJ, Craig JR (1978) Mineral chemistry of natural sulfides. Cambridge University Press, Cambridge
30. Huber C, Wächterhäuser G (1998) Science 281:670
31. Cody GD (2004) Ann Rev Earth Planet Sci 32:569
32. Anet FAL (2004) Curr Opin Chem Biol 8:654
33. Pereto J (2005) Int Microbiol 8:23
34. Trevors JT, Abel DL (2004) Cell Biol Int 28:729
35. Lindahl PA, Chang B (2001) Orig Life Evol Biosph 31:403
36. Grahame DA, Gencic S, Demoll E (2005) Arch Microbiol 184:32
37. Hedderich R (2004) J Bioenerg Biomembr 36:65
38. Vignais PM, Billoud B, Meyer J (2001) FEMS Microbiol Rev 25:455
39. Soboh B, Linder D, Hedderich R (2002) Eur J Biochem 269:5712
40. Volbeda A, Charon MH, Piras C, Hatchikian EC, Frey M, Fontecilla-Camps JC (1995) Nature 373:580
41. Volbeda A, Garcin E, Piras C, De Lacey AL, Fernandez VM, Hatchikian EC, Frey M, Fontecilla-Camps JC (1996) J Am Chem Soc 118:12989
42. Higuchi Y, Yagi T, Yasuoka N (1997) Structure 5:1671
43. Higuchi Y, Ogata H, Miki K, Yasuoka N, Yagi T (1999) Structure 7:549
44. Ogata H, Mizoguchi Y, Mizuno N, Miki K, Adachi S, Yasuoka N, Yagi T, Yamauchi O, Hirota S, Higuchi Y (2002) J Am Chem Soc 124:11628
45. Montet Y, Amara P, Volbeda A, Vernede X, Hatchikian EC, Field MJ, Frey M, Fontecilla-Camps JC (1997) Nat Struct Biol 4:523
46. Volbeda A, Montet Y, Vernède X, Hatchikian EC, Fontecilla-Camps JC (2002) Int J Hydrogen Energy 27:1449

47. Matias PM, Soares CM, Saraiva LM, Coelho R, Morais J, Le Gall J, Carrondo MA (2001) J Biol Inorg Chem 6:63
48. Frey M, Fontecilla-Camps JC, Volbeda A (2001) Ni,Fe-Hydrogenases. In: Messerschmidt A, Huber R, Poulos T, Wieghardt K (eds) Handbook of Metalloproteins. Wiley, New York, p 880
49. Garcin E, Vernede X, Hatchikian EC, Volbeda A, Frey M, Fontecilla-Camps JC (1999) Structure 7:557
50. Volbeda A, Martin L, Cavazza C, Matho M, Faber BW, Roseboom W, Albracht SP, Garcin E, Rousset M, Fontecilla-Camps JC (2005) J Biol Inorg Chem 10:239
51. Ogata H, Hirota S, Nakahara A, Komori H, Shibata N, Kato T, Kano K, Higuchi Y (2005) Structure 13:1635
52. Bagley KA, Duin EC, Roseboom W, Albracht SPJ, Woodruff WH (1995) Biochemistry 34:5527
53. Happe RP, Roseboom W, Pierik AJ, Albracht SPJ, Bagley KA (1997) Nature 385:126
54. De Lacey AL, Hatchikian EC, Volbeda A, Frey M, Fontecilla-Camps JC, Fernandez VM (1997) J Am Chem Soc 119:7181
55. Pierik AJ, Roseboom W, Happe RP, Bagley KA, Albracht SPJ (1999) J Biol Chem 274:3331
56. Pierik AJ, Hulstein M, Hagen WR, Albracht SPJ (1998) Eur J Biochem 258:572
57. Peters JW, Lanzilotta WN, Lemon BJ, Seefeldt LC (1998) Science 282:1853
58. Nicolet Y, De Lacey AL, Vernède X, Fernandez VM, Hatchikian EC, Fontecilla-Camps JC (2001) J Am Chem Soc 123:1596
59. Lyon EJ, Shima S, Boecher R, Thauer RK, Grevels FW, Bill E, Roseboom W, Albracht SP (2004) J Am Chem Soc 126:14239
60. Brecht M, Van Gastel M, Buhrke T, Friedrich B, Lubitz W (2003) J Am Chem Soc 125:13075
61. Reissmann S, Hochleitner E, Wang H, Paschos A, Lottspeich F, Glass RS, Bock A (2003) Science 299:1067
62. Roseboom W, Blokesch M, Böck A, Albracht SP (2005) FEBS Lett 579:469
63. Vignais PM, Colbeau A (2004) Curr Issues Mol Biol 6:159
64. Albracht SPJ (1994) Biochim Biophys Acta 1188:167
65. Cammack R, Robson R, Frey M (eds) (2001) Hydrogen as a fuel. Learning from Nature. Taylor and Francis, London
66. Maroney MJ, Bryngelson PA (2001) J Biol Inorg Chem 6:453
67. Volbeda A, Fontecilla-Camps JC (2003) Dalton Trans 21:4030
68. Armstrong FA (2004) Curr Opin Chem Biol 8:133
69. Best SP (2005) Coord Chem Rev 249:1536
70. De Lacey AL, Fernandez VM, Rousset M (2005) Coord Chem Rev 249:1596
71. Vignais PM (2005) Coord Chem Rev 249:1677
72. Armstrong FA, Albracht SP (2005) Phil Trans Royal Soc A 363:937
73. Niu S, Thomson LM, Hall MB (1999) J Am Chem Soc 121:4000
74. Amara P, Volbeda A, Fontecilla-Camps JC, Field MJ (1999) J Am Chem Soc 121:4468
75. De Gioia L, Fantucci P, Guigliarelli B, Bertrand P (1999) Inorg Chem 38:2658
76. Siegbahn PEM, Blomberg MRA, Wirstam M, Crabtree RH (2001) J Biol Inorg Chem 6:460
77. Stadler C, De Lacey AL, Montet Y, Volbeda A, Fontecilla-Camps JC, Conesa JC, Fernandez VM (2002) Inorg Chem 41:4424
78. Stein M, Lubitz W (2004) J Inorg Biochem 98:862
79. Liu P, Rodriguez JA (2005) J Am Chem Soc 127:14871

80. Bleijlevens B, Van Broekhuizen FA, De Lacey AL, Roseboom W, Fernandez VM, Albracht SPJ (2004) J Biol Inorg Chem 9:743
81. Happe RP, Roseboom W, Albracht SPJ (1999) Eur J Biochem 259:602
82. De Lacey AL, Stadler C, Fernandez VM, Hatchikian EC, Fan H-J, Li S, Hall MB (2002) J Biol Inorg Chem 7:318
83. Dole F, Medina M, More C, Cammack R, Bertrand P, Guigliarelli B (1996) Biochemistry 35:16399
84. Evans DJ, Pickett CJ (2003) Chem Soc Rev 32:268
85. Artero V, Fontecave M (2005) Coord Chem Rev 249:1518
86. Bouwman E, Reedijk J (2005) Coord Chem Rev 249:1555
87. Dobbek H, Svetlitchnyi V, Gremer L, Huber R, Meyer O (2001) Science 293:1281
88. Dobbek H, Svetlitchnyi V, Liss J, Meyer O (2004) J Am Chem Soc 126:5382
89. Drennan CL, Heo J, Sintchak MD, Schreiter F, Ludden PW (2001) Proc Natl Acad Sci USA 98:11973
90. Doukov TI, Iverson TM, Seravalli J, Ragsdale SW, Drennan CL (2002) Science 298:567
91. Darnault C, Volbeda A, Kim EJ, Legrand P, Vernède X, Lindahl PA, Fontecilla-Camps JC (2003) Nat Struct Biol 10:271
92. Kim EJ, Feng J, Bramlett MR, Lindahl PA (2004) Biochemistry 43:5728
93. Hu Z, Spangler NJ, Anderson ME, Xia JQ, Ludden PW, Lindahl PA, Münck E (1996) J Am Chem Soc 118:830
94. Volbeda A, Fontecilla-Camps JC (2005) Dalton Trans 21:3443
95. Feng J, Lindahl PA (2004) J Am Chem Soc 126:9094
96. Ensign SA (1995) Biochemistry 34:5372
97. Lindahl PA (2002) Biochemistry 41:2097
98. Chen J, Huang S, Seravalli J, Gutzman H Jr, Swartz DJ, Ragsdale SW, Bagley KA (2003) Biochemistry 42:14822
99. Heo J, Staples CR, Halbieb CM, Ludden PW (2000) Biochemistry 39:7956
100. Anderson ME, Lindahl PA (1994) Biochemistry 33:8702
101. DeRose VJ, Telser J, Anderson ME, Lindahl PA, Hoffman BM (1998) J Am Chem Soc 120:8767
102. Seravalli J, Kumar M, Lu WP, Ragsdale SW (1995) Biochemistry 34:7879
103. Anderson ME, Lindahl PA (1996) Biochemistry 35:8371
104. Svetlitchnyi V, Dobbek H, Meyer-Klaucke W, Meins T, Thiele B, Romer P, Huber R, Meyer O (2004) Proc Natl Acad Sci USA 101:446
105. Volbeda A, Fontecilla-Camps JC (2004) J Biol Inorg Chem 9:525
106. Tan X, Loke HK, Fitch S, Lindahl PA (2005) J Am Chem Soc 127:5833
107. Seravalli J, Gu W, Tam A, Strauss E, Begley TP, Cramer SP, Ragsdale SW (2003) Proc Natl Acad Sci USA 100:3689
108. Golden ML, Rampersad MV, Reibenspies JH, Darensbourg MY (2003) Chem Commun (Camb) 7:1824
109. Krishnan R, Voo JK, Riordan CG, Zakharov L, Rheingold AL (2003) J Am Chem Soc 125:4422
110. Linck RC, Spahn CW, Rauchfuss TH, Wilson SR (2003) J Am Chem Soc 125:8700
111. Bramlett MR, Tan X, Lindahl PA (2003) J Am Chem Soc 125:9316
112. Seravalli J, Xiao Y, Gu W, Cramer SP, Antholine WE, Krymov V, Gerfen GJ, Ragsdale SW (2004) Biochemistry 43:3944
113. Gencic S, Grahame DA (2003) J Biol Chem 278:6101
114. Funk T, Gu W, Friedrich S, Wang H, Gencic S, Grahame DA, Cramer SP (2004) J Am Chem Soc 126:88

115. Russell WK, Stålhandske CMV, Xia J, Scott RA, Lindahl PA (1998) J Am Chem Soc 120:7502
116. Tan X, Bramlett MR, Lindahl PA (2004) J Am Chem Soc 126:5954
117. Barondeau DP, Kassmann CJ, Bruns CK, Tainer JA, Getzoff ED (2004) Biochemistry 43:8038
118. Wuerges J, Lee JW, Yim YI, Yim HS, Kang SO, Carugo KD (2004) Proc Natl Acad Sci USA 101:8569
119. Hegg EL (2004) Acc Chem Res 37:775
120. Riordan CG (2004) J Biol Inorg Chem 9:542
121. Evans DJ (2005) Coord Chem Rev 249:1582
122. Harrop TC, Mascharak PK (2005) Coord Chem Rev 249:3007
123. Fan C, Gorst CM, Ragsdale SW, Hoffman BM (1991) Biochemistry 30:431
124. Heo J, Skjeldal L, Staples CR, Ludden PW (2002) J Biol Inorg Chem 7:810
125. Nicolet Y, Lemon BJ, Fontecilla-Camps JC, Peters JW (2000) Trends Biochem Sci 25:138
126. Shima S, Lyon EJ, Thauer RK, Mienert B, Bill E (2005) J Am Chem Soc 43:10430
127. Coremans JMCC, Van Garderen CJ, Albracht SPJ (1992) Biochim Biophys Acta 1119:148
128. Barondeau DP, Roberts LM, Lindahl PA (1994) J Am Chem Soc 116:3442
129. Pershad HR, Duff JL, Heering HA, Duin EC, Albracht SPJ, Armstrong FA (1999) Biochemistry 38:8992
130. Dementin S, Burlat B, De Lacey AL, Pardo A, Adryanczyk-Perrier G, Guigliarelli B, Fernandez VM, Rousset M (2004) J Biol Chem 279:10508
131. Jones AK, Sillery E, Albracht SPJ, Armstrong FA (2002) Chem Commun 866
132. Seravalli J, Kumar M, Lu W-P, Ragsdale SW (1997) Biochemistry 36:11241
133. Gu W, Seravalli J, Ragsdale SW, Cramer SP (2004) Biochemistry 43:9029
134. Menon S, Ragsdale SW (1996) Biochemistry 35:15814
135. Lu ZL, Crabtree RH (1995) J Am Chem Soc 117:3994
136. Forster D (1976) J Am Chem Soc 98:846
137. Seravalli J, Kumar M, Ragsdale SW (2002) Biochemistry 41:1807
138. George SJ, Seravalli J, Ragsdale SW (2005) J Am Chem Soc 127:13500
139. Grahame DA, Khangulov S, DeMoll E (1996) Biochemistry 35:593
140. Barondeau DP, Lindahl PA (1997) J Am Chem Soc 119:3959
141. Tan XS, Sewell C, Yang Q, Lindahl PA (2003) J Am Chem Soc 125:318
142. Lindahl PA (2004) J Biol Inorg Chem 9:516
143. Schenker RP, Brunold TC (2003) J Am Chem Soc 125:13962
144. Amara P, Volbeda A, Fontecilla-Camps JC, Field MJ (2005) J Am Chem Soc 127:2776
145. Maynard EL, Lindahl PA (2001) Biochemistry 40:13262
146. Thauer RK, Jungermann K, Decker K (1977) Bacteriol Rev 41:100
147. Schulte MD, Rogers KL (2004) Geochim Cosmochim Acta 68:1087
148. George SJ, Kurkin S, Thorneley RN, Albracht SPJ (2004) Biochemistry 43:6808

Top Organomet Chem (2006) 17: 83–122
DOI 10.1007/3418_006

Published online: 12 April 2006

Carbene Complexes of Heme Proteins and Iron Porphyrin Models

Gérard Simonneaux (✉) · Paul Le Maux

Laboratoire de Sciences Chimiques de Rennes, UMR CNRS 6226, Université de Rennes 1, 35042 Rennes cedex, France
gerard.simonneaux@univ-rennes1.fr

Abstract The possible formation of carbene complexes of cytochromes P450 enzymes in various metabolisms of xenobiotics is first described. In view of these observations, the organometallic chemistry of iron porphyrins, serving as a basis for elucidating the mechanism of these enzymes is then presented, emphasizing the biological relevance. The chemistry of metalloporphyrin models has been extended to ruthenium and osmium complexes, both metals belonging to the iron triads. This review also illustrates that metalloporphyrins are versatile organometallic catalysts that can be used in many different reactions involving carbene transfer. Their development will be considerably widened in the future, with possible application in asymmetric catalysis.

Keywords Carbene · Catalysis · Cytochrome P450 · Iron porphyrin · Ruthenium

Abbreviations

$(TPP)H_2$	tetraphenylporphyrin
$(TTP)H_2$	tetratolylporphyrin
$(TPFPP)H_2$	tetrapentafluorophenylporphyrin
$(T\text{-}p\text{-}ClPP)H_2$	tetra-*p*-chlorophenylporphyrin
$(TMP)H_2$	tetramesitylporphyrin
$(OEP)H_2$	octaethylporphyrin
$(T\text{-}2,6\text{-}ClPP)H_2$	tetra-2,6-dichlorophenylporphyrin
$(T\text{-}p\text{-}FPP)H_2$	tetra-*p*-fluorophenylporphyrin

1 Introduction

The chemistry of coenzyme B12 and related systems containing the cobalt-carbon bond was the first major organometallic chemistry of biological importance to be recognized [1, 2]. The importance of organometallic chemistry of heme proteins and iron porphyrins was recognized later in the 1970–1980s. Interest in organometallic chemistry of iron porphyrins was directly related to the observation that alkyl (or aryl) and carbene iron complexes are detected during the metabolism of xenobiotics by cytochrome P450 enzymes [3–5]. Thus, possible formation of carbene complexes of cytochrome P450 enzymes was suggested in reductive metabolism of polyhalogenated compounds [3, 6], oxidative metabolism of benzodioxole derivatives [7] and in propylene epoxidation [8]. In a more general way, mimicry of xenobiotic interaction with P450 enzymes has been the goal of oxidation reactions catalyzed by metalloporphyrins for at least 25 years [9–11].

Preferably, the participation of all different types of iron-carbon bonds in heme protein reactions, such as σ-alkyl, σ-aryl and carbene metal bonds, should be first presented among the general framework of organometallic reactions related to the heme protein area. Such a comprehensive approach to the subject is well beyond the scope of this review which is limited to examples of carbene complexes. A multivolume work on porphyrin chemistry, including some chapters on heme protein chemistry, has been recently published, and these volumes contain several chapters on the organometallic chemistry of metalloporphyrins [12]. However, an overview related to the organometallic chemistry of heme proteins, including σ-alkyl or σ-aryl iron complexes when necessary, will be presented first since these complexes are often intimately related to carbene complexes of metalloporphyrins. Although there have been relatively few developments in this area over the last 5 years, it may serve as an excellent basis for the discovery of new organometallic reactions catalyzed by metalloporphyrins. These catalytic reactions will be described in the following section and restricted to reactions of complexes of the iron triad elements.

This review is subdivided into four parts which correspond to the major aspects of carbene chemistry involving heme proteins and iron porphyrin models (and iron triad: Fe, Ru and Os). The review is organized as follows: the first part deals with metal-carbene bond formation and reacting properties of heme proteins together with an overview of older results; the second part is devoted to methods of synthesis and structure of metal-carbon bonded complexes in iron porphyrins, and the reactivity of these synthetic metalloporphyrins. The third part describes carbene complexes of ruthenium and osmium porphyrins. The interest in ruthenium chemistry was inspired by the periodic relationship of ruthenium to iron and the possibility of preparing stable carbene ruthenium derivatives. The fourth part reviews various reactions involving carbene transfers catalyzed by metal porphyrins which are presently under extensive development. Finally, the authors will tentatively propose possible developments in the organometallic chemistry of heme proteins, in particular in the use of heme proteins as catalysts for organometallic reactions. We suggest a name for this new area: bioorganometallic catalysis by heme proteins.

One family of derivatives that will not be treated in this review, even though they contain metal-carbon bonds are carbonyl and cyanide (or isocyanide) complexes. These complexes are more relevant to coordination chemistry than organometallic chemistry.

2 Overview: Carbenes as Ligands to Heme Proteins

2.1 Cytochrome P450

The great diversity of reactions catalyzed by cytochromes P450 is now well-documented [9, 13]. Concerning the cytochrome P450 field, a multi-author work edited by Ortiz de Montellano was published in 1986 [14] and progress during the decade 1986–1995 was reported in a second volume in 1995 [15]. Most of the cytochrome P450 reactions are oxidations. However, it has long been recognized that cytochrome P450 can catalyze the reduction of polyhalogenated compounds. Thus, the reduced form of cytochrome P450 can be an extremely efficient reducing agent and transfer an electron, not only to dioxygen, but to other substrates that have electron-accepting properties such as halogenated compounds. Carbon tetrachloride, for example, produced a stable carbene complex via a reductive dehalogenation [6]. This discovery represents the starting point of the bioorganometallic chemistry of heme proteins. Thus, although carbenes are not natural substrates in biological systems, their possible formation with heme proteins has been extensively studied. Initial experiments were carried out in the 1970–1980s, and several short reviews on carbene formations during interactions of heme proteins with xenobiotics are available [4, 14, 16–18]. Many more specialized reviews on organoiron derivatives of heme protein have been reported since then, and attention will focus herein on new insights concerning recent reactions with drugs. Thus, the literature before 1985 will be summarized first and more recent results will then be highlighted.

Evidence was first presented in 1974 and then in 1977 for microsomal cytochrome P450 carbene complexes by Ullrich, Mansuy and coworkers [3,6]. Carbon tetrachloride is a hepatotoxic molecule, resulting in the formation of trichloromethyl radicals which are responsible for lethal cellular damage [19]. Initially, the concept of the toxicity of carbon tetrachloride was discussed in a paper by Ullrich and Schnabel in 1973 on carbanion species as ligands of cytochrome P450 [20]. Later, it was proposed that a two-step reduction of CCl_4 to the CCl_3^- anion would liberate a chloride ion to yield the highly reactive dichlorocarbene (Eqs. 1–4) [6].

$$R^1R^2CX_2 + (P)Fe^{II} \xrightarrow[-2X^-]{+2e^-} [R^1R^2C \rightarrow (P)Fe^{II}] \quad (1)$$

$$R^1R^2CX_2 + (P)Fe^{II} \xrightarrow{-X^-} R^1R^2C^{\cdot}X + (P)Fe^{III} \quad (2)$$

$$R^1R^2C^{\cdot}X + (P)Fe^{II} \longrightarrow [(P)Fe^{III} - CXR^1R^2] \quad (3)$$

$$(P)Fe^{III} - CXR^1R^2 \xrightarrow[-X^-]{+e^-} [(P)Fe^{II} \leftarrow CR^1R^2] \quad (4)$$

Carbon monoxide and chloroform, two other metabolites that result from the reduction of CCl_4, were detected in incubations that contained P450 and dithionite or in complete systems (P450 and reductase) (Eq. 5) [6,21]. The metabolism of carbon tetrachloride and its relationship to lipid peroxidation has been investigated in hepatic microsomes and in reconstituted monoxygenase systems [21]. The results, obtained using purified enzymes, demonstrate that cytochrome P450 can catalyze both the one- and the two-electron reductions of CCl_4. It seems also from these studies that the cytochrome P450-mediated reduction of CCl_4 and CCl_4-induced lipid peroxidation are independent reactions.

The mechanism of reductive dehalogenation of polyhalogenated compounds by microsomal cytochrome P450 has been studied in detail [22]. The main products of the in vitro metabolism of hexa- and pentachloroethane were tetra- and trichloroethene, respectively. In this case, the reductive dehalogenation probably proceeds by two sequential one electron reductions forming first a radical and then a carbanion. The carbanion may undergo

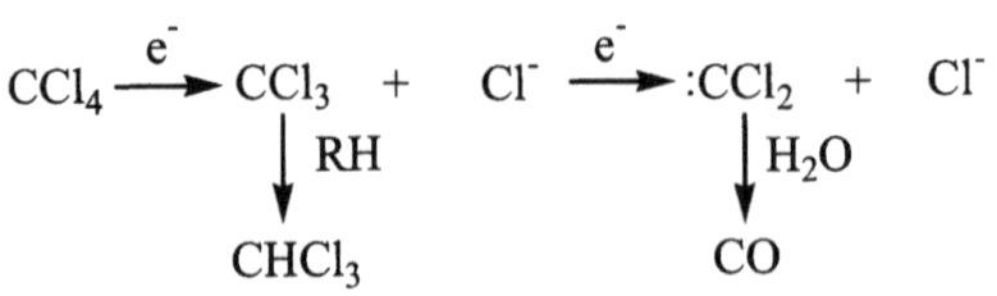

(5)

protonation, alpha or beta elimination forming a monohalogenated alkane, a carbene or an olefin, respectively (Scheme 1) [22].

Since halothane ($CF_3CHClBr$) is one of the most widely used polyhalogenated anaesthetics, a possible formation of carbene species during the interaction of halothane and cytochrome P450 was also proposed by analogy with the results obtained with CCl_4 (Eq. 6). This hypothesis was tested by producing trifluoromethyl carbene chemically from a diazo derivative and by then studying its interaction with rat liver cytochrome P450 [3]. The similarity of the halothane-induced difference spectrum with that obtained from trifluoro diazoethane suggested the formation of the corresponding carbenoid complex with halothane ($\lambda_{max} = 470$ nm).

$$CF_3CHClBr \xrightarrow[-Cl^-,-Br^-]{+2e^-} CF_3\bar{C}H \qquad (6)$$

Other cytochrome P450-carbene complexes were found under anaerobic reducing conditions with various polyhalogenated methanes such as CBr_4, CCl_4, CCl_3F, CCl_3Br, CCl_3CN, CHI_3, $CHBr_3$, and $CHCl_3$, showing absorption bands between 450 and 480 nm [6, 23]. Carbon monoxide was detected as a metabolic product. Carbon monoxide is a known hydrolysis prod-

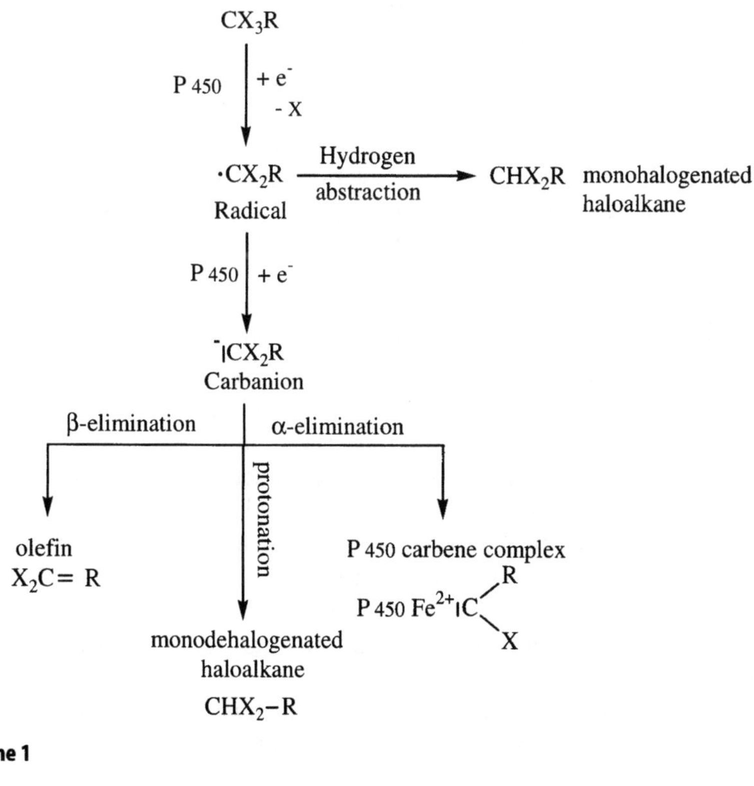

Scheme 1

uct of dihalogeno carbenes [24,25]. Comparison of cytochrome P450 complex formation using liver microsomal preparations from phenobarbital and 3,4-benzopyrene treated rats showed differences that could be accounted for by a decreasing stability of the halogenomethane complex when the 3,4-benzopyrene-induced form is used [6].

Complexes in which the halocarbon is σ-bonded to the ferric prosthetic heme groups that are formed in preference to carbene complexes were also suggested [26–28]. The oxidation of heme proteins by alkyl halides was studied by Castro and coworkers [29,30]. More recent studies [31], under anaerobic reducing conditions with bacterial cytochrome $P450_{CAM}$, confirm previous results obtained by the groups of Ullrich and Mansuy [3,6]. It was also shown that a purified reconstituted cytochrome P450-containing system is capable of reductive dehalogenation of CCl_4. It is now well known that halogenated hydrocarbons interact with models of cytochrome P450 to form iron carbene [32,33] or σ-bonded alkyl derivatives [26–28].

Cytochrome P450 inactivation was observed during reductive metabolism of hydrochlorofluorocarbons, which have been developed as candidate substitutes for the ozone-depleting chlorofluorocarbons [34]. Rat liver microsomes were used providing indirect evidence for the involvement of both P4502E1 and P4502B1/2. The reaction with 1,1-dichloro-2,2,2-trifluoroethane is prevented by both carbene and free-radical scavengers, providing indirect evidence of a possible role of reactive carbene species in the mechanism. Indeed, various other hydrochlorofluorocarbons are biotransformed by cytochrome P450 to oxygenated products such as 1,1-dichloro-1-fluoroethane [35]. Two theoretical investigations of the anaerobic reduction of halogenated alkanes by cytochrome P450 were also reported in order to predict the reactivity of the substrates [36,37]. An anaerobic reduction cycle for polyhalogenated substrates has been proposed [36]. The strengths of C – H bonds in halogenated methanes were theoretically calculated to correlate to the activity of the radical produced in anaerobic reduction.

Bacterial cytochrome P450 has been used as an excellent model to better understand bacterial reductive dehalogenation biochemistry [31,38]. Thus, the binding of halogenated pollutants to cytochrome $P450_{CAM}$ has been investigated. Hexachloroethane was found to bind more tightly to Fe(III)-$P450_{CAM}$ than the physiological substrate camphor [31]. In this study, it was found that the enzyme catalyzed a single turnover stoichiometric reduction of $CFCl_3$ to carbon monoxide, indicating a carbene intermediate in the reaction pathway [31]. A similar carbene pathway to CO has been proposed during reduction of $CFCl_3$ by cobalamins or a methanogenic bacterium [39]. It is also known that fluorine substituents stabilize carbenes [40]. However, the reported data do not discriminate between potential free or iron-bound heme carbene intermediates.

The 1,3-benzodioxole derivatives are oxidatively metabolized by cytochrome P450 monooxygenases with formation of very stable complexes of this cytochrome in the ferrous state, characterized by a Soret peak at 455 nm [41, 42]. Direct evidence for the presence of 1,3-benzodioxol-2-carbene complexes of cytochrome P450 came from model studies (vide infra) [7, 10, 43]. Structure-activity relationships in the interaction of alkoxymethylenedioxybenzene derivatives with microsomal mixed-function oxidase in vivo have been reported [44] and the mechanism of this reaction has also been discussed in a review by Ortiz de Montellano (Scheme 2) [5].

The metabolism of aryldioxymethylene compounds to catechol, carbon monoxide and formic acid is consistent with hydroxylation of the carbene complex (Scheme 3) [5].

The interactions of related derivatives, methylenedioxyphenyl HIV protease inhibitors with ferrous P450 have been examined [45]. It has also been proposed that a carbene is the reactive intermediate generated from these derivatives, which coordinates to the prosthetic heme of P450. Complex formation is reduced when the substituent on the methylenedioxyphenyl moiety is an electron-withdrawing group which destabilizes the carbene iron complex, whereas, if the substituent is electron-donating, complex formation is increased and stabilized [45]. The electron-donating stabilization of the metabolite-P450 complex was previously observed with other similar derivatives [46, 47]. Catechol formation of the aryldioxymethylene derivatives is now well-documented. P450 enzymes mediate the hydroxylation of the methylene carbon of many drugs, and resulting hydroxylated inter-

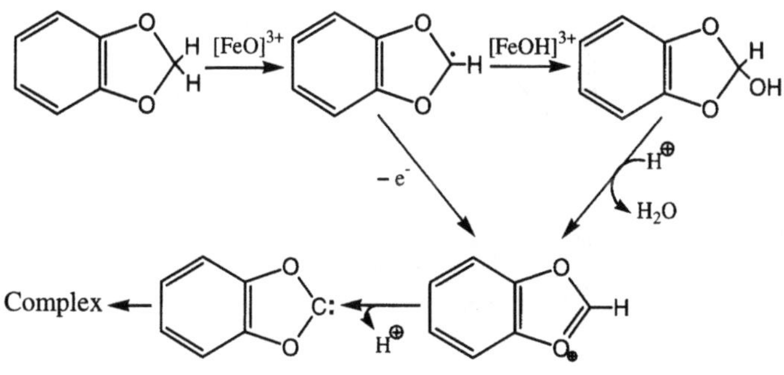

Scheme 2

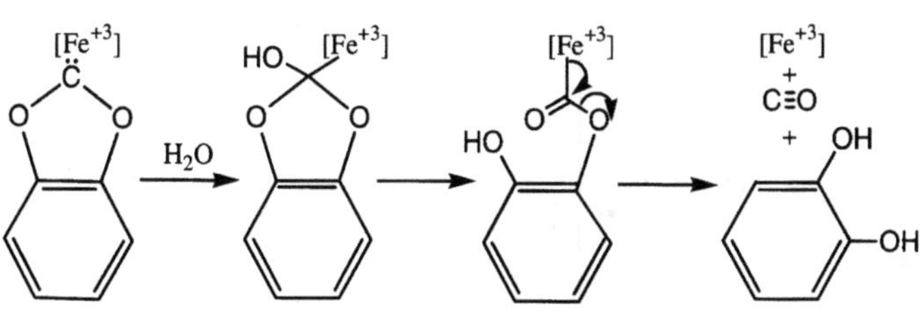

Scheme 3

mediates can undergo hydrolysis to yield the catechol intermediate. The metabolism of dimethyl-4,4′-dimethoxy-5,5′,6,6′-dimethylene dioxybiphenyl-2,2′-dicarboxylate, a hepatic drug, by cytochrome P450 is a recent example of such behavior [48]. Regiochemical differences in cytochrome P450 isozymes responsible for the oxidation of methylenedioxyphenyl groups by rabbit liver have been observed [49]. In particular, it was suggested that complex formation with methylenedioxyamphetamine was not due to the carbene pathway involving the methylene dioxy group but was due to oxidation of the amino group.

The metabolism of sydnones, which are a pharmacologically interesting class of drugs [50], has been shown to be a mechanism-based inactivator of microsomal P450 (Scheme 4) [5]. Enzymatic destruction is accompanied by the formation of *N*-vinylprotoporphyrin IX. It was first suggested that intermediate formation of diazo compounds (Scheme 5) [51] from sydnone metabolism gives bridged Fe – C – N iron carbene complexes [5]. To more define the mechanism of these reactions, other sydnone substrates that do not have a leaving group were also examined in order to explain the formation of the *N*-alkyl heme adduct [52]. Reaction of the same diazoalkane with iron-porphyrin models confirms the formation of the carbene complexes as precursors of the *N*-alkyl porphyrins [53].

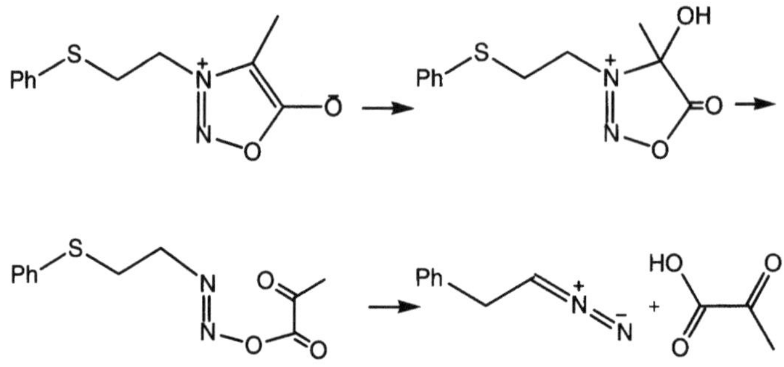

Scheme 4

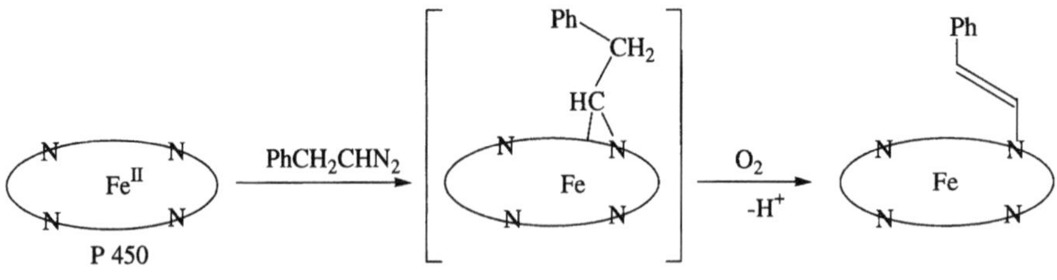

Scheme 5

2.2 Myoglobin and Hemoglobin

The oxidation of two other heme proteins by alkyl halides was studied by Castro and coworkers [54]. The oxidation of deoxymyoglobin and deoxyhemoglobin by bromomalononitrile has been reported as yielding metmyoglobin, methemoglobin and malononitrile [54, 55]. The reductive metabolism of $BrCCl_3$ by ferrous deoxymyoglobin leads to the formation of three major modified heme products (Scheme 6) and a protein-bound heme adduct which were identified [56]. All of these metabolites appear to result from the initial regiospecific attack of the trichloromethyl radical on the vinyl group of the heme.

More recently, the reductive activation of halothane ($CF_3CHBrCl$), which is a hepatotoxic anaesthetic molecule, by human hemoglobin results in the modification of the prosthetic heme [57]. The inhibition of the reaction by adding exogeneous CO or the spin trapping agent *N*-*t*-butyl-*α*-phenyl nitrone to the incubation mixture indicated that (i) a reduced and free heme iron is required by Hb to activate the halogenated substrate; and (ii) the formation of free radical species is responsible for Hb inactivation. However, no carbene species were detected in these reactions. The mechanism is shown in Scheme 7.

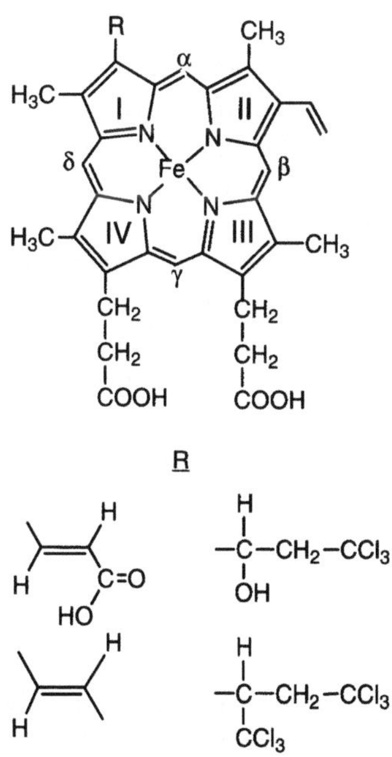

Scheme 6

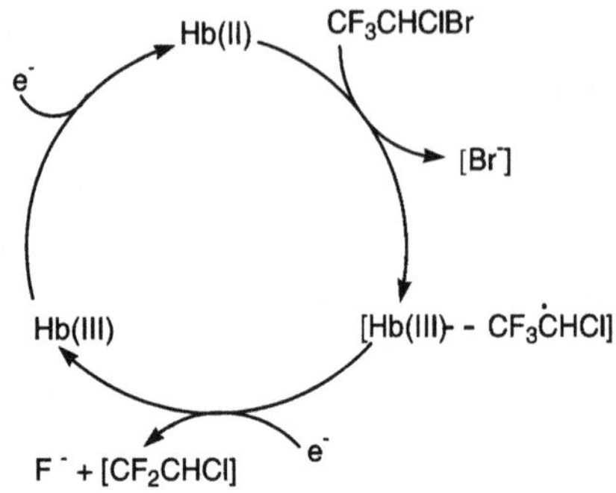

Scheme 7

Finally, a large amount of work has been accomplished over the years on the synthesis and characterization of the metal-carbene group in carbene heme proteins. Evidence of carbene formation during the metabolism of various drugs is still increasing. For example, it is probable that the availability of new drugs [58] such as recent HIV protease inhibitors [45] or other drugs [48] will generate more discoveries involving new aspects of the carbene chemistry relevant to biological systems.

3 Carbenes as Ligands to Iron Porphyrins

The desire to mimic the heme protein enzymes like cytochrome P450 led to a large development of research involving synthetic metalloporphyrins as models of the active sites [12]. This area continues to be very attractive and is still mainly based on the catalytic oxidation reactions. This fact is related to the presence of high valent oxoiron species, which are important intermediates in numerous catalytic cycles of cytochromes P450 and peroxidases [59]. The carbon analogues of oxoiron porphyrins, namely, iron porphyrin carbene complexes, have also received much attention [32], due mainly to the possible formation of carbene complexes of cytochromes P450 enzymes (vide supra) and recently, there was also a renewal of interest in reactions involving metal-carbon bonds, in particular carbene-transfer reactions [60, 61] (see the next part, for catalytic reactions).

3.1 General

It has been well known for several decades that divalent carbon species of the type $:CR^1R^2$, where one of the two substituents at the carbene carbon atom is bonded via a heteroatom (R^1 = alkyl, R^2 = OR, NR_2) exhibit σ-donor and π-acceptor properties upon binding to transition metals [62]. In general, a "double bond", between the metal and the carbon results, depending somewhat on the nature of the R^1 and R^2 groups. More generally, the literature discriminates between electrophilic (Fisher) carbenes [62] and nucleophilic (Schrock) carbenes [63, 64].

Comprehensive reviews on metal carbene compounds in organometallic chemistry are available [64–71] and this general topic will not be discussed here. We shall rather focus on the metalloporphyrin area.

Two different metalloporphyrin carbene complexes have been reported. These are the axially symmetric complex with multiple metal-carbon order and the metal-nitrogen inserted complex (Scheme 8). The relative stability of the two types, A and B, has been previously discussed [72]. A carbene fragment in a metalloporphyrin should insert in a M – N bond when the molecule has a d^8 electronic configuration and/or the d orbitals of the central metal are lowered in energy [72]. Occupancy of two electrons in the M-C(carbene) π^* level seems to be the reason that the d^8 molecule tends towards geometry B rather than A (Scheme 8). Considering the strong σ-donor and π-acceptor character of a carbene ligand, it is also expected that the addition of σ-donor or a π-acceptor would weaken the M = C bond, thus destabilizing structure A.

Scheme 8

The formalism generally adopted for late-transition-metal complexes is a formally neutral carbene fragment bonded to M(II) in the group 6 compound [73], but carbene complexes have been also described as M(IV) complexes on the basis of Môssbauer results [74, 75]. In a theoretical study on the structure of iron carbene derivatives, $(TPP)FeCCl_2$ was considered as a d^6 electron system [72]. Carbene adducts of iron porphyrins have also been considered as carbon analogues of the porphyrin-iron-oxo species on the basis of similarity in the electronic spectra [16, 72, 76]. Although, it is generally accepted that metal-carbene carbon bonds in carbene complexes for cyclopropanation should be double-bonded, it has recently been suggested that the metal-carbon bond of some carbene complexes could be identified as a single bond [77, 78].

3.2 Syntheses

Mansuy and coworkers [32] reported the first isolated metalloporphyrin-carbene complex in 1977 and it was an iron porphyrin carbene complex. It was suggested by the authors that iron carbene complexes could be involved in the metabolism of xenobiotics. Thus, the five-coordinated (TPP)Fe(CCl_2) was prepared as a purple red solid by treatment of (TPP)FeCl with CCl_4 in the presence of an excess of iron powder (90% yield) [32]. It must be emphasized that the possible formation of carbene complexes after the reduction of iron deuteroporphyrin in CCl_4 was initially proposed by Brault and coworkers [79]. An original method of reduction by aqueous sodium dithionite solution was described but, under these conditions, it was not possible to isolate any complex although the reaction leads to a compound of unusually good stability towards air.

A later crystal structure of a related six-coordinate complex, (TPP)Fe(CCl_2)(H_2O), confirmed the formation of such a CCl_2 complex which was also the first example of a dihalogenated carbene complex of a transition metal [33]. In this complex, the overall deviations from the planarity of the porphyrin core are very small (0.03 Å) and the iron atom is not significantly displaced from this plane. The average length of the equivalent Fe – N bonds (1.984(4) Å) and the distances within the porphyrin core are in good agreement with values reported for other low-spin iron porphyrins [80, 81]. Resonance Raman spectra of the same carbene complexes have also been reported [82]. This synthesis was an opening route to other carbene complexes since various polyhalogenated compounds, RCX_3, react similarly with iron porphyrins leading to the corresponding porphyrin iron carbene complexes. A large series of iron carbene porphyrins were synthesized, starting from (TPP)FeCl and polyhalogenated compounds in the presence of an excess of reducing agent. Thus, the carbenes, CCl_2 [32], CBr_2 [4], CF_2 [4], CFCl [4], CFBr [4], and CClCN [83] have been attached to the iron atom by reaction with CCl_4, CBr_4, CF_2Br_2, $CFCl_3$, $CFBr_3$, and CCl_3CN, respectively.

The stable complex formed during the reaction between DDT (2,2-bis (*p*-chlorophenyl)-1,1,1-trichloroethane), a widely used insecticide, and iron porphyrins under reducing conditions is a nice example of such carbene formation (Eq. 7). Actually, these carbene complexes were the first reported vinylidene carbene complexes of an iron porphyrin [84]. The syntheses were then extended to other chlorocarbenes having an electron-withdrawing group on the carbene carbon atom such as CClX yielding (TPP)Fe(CClX) (X = CO_2Et) [83]. However, these complexes were less stable than (TPP)Fe(CCl_2), due to an increase in reactivity towards nucleophiles.

$$(TPP)Fe + R_2CHCCl_3 \xrightarrow[-2Cl^-]{+2e^-} (TPP)Fe[C=C(R)_2] \quad R = C_6H_4Cl \tag{7}$$

Characterization of the (TPP)Fe(1,3-benzodioxol-2-carbene) complex (Eq. 8) was also reported, giving indirect evidence for the presence of this carbene as a ligand in the benzodioxole-derived cytochrome P450 complex (Eq. 9) [7, 16].

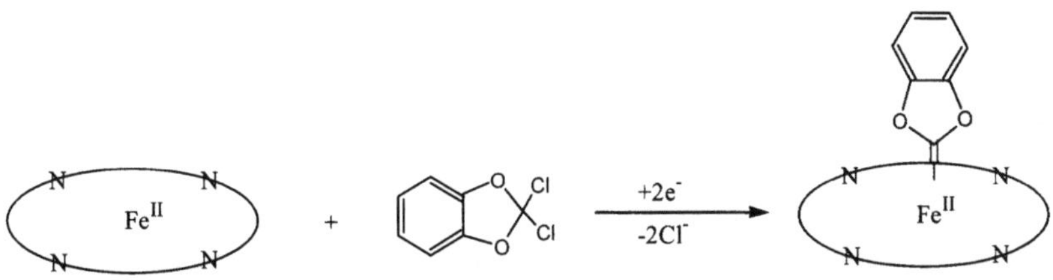

(8)

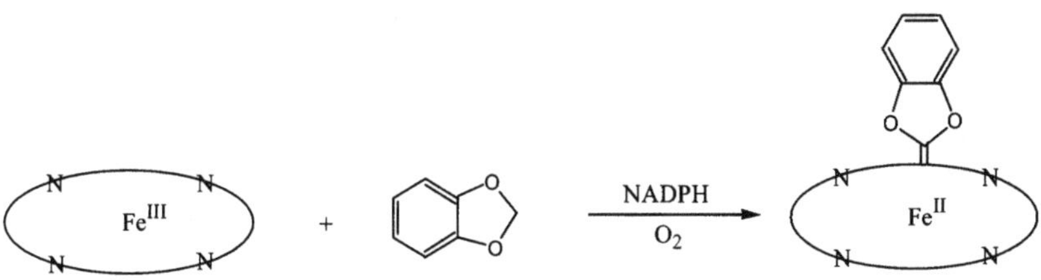

(9)

Fe(TPP)(C(Cl)SeCH$_2$C$_6$H$_5$) [85] and Fe(TPP)(C(Cl)SR) (R = C$_6$H$_5$, CH$_2$C$_6$H$_5$) [86] were also prepared as precursors of the selenocarbonyl and thiocarbonyl complexes, respectively (Eqs. 10, 11).

$$(TPP)Fe + C_6H_5CH_2SeCCl_3 \xrightarrow{Na_2S_2O_4} (TPP)Fe[C(Cl)SeCH_2C_6H_5] \xrightarrow{FeCl_2} (TPP)Fe(CSe) + C_6H_5CH_2Cl \tag{10}$$

$$(TPP)Fe + C_6H_5CH_2SCCl_3 \xrightarrow{Na_2S_2O_4} (TPP)Fe[C(Cl)SCH_2C_6H_5] \xrightarrow{FeCl_2} (TPP)Fe(CS) + C_6H_5CH_2Cl \tag{11}$$

A μ-carbido dimeric complex, [(TPP)Fe]$_2$C is formed upon reaction with carbon tetraiodide (Eq. 12) [87]. This was the first example of a transition-metal complex of the type M = C = M involving a formally dicarbenic carbon atom ligand bridging two transition metals.

$$(TPP)Fe + CI_4 \xrightarrow[-4I^-]{+4e^-} (TPP)Fe=C=Fe(TPP) \tag{12}$$

The X-ray structure of this complex was reported by Bottomley and coworkers [88]. Remarkably, the electronic structure of such a complex was predicted before its preparation [72]. Reduction of trichloromethyltrimethylsilane by iron(II) tetraarylporphyrins in the presence of a reducing agent also leads to the carbide complex [89]. This surprising result can be explained by the involvement of an unstable α-silylcarbene ferroporphyrin complex. A heterometallic μ_2-carbido complex was isolated from the reaction of a dichloro carbene iron porphyrin and pentacarbonylrhenate (Eq. 13) [90]. The X-ray structure of this trinuclear complex, $[(TPP)Fe=C=Re(CO)_4Re(CO)_5]$, shows a 1,3-dimetalla-allene system.

$$(TPP)Fe(CCl_2) \xrightarrow[-CO,-2Cl^-]{2Re(CO)_5^-} (TPP)Fe=C=Re(CO)_4(Re(CO))_5 \qquad (13)$$

Bridged carbene complexes were also reported in the iron series. Thus, a bridged carbene complex with a vinylidene group inserted into an Fe – N bond of (TPP)FeCl was independently reported by Mansuy and coworkers [76, 91] and Balch, La Mar and coworkers [92, 93]. An X-ray analysis of the complex $(TPP)Fe[C{=}C(p\text{-}ClC_6H_4)_2]Cl$ was reported by two different groups [76, 94]. In this compound, the metal is five-coordinated by three of the four pyrrole nitrogen atoms, a chlorine atom and the carbon of the vinylidene group. A 1.914(7) Å (Fe – C) distance and 1.387(6) Å (Fe – N) distance was found, respectively, for the carbene inserted group. The four pyrrole nitrogen atoms are approximately coplanar and the iron atom is displaced out of this plane by 0.3 Å. Magnetic susceptibility, electron spin resonance and Môssbauer spectroscopic studies indicate that this complex is described as an iron(III) complex with an $S = 3/2$ ground state [95].

The decomposition of diazo compounds by transition metal complexes is one of the best methods in chemistry for new synthetic methodologies [70]. This method can be used for catalytic reactions or for synthesis of carbene complexes with iron porphyrins. Reaction of the diazocompound $PhCH_2CHN_2$ with (T-*p*-ClPP)Fe also led to a bridged carbene complex with a $PhCH_2CH$ moiety inserted between the iron and a pyrrole nitrogen atom (Scheme 9) [53]. Reduction of this intermediate spin complex gives the corresponding diamagnetic axial carbene complex. This reaction also confirms the intermediate formation of a bridged carbene species upon reaction of $PhCH_2CHN_2$ with cytochrome P450, which was previously reported by Ortiz de Montellano and coworkers [52, 96]. Similar intermediates were also suggested during the inactivation of sydnones by P450 [96].

Iron-porphyrin model reactions have been reported for several steps of metabolic oxidation of sydnones by cytochrome P450 [52, 96]. Thus, diazoketones react with iron(II) porphyrins to give iron carbene complexes and then *N*-alkylporphyrins after a one electron oxidation. In this case, a migration of the carbene moiety to the pyrrole nitrogen atom is observed (Scheme 10) [97]. Similar results were obtained with $PhCH_2CHN_2$ [53].

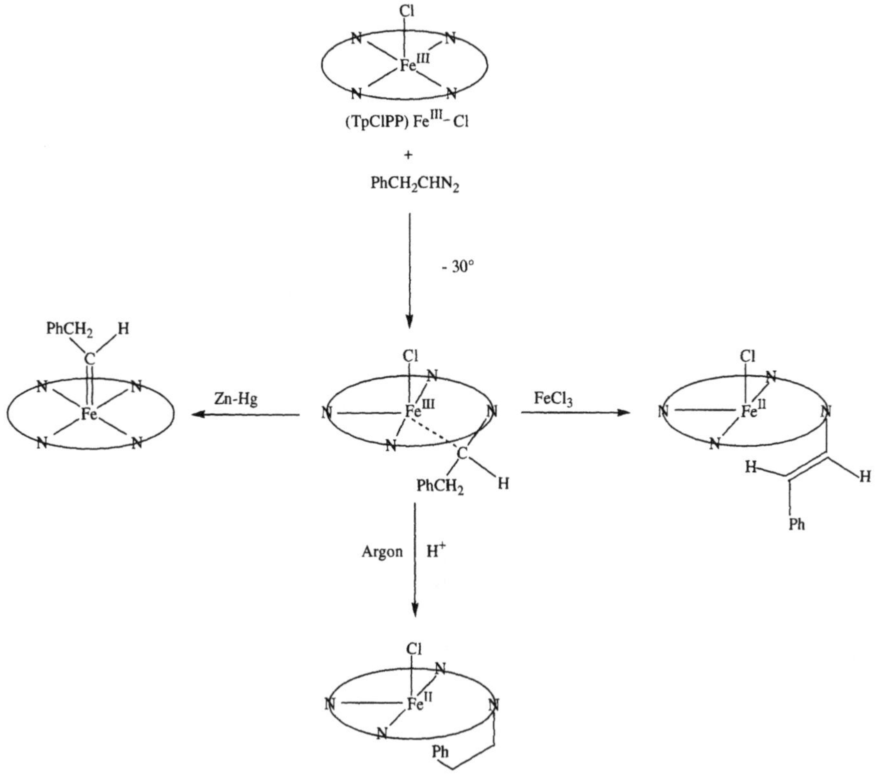

Scheme 9

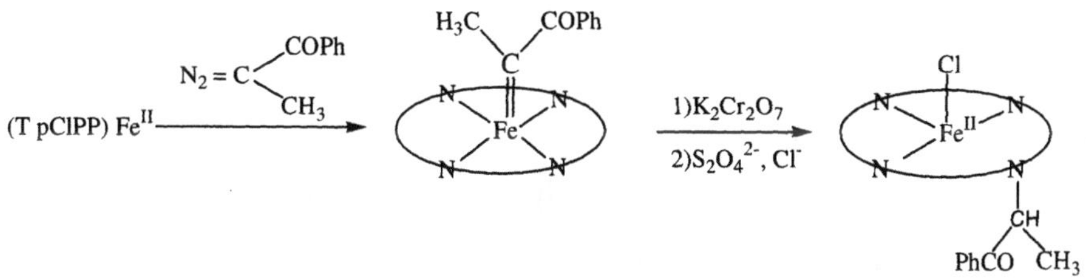

Scheme 10

On treatment of iron(II) porphyrin complexes with reactive diazoreagents, such as mesityldiazomethane and trimethylsilyldiazomethane as the carbene sources, the carbene complexes were observed by ^{1}H NMR, although the sample was contaminated by ferrous porphyrin [60]. This result is particularly important since these carbene complexes are probably involved in the cyclopropanation reaction.

Despite the report of a large number of iron porphyrin carbene complexes in the literature, there are extremely few iron porphyrins bearing non heteroatom-stabilized carbenes [53, 60]. Recently, Che and coworkers re-

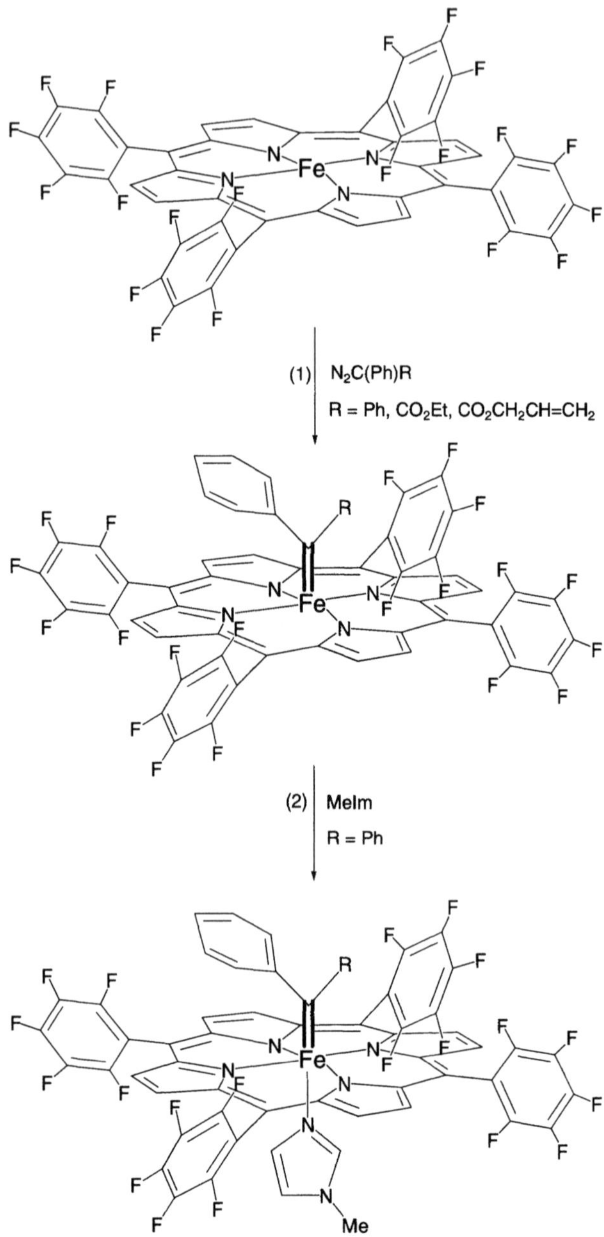

Scheme 11

ported that phenyl-substituted aryl- or (alkoxycarbonyl)carbene complexes of iron meso-tetrakis(pentafluorophenyl)porphyrins (Scheme 11) are sufficiently stabilized to be isolated and fully characterized by NMR and X-ray crystallography [61]. X-ray crystal structures reveal $Fe=CPh_2$ bond lengths of 1.767 and 1.827 Å for the pentacoordinated and hexacoordinated complexes, respectively (Table 1). Reactivity of these metallocarbenes towards carbon-carbon double bonds and C – H insertion was remarkably noted (vide infra).

3.3 X-ray Structures and Spectroscopic Characterizations

From the reactions of various carbene precursors, it appears that the nature of the adduct depends on the nature of the coordinating metal and even on the nature of the porphyrin. Two main binding modes are possible as shown in Scheme 8. The carbene is an axial metalloporphyrin ligand or is inserted into a nitrogen-metal bond. These two structural types have been established by single-crystal X-ray crystallography. Structural data for carbene metal porphyrin complexes are summarized in Tables 1 and 2. Although a large series of carbene complexes of metalloporphyrins have been synthesized, only a few of the compounds have been characterized by single-crystal X-ray diffraction analysis. For the compounds of group 6 (Fe, Ru and Os), there is a complete series. Two papers dealing with the theoretical interaction of metalloporphyrins with carbenes to give axial-metal or nitrogen-metal inserted complexes have been published [72, 73].

Table 1 X-ray structural data for carbene and carbide complexes of iron porphyrin axial carbenes

Complex	Molecular parameters [a]		Refs.
(TPP)Fe(CCl_2)(H_2O)	Fe – C	1.83(3)	[33]
	Fe – N [b]	1.984(4)	
	Fe – O	2.13(3)	
(TPFPP)Fe(CPh_2)	Fe – C	1.767(3)	[61]
	Fe – N [b]	1.966(3)	
(TPFPP)Fe(CPh_2)(MeIm)	Fe – C	1.827(5)	[61]
	Fe – N [b]	1.973(4)	
[(TPP)Fe]$_2$C	Fe – C	1.675(4)	[88]
	Fe – N	1.980(4)	
(TPP)FeCRe(CO)$_4$Re(CO)$_5$	Fe – C	1.605(13)	[90]
	Fe – N	1.982(10)	
	Re – C	1.957(12)	

[a] Bond lengths are reported in Å (errors not reported)
[b] average value

Table 2 X-ray structural data for M – N inserted carbene complexes of metalloporphyrins

M – N Inserted Carbene Complex	Molecular parameters [a]		Refs.
(TPP)Ni[CH(CO_2Et)]	Ni – C	1.905(4)	[98]
	Ni – N	1.911(3)	
	Ni – N	1.910(3)	
	Ni – N	1.905(4)	
(TPP)Fe[C = C(*p*-ClC_6H_4)$_2$]Cl	Fe – C	1.914(7)	[76]
	Fe – N	1.990(5)	
	Fe – Cl	2.290(2)	
(TPP)Fe[C = C(*p*-ClC_6H_4)$_2$]Cl	Fe – C	1.921(5)	[92]
	Fe – N	2.002(4)	
	Fe – N	1.991(4)	
	Fe – N	1.985(4)	
	Fe – Cl	2.299(1)	
(OEP)Co(CHCO_2Et)$_2$(NO_3^-)	Co – C	1.98(1)	[108]
	Co – C	2.00(1)	
	Co – N	1.90(1)	
	Co – N	1.93(1)	

[a] Bond lengths are reported in Å (errors not reported)

A structural comparison of halocarbene iron complex and diphenyl carbene complexes has recently been reported [61]. First, a small ruffling distortion of the porphyrin ring (0.138 Å from the mean plane of 24 component atoms) is observed with the diphenyl iron carbene complexes whereas the porphyrin ring of the dichloro carbene complex (TPP)Fe(CCl_2)(H_2O) is essentially planar. More importantly, the Fe – C(carbene) distance of 1.767(3) Å in the pentacoordinate complex [61] is considerably shorter than that of (1.83(3) Å) for hexacoordinate (TPP)Fe(CCl_2)(H_2O). This short distance has been related to the stability of the diphenyl carbene complex. In contrast, the (TPFPP)Fe(CPh_2)(MeIm) complex exhibits a long Fe – C(carbene) distance (1.827(5) Å) [61] which was attributed to the influence of a trans donor ligand on the metal-carbene bond.

Turning to the metal carbene complexes where the carbene is inserted into one of the metal-nitrogen bonds, structural data are also available [76, 92, 98]. In these complexes (Ni and Fe), the porphyrin macrocycle is largely distorted and the metal is bound to three of the four pyrrole nitrogen atoms and the carbene carbon (and chloride in the iron complex) (Table 2).

Hyperporphyrin spectra, showing a "split Soret" band for carbene complexes of metalloporphyrins have been predicted [99]. Actually, addition of alkylthiolates to (TPP)Fe(II)[C = C(*p*-Cl – C_6H_4)$_2$] resulted in the formation of an hyperporphyrin spectrum with Soret peaks at $\lambda = 385$ and 461 nm [100]. The position of the red-shifted Soret peak in the iron complex is very close to that reported for carbene complexes of cytochrome P450 [3,6] and confirms the ligation of a cysteinate axial ligand *trans* to the carbene in the cytochrome P450 complex. Such a result is characteristic of sulfur binding to metalloporphyrins containing a strong iron-carbon bond. This was first reported for carbon monoxide complexation to cytochrome P450Fe(II) [101].

The ^{13}C chemical shifts of the carbene complex vary from 210 to 385 ppm in the iron series. These large differences have been interpreted by taking into account the more electrophilic nature of the carbene atom due to a possible stabilization of a positive charge by an alkyl group in some cases [102]. An increase of the chemical shift is also noted in going from the pentacoordinate complex to the hexacoordinate adduct. For example, the ^{13}C chemical shift of the diphenyl carbene increases from 358.9 to 385.4 ppm with iron porphyrins [61].

Recently, FT-IR and theoretical analysis of iron porphyrin carbene complexes revealed a possible paramagnetic intermediate carbene state during the reaction of ethyl diazoacetate with Fe(TPP). Focusing on the $C=O$ stretching frequency in the carbene iron complex, a decrease of the value is observed, in comparison with diamagnetic carbene complexes since the radical on the C(carbene) delocalizes over the α-carbonyl group [78]. All isolated axial iron carbene complexes are, however, diamagnetic.

3.4 Reactivity

Addition of alkylthiolates to porphyrin-iron-carbene complexes immediately gave new complexes characterized by their hyperporphyrin spectra [100]. Similar spectra were obtained with the analogous iron thiocarbonyl complexes [103], suggesting that these complexes have similar electronic structures.

The reversible one-electron oxidation of the vinylidene complex (TPP)Fe($C=CAr_2$) induces the carbene ligand to adopt a bridging structure [76,95, 104] giving a complex with an intermediate spin iron(III) [95]. Further oxidation results in the formation of a *N,N′*-4-vinylidene-bridged porphyrin [105].

The dichlorocarbene complex (TPP)Fe(CCl_2) is a useful synthetic intermediate forming isocyanide complexes upon reaction with primary amines (Eq. 14) [106].

$$(TPP)Fe(CCl_2) + 2RNH_2 \longrightarrow (TPP)Fe(RNH_2)(RNC) + 2HCl \qquad (14)$$

A possible application to the synthesis of labeled ^{13}C (or ^{14}C) isocyanides using $^{13}CCl_4$ (or $^{14}CCl_4$) was suggested. Later, a kinetic investigation established two alternate mechanisms for the reaction of primary amines with (TPP)Fe(CCl_2) [107]. Both the formation of the mixed-ligated complex

$(TPP)Fe(CNR)(RNH_2)$ and the bis(amino) complex $(TPP)Fe(RNH_2)_2$ was detected depending of the basicity or the steric hindrance of the primary amine (RNH_2).

An unusual intramolecular rearrangement of an iron carbene complex yielding a *N,N′-cis* bridged tetraarylporphyrin was reported [105]. This rearrangement results from oxidation with $FeCl_3$ of the corresponding vinylidene iron complex obtained from reaction of DDT with tetraanisylporphyrin iron chloride under reducing conditions. An intermediate in which the carbene has been inserted into an iron-nitrogen bond was also suggested by analogy with cobalt porphyrin carbene chemistry [108, 109].

Iron porphyrin carbenes and vinylidenes are photoactive and possess a unique photochemistry since the mechanism of the photochemical reaction suggests the liberation of "free" carbene species in solution [110, 111]. These free carbenes can react with olefins to form cyclopropanes (Eq. 15). The photochemical generation of the free carbene fragment from a transition metal carbene complex has not been previously observed [112, 113]. Although the photochemistry of both Fischer and Schrock-type carbene has been investigated, no examples of homolytic carbene dissociation have yet been found. In the case of the metalloporphyrin carbene complexes, the lack of other coordinatively labile species and the stability of the resulting fragment both contribute to the reactivity of the iron-carbon double bond. Thus, this photochemical behavior is quite different to that previously observed with other classes of carbene complexes [113, 114].

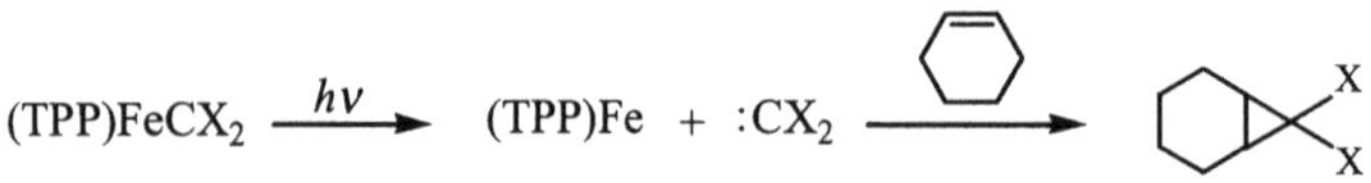

(15)

However, recently, isolated and fully characterized iron porphyrin carbene complexes such as $(TPFPP)Fe(CPh_2)(MeIm)$ [61], have been found to react with hydrocarbons to form C – H insertion products. For example, they can undergo carbon atom transfer into benzylic C – H bonds of cumene or into allylic C – H bonds of cyclohexene. These complexes can also cyclopropanate alkenes without photolysis and undergo intermolecular carbon atom transfer into saturated C – H bonds.

The reductive electrochemistry of iron-carbene porphyrins has been investigated in aprotic solvents [115]. With the vinylidene complex, there is a $2e + H^+$ reduction of the ligand leading to the formation of the corresponding iron(II) vinyl complex. The energies required to reduce by two electrons the other carbene complexes are quite similar [115]. The dichlorocarbene complex is an exception because the reduction is facilitated by the extreme instability of the one-electron intermediate. Formation of σ-alkyl iron(III) porphyrins has been confirmed by independent synthesis [116]. The σ-alkyl iron(III) porphyrins can then be obtained by a one-electron reoxidation reaction. These results are based on an electrochemical investigation of the reduction of carbene complexes in aprotic media [116]. The electrochemistry of organometallic iron porphyrin complexes has been reported by Guilard et al. [117].

Reaction of trimethylsilyldiazomethane with $[(OEP)Fe^{III}](ClO_4)$ gave an *N,N'*-ethano-bridged porphyrin via a possible bridged carbene complex [118]. The chemistry of the naturally occurring iron porphyrins and other iron, cobalt, and ruthenium porphyrins has been reviewed by Setsune and Dolphin [18]. The chemical properties of carbene iron porphyrins and their relationships with *N*-alkyliron(III) porphyrins and σ-alkyl iron(III) porphyrins have been discussed in this review. Thus, the organometallic chemistry of iron porphyrins may serve as an excellent basis for elucidating the mechanism of the catalytic reactions in which the cytochromes P450 play an important role, in particular towards the formation of carbene iron species.

Carbene intermediates have also been identified as products of the reduction of polyhalogenated methanes by iron porphyrins in the presence of cysteine [119]. Such a process seems to be of particular interest because of its potential applicability in the treatment of wastes as well as in remediation approaches to removing polyhalogenated methanes from contaminated soils.

4 Carbenes as Ligands to Ruthenium and Osmium Porphyrins

4.1 Ruthenium

The first ruthenium porphyrin carbene complex was reported by Balch and coworkers [120] by metallation of an *N,N'*-vinyl-bridged porphyrin [105, 121] with $Ru_3(CO)_{12}$ (Scheme 12). In this reaction, both of the C – N bonds (vinyl) were broken. Surprisingly, this reaction also yields two ruthenium(II) dicarbonyl complexes in which the *N,N'*-vinyl bridge remains intact, but the ruthenium has been inserted into a pyrrole C – N bond [122, 123]. Upon heating, these two complexes are converted to the axial ruthenium carbene complex.

Cleavage of a ruthenium dimer such as $[(TTP)Ru]_2$ [124] upon treatment with diazoalkanes and diazoesters affords the corresponding carbene complexes, $(TTP)Ru(CHCH_3)$ and $(TTP)Ru(CHCO_2CH_2CH_3)$, respectively (Scheme 13) [125, 126]. These carbene complexes were the first such metalloporphyrin species to contain a proton on the carbene carbon atom. Unfortunately, the methylene carbene complex was not detected when diazomethane was used as the reagent. Instead, the ethylene complex was formed.

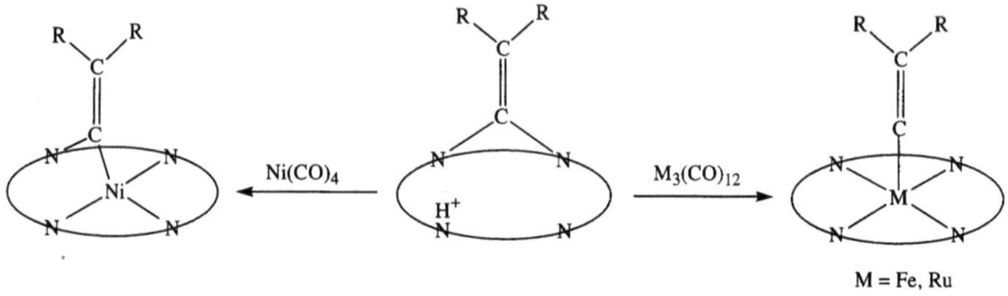

Scheme 12

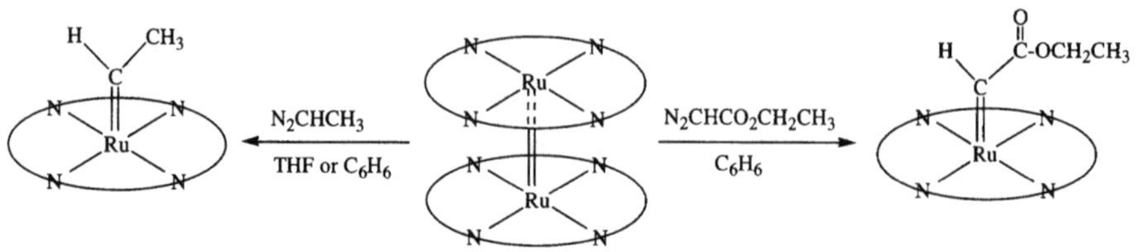

Scheme 13

The carbene complexes were also prepared by interaction of geminal dihalides with zero-valent ruthenium porphyrins such as $K_2[(TTP)Ru]$ by the same authors (Eq. 16) [125].

$$K_2[(TPP)Ru] + CCl_2CHR \longrightarrow (TPP)Ru(CHR) + 2KCL \quad (16)$$
$$R = CH_3\,, Si(CH_3)_3$$

It was later demonstrated that decomposition of a dialkyl ruthenium(IV) porphyrin also yields the carbene species $(TPP)Ru(CHCH_3)$ due to α-abstraction [127]. More recently [128], it was reported that addition of ethyl diazomethyl acetate to the 14 electron species, (TMP)Ru [129], also affords $(TMP)Ru(CHCO_2CH_2CH_3)$ in a more classical route (Eq. 17).

$$(TMP)Ru + N_2CHCO_2Et \longrightarrow (TMP)Ru(CHCO_2Et) + N_2 \quad (17)$$

The reaction of (TMP)Ru with ethyne to produce a μ-biscarbene complex $[(TMP)Ru]_2(\mu\text{-}C_2H_2)$ was reported by Rajapakse et al. (Eq. 18) [130]. It is worth noting that formation of the carbene complex appears to require at least laboratory light to proceed, the C_2H_2 reaction being stopped in the dark. This complex was characterized by ^1H and ^{13}C NMR ($\delta = 263.8$ ppm). A possible formation of a π-acetylene ruthenium complex was ruled out on the basis of the spectroscopic data and a comparison with a rhodium complex $[(OEP)Rh]_2(\mu\text{-}C_2H_2)$ which was formulated as a $Rh-CH=CH-Rh$ unit [131]. In contrast to C_2H_2 itself, PhCCPh and PhCCH form 1 : 1 complexes with (TMP)Ru [130]. Similar acetylene rearrangements in the reaction with ruthenium porphyrinogen have been recently reported by Floriani and coworkers [132].

$$2(TMP)Ru(N_2) + C_2H_2 \xrightarrow{C_6H_6} [(TMP)Ru]_2(\mu\text{-}C_2H_2) \quad (18)$$

The first X-ray structure of a ruthenium porphyrin carbene complex was reported by Simonneaux and coworkers (Fig. 1) [133]. To stabilize the ruthenium carbene complex, ethyl diazomalonate was used instead of ethyl diazomethyl acetate, as was previously reported in the bis(oxazolinyl)pyridine (pybox) series [134]. The presence of this complex as an intermediate in cyclopropanation was also discussed in relation with stoichiometric transfer to alkenes (vide infra) [135].

Figure 1 illustrates the molecular structure of (TPP)Ru(C(CO_2Et)$_2$) (MeOH) [133]. The coordination sphere of the ruthenium atom consists of four pyrrole nitrogen atoms, one carbon atom from the carbene ligand and one oxygen atom from the methanol group. As expected for a six coordinate complex, the porphyrin ligand is nearly planar. However, the ruthenium atom is slightly out of the mean porphyrin plane 0.12 Å towards the carbene ligand. A similar situation was previously observed for an osmium(II) porphyrin carbene complex, (TTP)Os($CHSiMe_3$)(THF) [136] and a rhodium(III) porphyrin carbene complex, (TPP)Rh[C($NHCH_2Ph$)$_2$]($CNCH_2Ph$)PF_6 [137]. The geometry of the coordination sphere is octahedral. The carbene fragment is slightly distorted since the angle is 112.2(7)° which is significantly smaller than the

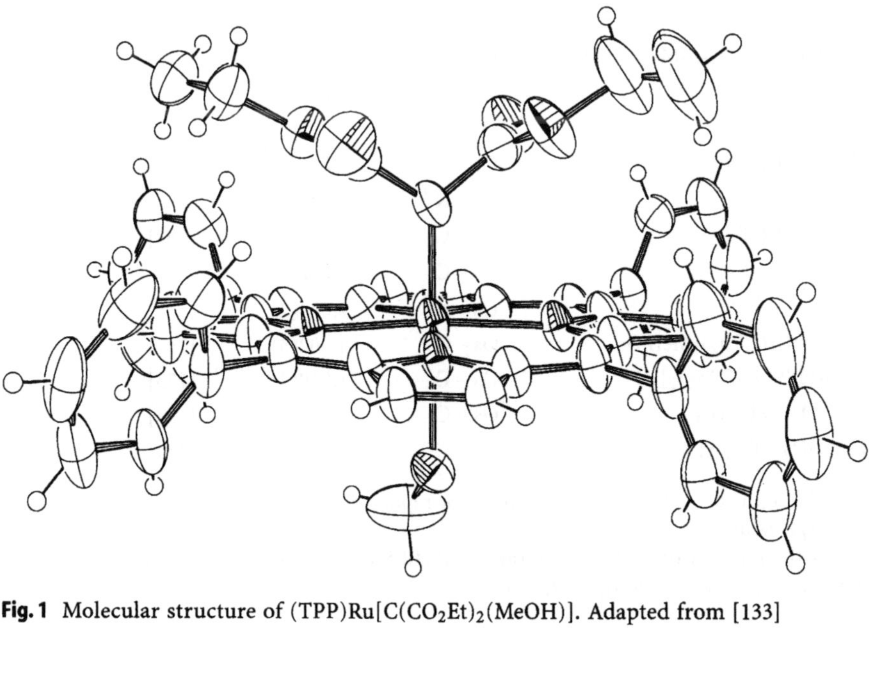

Fig. 1 Molecular structure of (TPP)Ru[C(CO_2Et)$_2$(MeOH)]. Adapted from [133]

Table 3 X-ray structural data for carbene and carbide complexes of ruthenium and osmium porphyrin carbenes

Complex	Molecular parameters [a]		Refs.
(TPP)Ru[C(CO_2Et)$_2$](THF)	Ru – C	1.829(9)	[133]
	Ru – N [b]	2.046(6)	
	Ru – O	2.293(6)	
(P*)Ru(CPh_2) [c]	Ru – C	1.860(6)	[140]
	Ru – N [b]	2.044(5)	
(P*)Ru(C(Ph)($CO_2CH_2CH=CH_2$)) [c]	Ru – C	1.847(3)	[140]
	Ru – N [b]	2.037(2)	
(TPFPP)Ru [C(*p*-C_6H_4OMe)$_2$]	Ru – C	1.854(4)	[141]
(TPFPP)Ru(CPh_2)(MeOH)	Ru – C	1.853(3)	[141]
(TPFPP)Ru(CPh_2)(EtSH)	Ru – C	1.858(5) [b]	[141]
(TPFPP)Ru (CPh_2)(MeIm)	Ru – C	1.876(3)	[141]
(TPFPP)Ru (CPh_2)($OPPh_3$)	Ru – C	1.853(3)	[141]
(TPFPP)Ru [C(Ph)CO_2Me](MeOH)	Ru – C	1.850(3)	[141]
(TPFPP)Ru [C(Ph)CO_2Et](MeOH)	Ru – C	1.868(3)	[141]
(TPFPP)Ru[C(Ph)CO_2R](MeOH) [d]	Ru – C	1.806(3)	[141]
(T-3,4,5-MeOPP)Ru(CPh_2)	Ru – C	1.866(7)	[141]
(T-2,6-ClPP) Ru(CPh_2)	Ru – C	1.859(5)	[141]
(TTP)Os($CHSiMe_3$)(THF)	Os – C	1.79(2)	[136]
	Os – N [c]	2.034(4)	
	Os – O	2.328(3)	
(TTP)Os[C(C_6H_4-*p*-Me_3)$_2$](THF)	Os – C	1.865(5)	[136]
	Os – N [b]	2.050(7)	
	Os – O	2.328(3)	
(TPFPP)Os(CPh_2)	Os – C	1.870(2)	[152]
	Os – N [b]	2.041(2)	
(TPFPP)Os(CPh_2)$_2$	Os – C	2.035(2)	[152]
	Os – C	2.027(3)	
	Os – N [b]	2.044(2)	
(TPFPP)Os(CPh_2)(MeIm)	Os – C	1.902(3)	[141]
[(T-*p*-FPP)Os(CPh_2)$_2$]$_2$O	Os – C	1.910(4)	[153]
(TTP)Os(CPh_2)(Py)	Os – C	1.903(7)	[153]
[(OEP)Os(CPh_2)$_2$]$_2$O	Os – C	1.930(8)	[154]

[a] Bond lengths are reported in Å (errors not reported)
[b] average value
[c] P* is the dianion of 5,10,15,20-tetrakis-[(1S,4R,5R,8S)-1,2,3,4,5,6,7,8-octahydro-1,4:5,8-dimethano anthracene-9-yl]porphyrin
[d] R is $CH=CH_2$

120° angle for ideal sp^2 hybridization as has been previously reported for an osmium porphyrin carbene complex [136]. Steric interactions between the two ethoxy groups of the carbene fragment and the phenyl groups of the porphyrin may explain this result. The Ru – C distance of 1.829(9) Å is slightly shorter than ruthenium-carbon double bonds reported for other nonporphyrin molecular structures of ruthenium carbene complexes. For example, the ruthenium-carbon distances in $RuCl_2(pybox)[C(CO_2Me)_2]$ (pybox: bis(oxazolinyl)pyridine) [134], $Ru(Cl_2)(PPh_3)(CH-CH=CPh_2)$ [138] and CpRuI(CO)[C(OEt)Ph] [139] are 1.880(7), 1.887(7) and 1.997(52) Å, respectively. For comparison, the M = C distances of the other carbene complexes that have been characterized by single-crystal X-ray diffraction analysis are reported in Table 3.

Two X-ray structures of chiral ruthenium carbene complexes were reported. $(P^*)Ru(CPh_2)$ and $(P^*)Ru(C(Ph)(CO_2CH_2CH=CH_2))$ were obtained by reaction of the chiral carbonyl ruthenium complex (P*)Ru(CO) (EtOH) (P* = 5,10,15,20-tetrakis-[(1S,4R,5R,8S)-1,2,3,4,5,6,7,8-octahydro-1,4 : 5,8-dimethanoanthracene-9-yl]porphyrin dianion) with the corresponding diazomethyl derivatives (Fig. 2) [140]. Both complexes contain a five coordinate ruthenium atom that is situated in a slightly distorted square-pyramidal coordination sphere with the carbene C atom at the vertex site. For the diphenyl carbene complex, the Ru – C distance is 1.860(6) Å and the ruthenium atom is displaced from the mean plane of the four pyrrole nitrogen atoms towards the carbene C atom by 0.19 Å. For $(P^*)Ru[C(Ph)(CO_2CH_2CH=CH_2)]$, the Ru – C distance is 1.847(3) Å (Table 3) and the ruthenium atom is displaced from the mean plane of the four pyrrole nitrogen atoms towards the carbene C atom by 0.22 Å. Five coordinate carbenes and their solvent adducts or even bis-carbene species were suggested as intermediates in the cyclopropanation reactions (vide infra) [140].

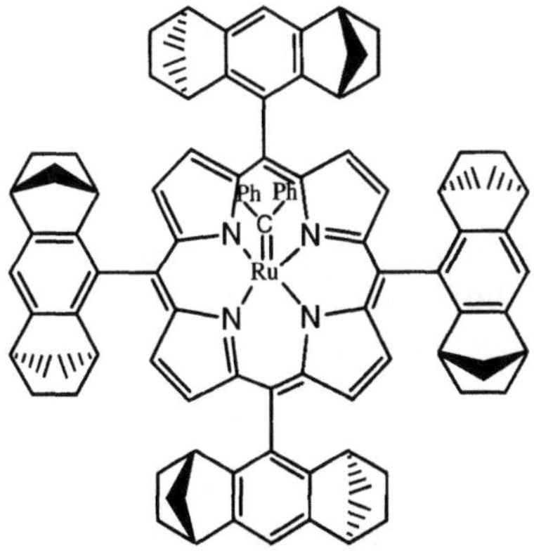

Fig. 2 Schematic structure of the chiral complex showing the Ru-carbene bond. Adapted from [140]

Very recently, several five and six-coordinate ruthenium monocarbene complexes have also been structurally characterized by the groups of Che [141] and Miyamoto [142–144]. X-ray crystal structure determinations revealed Ru = C distances of 1.806(3)–1.876(3) Å, comparable to those in the previously reported analogues (Table 3). In one case, (TPFPP)Ru[C(p-$C_6H_4OMe)_2$], the complex is a one-dimensional coordination polymer [141]. In this particular complex, the carbene groups function as bridges through the carbene carbon atom and one of the two methoxy oxygen atoms. Many of these complexes were also characterized by ^1H and ^{13}C NMR [141].

The isolation of a diamagnetic bridging methylene complex [(OEP-N-μ-CH_2) Ru(CH_3)](BF_4) from decomposition of [(OEP – N – CH_3)Ru(CH_3)](BF_4) was also possible. This complex has been characterized by ^1H NMR and partially by an X-ray structure [145]. Unfortunately, reduction of this complex did not result in formation of an axial methylene carbene complex as was postulated by James and Dolphin [146]. Although M = CH_2 species have been prepared [147, 148], similar metalloporphyrin complexes are not yet known. Ruthenium carbene complexes which are involved in catalytic reactions will be discussed below.

4.2 Osmium

A series of osmium *meso*-tetra-*p*-tolyl-porphyrin carbene complexes, (TTP) Os(CRR′)(R,R′ = *p*-tolyl; R = H, R′ = $SiMe_3$ or CO_2Et) were first prepared by Woo and Smith [149] by treating [$(TTP)_2Os$] with the appropriate diazoalkanes (Eq. 19). As an indirect method, one of these carbene complexes can also be prepared by reaction of the silylene complex (TTP)Os($SiEt_2$)(THF) [150] with di-*p*-tolyl diazomethane.

$$1/2[(TPP)Os]_2 + N_2CRR' \rightarrow (TPP)Os(CRR') + N_2 \qquad (19)$$

$$R = R' = p\text{-}C_6H_4CH_3$$

$$R = H, \quad R' = SiMe_3$$

$$R = H, \quad R' = CO_2Et$$

Addition of four-substituted pyridine derivatives to (TTP)Os($CHCO_2Et$) affords stable osmium ylide complexes (Scheme 14) [151]. The same group also reported the molecular structure of the two carbene complexes (TTP)Os ($CHSiMe_3$)(THF) and (TTP)Os[C(C_6H_4-*p*-Me$)_2$](THF) [136]. As expected for a six-coordinate complex, the porphyrin in (TTP)Os[C(C_6H_4-*p*-Me$)_2$](THF) is nearly planar but the osmium atom is slightly out of the mean porphyrin plane 0.14 Å toward the carbene atom. The Os – C (carbene) distance is 1.865(5) Å and the carbene ligand is slightly distorted since the angle formed by the carbene carbon atom and the two adjacent carbon atoms is 113.0(4)°,

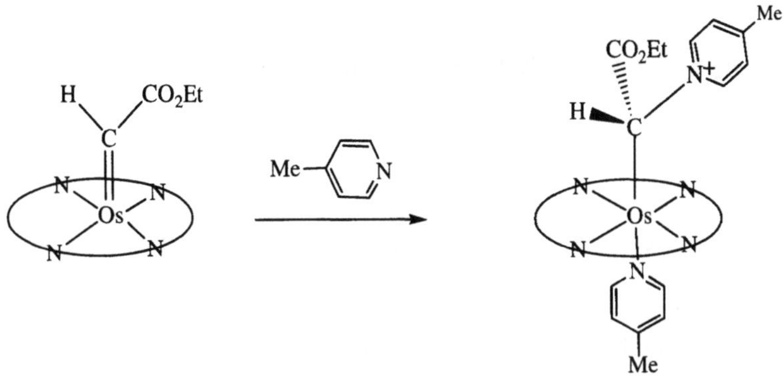

Scheme 14

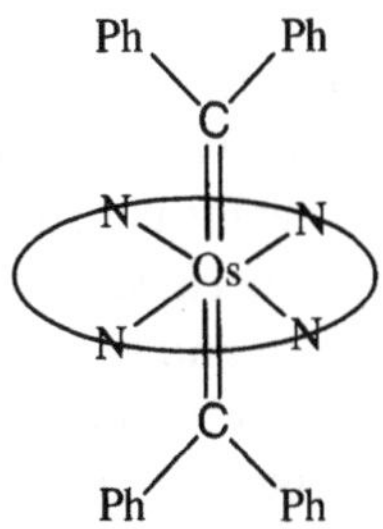

Fig. 3 Schematic structure of (TPFPP)Os(CPh_2)$_2$ showing the two Os–carbene bonds. Adapted from [152]

a value which is significantly smaller than the 120° angle for ideal sp^2 hybridization. Similar data were obtained with (TTP)Os($CHSiMe_3$)(THF) [136]. The presence of a *trans* bis-carbene species was also mentioned but the complex was contaminated with the mono carbene species.

Che and coworkers [152] were able to isolate and characterize a pure bis-carbene (TPFPP)Os(CPh_2)$_2$ (Fig. 3). The bis-carbene species represents the first structurally characterized *trans*-bis-carbene metal complex whose carbene groups are not stabilized by heteroatoms. The related pentacoordinated mono-carbene complex was also prepared and characterized by an X-ray structure. A comparison of the reactivity of these complexes with olefins suggests that the bis-carbene species acts as an intermediate in cyclopropanation. Thus, the inertness of the mono-carbene complex towards stoichiometric styrene cyclopropanation and the observation of an efficient cyclopropanation of styrene in the presence of the bis-carbene complex as a catalyst support this suggestion [152]. A recent X-ray structure determination for (TPFPP)Os(CPh_2)(MeIm) revealed an Os = C distance of 1.902(3) Å (Table 3) [141]. Recently, Che and coworkers [153] and Miyamoto and coworkers [154] reported oxo-bridged carbene complexes of osmium porphyrins (see Table 3). These compounds are rare examples of oxo-binuclear carbene complexes.

5 Metalloporphyrins as Catalysts for Carbene Transfer

5.1 Cyclopropanation

Enantioselective carbene transfer to olefins is an important area of asymmetric synthesis [134, 155–157]. Recent growth in the area of transition metal porphyrin chemistry has, in part, been driven by the increased interest associated with metal-catalyzed cyclopropanation, in particular with rhodium porphyrins [158–166].

The use of metalloporphyrins as cyclopropanation catalysts originated with Callot who reported that (TPP)RhI provided a *cis* preference for the cyclopropanation of styrene with ethyl diazoacetate (Scheme 15) [158, 159]. This was quite unexpected because cyclopropanation of *cis*-olefin using diazoesters and metal derivatives as catalysts usually gives the *trans* cyclopropyl ester as the major product [167]. The *cis*-selectivity increased with the size of the substituents at the *meso* position and suggested a preferential direction of approach of the alkene towards a rhodium carbene complex [159].

Examples involving osmium [168, 169] and iron [170] porphyrins as catalysts have also been reported but the catalysts mainly provide a *trans* product. A mechanism for iron porphyrin-catalyzed cyclopropanation was proposed by Kodadek and coworkers [170]. A transition state for carbene transfer which is reached later than in the rhodium porphyrin catalyzed reaction is suggested. In this case, the olefin is parallel to the metallocarbene and significant bonding has occurred. This geometry explains why 1,2 disubstituted alkenes are poor substrates since there is a steric interaction between the porphyrin ring and the alkene [170].

The catalytic production of olefins, diethyl maleate and fumarate, from ethyl diazoacetate has been reported with osmium [149] and ruthenium [128] porphyrins. Despite the periodic relationship of ruthenium to iron and osmium and the syntheses of different carbene complexes of ruthenium porphyrins, developed by Collman et al. [125–127], it is only very recently that cyclopropanation [135, 171] and ethyl diazoacetate insertion into heteroatom bond reactions [172] were observed using ruthenium porphyrins as catalysts. The details of the catalytic reaction of diazoesters with simple olefins catalyzed with ruthenium porphyrins have been reported [173]. Product yields,

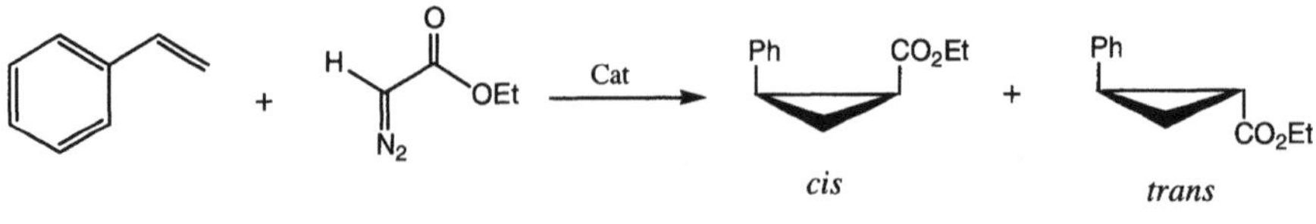

Scheme 15

stereoselectivities and regioselectivities for ruthenium porphyrin-catalyzed cyclopropanation reactions of ethyl diazoacetate with styrene derivatives are compared with observed stereoselectivities for cyclopropanation reactions catalyzed with other metalloporphyrin catalysts. Linear correlations are observed when the rates for competitive cyclopropanation or product stereoisomer ratio are plotted against Hammet constants of various ring-substituted groups on the styrenes [173]. Isomeric distributions for the cyclopropanation of isoprene and 1,3-pentadiene with ethyl diazoacetate and competition studies of the cyclopropanation have also been reported. All of these results agree with a major electronic and steric influence on both the regiochemical and stereochemical control in the catalytic cyclopropanation reactions [173]. Recently, it was discovered that ruthenium porphyrins catalyze effective cyclization of γ-alkoxy-α-diazo-β-ketoesters to form 1,3-dioxolanes selectively [174]. Reaction of (TTP)Ru(CO) with a diazo ketoester affords a ruthenium-carbene complex which has been isolated [174].

There has also been a renewal of interest in reactions catalyzed by ruthenium(II) porphyrin complexes, simultaneously with the development of new chiral ruthenium porphyrins [175–178]. Although these reactions focus mainly on asymmetric epoxidation of olefins [179, 180], in some cases asymmetric cyclopropanations were very successful. As a recent example, the intermolecular cyclopropanation of styrene and its derivatives with ethyl diazoacetate afforded the corresponding cyclopropyl esters in up to 98% ee with high *trans/cis* ratios of up to 36 and extremely high catalyst turnovers of up to 1.1×10^4 [140]. The structure of the metalloporphyrin is given in Fig. 2. Asymmetric intramolecular cyclopropanations were also reported with the same catalyst [140]. In this case, the decomposition of a series of allylic diazoacetates afforded the cyclopropyl lactones in up to 85% ee. Both the inter- and intramolecular cyclopropanation were proposed to proceed via a reactive chiral ruthenium carbene intermediate. The enantioselectivities in these processes were rationalized on the basis of the X-ray crystal structures of closely related stable chiral carbene complexes obtained from the reaction of the chiral complex with N_2CPh_2 and $N_2C(Ph)CO_2CH_2CH=CH_2$.

Frauenkron and Berkessel [181], and Che et al. [171], independently reported that the ruthenium complex of the same chiral porphyrin, can be used to catalyze the cyclopropanation of styrene. The synthesis of this chiral porphyrin was previously reported by Halterman and Jan [176]. This reaction is particularly interesting since the enantiomeric excesses are quite high (90%). Surprisingly, changing the solvent from 1,2-dichloroethane to benzene resulted in an inversion of the absolute configuration of the major enantiomer for the *cis*-cyclopropane and no change for the *trans*-cyclopropane [181].

Gross et al. [182] described asymmetric cyclopropanation of styrene by an enantiopure carbenoid by ruthenium porphyrins as a catalyst. A comparison with the classical approach, chiral porphyrin and nonchiral carbenoid, provides significant insight into the mechanistic aspects of these reactions.

Using different metal porphyrin complexes (Ru, Fe, Os, and Rh), the authors clearly demonstrate that the absolute configuration of the major diastereomer is related not to the metal but rather to the structure of the porphyrin.

Further insight into the mechanism of osmium(II) porphyrin-catalyzed cyclopropanation of alkenes by diazoalkanes was reported by Woo and coworkers [169]. A mono-carbene complex, (TTP)Os($CHCO_2Et$), has been isolated but is not the catalytically active species. An electron-withdrawing ligand *trans* to the carbene activates the carbon fragment towards transfer to an olefin. Substrate reactivity profiles and labeling studies are consistent with a *trans*-osmium(II) bis-carbene species as the active catalyst [169].

Recently, work in the laboratories of Che [183] found that tosylhydrazone derivatives could be used as a carbene source for cyclopropanation catalyzed by ruthenium porphyrins. The practical utility of these ruthenium catalysts was illustrated in the synthesis of a potent HIV-reverse transcriptase inhibitor.

Finally, asymmetric heterogeneous carbene transfer catalyzed by polymers bearing optically active metalloporphyrins have been recently reported by two groups [184–186]. Thus, it was reported that anodic oxidation of ruthenium complexes of various spirobifluorenylporphyrins [187] and tetrafluorenylporphyrins [188] leads to the coating of the working electrode by insoluble optically active films. They were characterized by electrochemical behavior and physicochemical properties. After removing from the electrode, the ruthenium-complexed polymers were evaluated as catalysts for the cyclopropanation of olefins by ethyl diazocetate. Extension of this method to

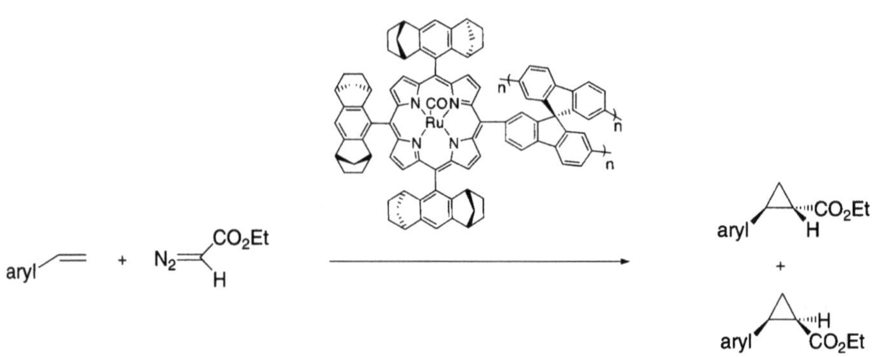

Scheme 16

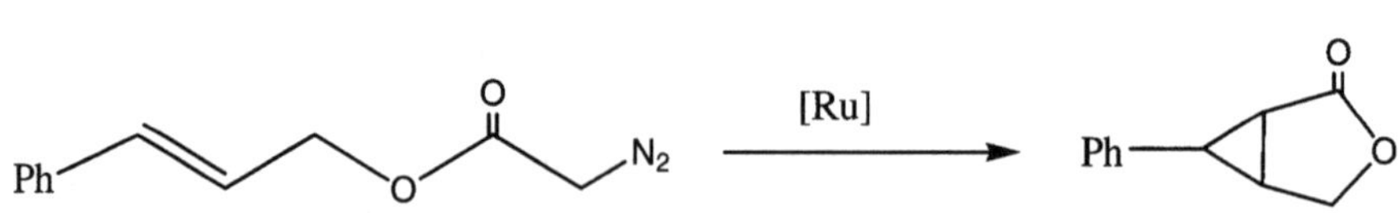

Scheme 17

optically active ruthenium porphyrins have also been recently reported [184]. It was decided to target porphyrins bearing spirobifluorenyl groups to allow polymerization and chiral groups for asymmetric induction, as monomers (Scheme 16). As a second example of carbene transfer catalyzed by chiral ruthenium porphyrin polymers, intramolecular cyclopropanation of *trans*-cinnamyl diazoacetate (Scheme 17) was found to proceed with good enantioselectivity (85%) and high product turnover numbers [186].

5.2 Insertion

Rhodium(III) porphyrins are known to catalyze the insertion of carbethoxycarbenes from ethyl diazoacetate into the C – H bonds of saturated compounds with yields up to 20–25% corresponding to a large increase of the primary/secondary selectivity [189]. In this case the substrates (C_6 to C_{12} *n*-alkanes) were used as solvents. The rhodium porphyrins, (TPP)RhI, (TMP)RhI and (OEP)RhI efficiently catalyze carbene insertion in O – H bonds, leading to ethers by using ethyl diazoacetate under mild conditions [190]. Using (TMP)RhI as the catalyst, a stereoselective insertion reaction was observed with the order of primary > secondary > tertiary for various alcohols.

The ruthenium porphyrins, (TPP)RuCO and (TMP)RuCO catalyze carbene insertion into S – H bonds, leading to dialkyl and alkyl aryl sulfides using ethyl diazoacetate under mild conditions. The insertion process is regiospecific since dithiothreitol reacts to give the S – H insertion product without any trace of the ether compound (Scheme 18) [172]. With a homochiral porphyrin ruthenium complex, asymmetric insertions were obtained but with low enantioselectivities [191].

The ruthenium porphyrins, (TPP)Ru(CO) and (TMP)Ru(CO) also catalyze carbene insertion into N – H bonds [172]. Thus, the complex (TMP)Ru(CO)

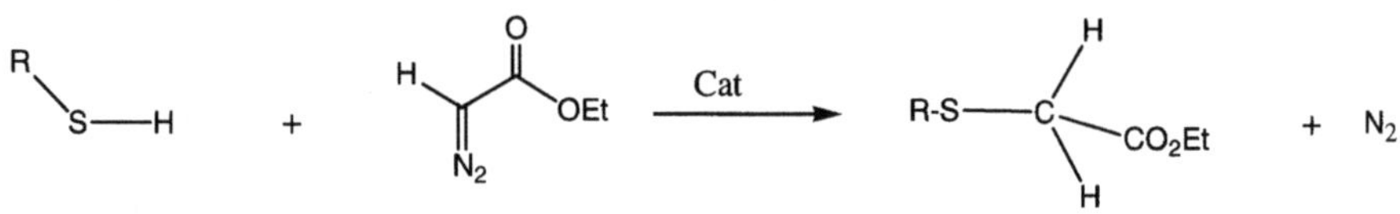

Scheme 18

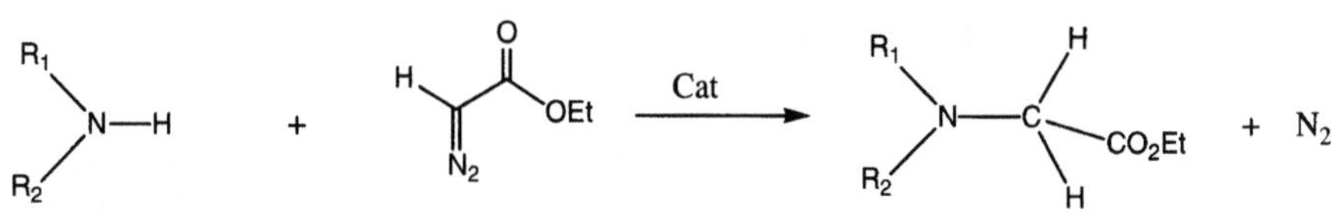

Scheme 19

reacts with ethyl diazoacetate in the presence of alkyl and aromatic amines to give the corresponding *N*-substituted glycine ethyl esters (Scheme 19). Both primary and secondary amines react with EDA but it is necessary to add simultaneously the diazo ester and the substrate into the solution to avoid too large an excess of amine in the presence of the catalyst. The nucleophilic amines clearly coordinate to the ruthenium and, to a certain extent, poison the catalyst.

The first ruthenium porphyrin-catalyzed intramolecular carbenoid C – H insertion to afford selectively *cis*-2,3-disubstituted-2,3-dihydroergocornine using tosylhydrazone salts as the carbene source was reported by Zheng et al. [192]. This general strategy was applied in natural product synthesis to provide a route to the total synthesis of racemic *epi*-conocarpan. Enantioselective synthesis of 2,3-dihydrobenzofurans was also achieved by a similar route using chiral ruthenium porphyrins as catalysts for this interesting carbon-carbon bond formation [193]. Recently, it was found that dinuclear μ-oxo osmium porphyrins are able to catalyze intermolecular carbene insertion into C – H bonds in cyclohexene [153].

5.3 Sigmatropic Rearrangements

The catalytic effectiveness of ruthenium porphyrins for ylide generation in reactions of ethyl diazoacetate and diisopropyl diazomethylphosphonate with some allylic substrates was described for the first time by Simonneaux and coworkers (Scheme 20) [194]. These reactions result in products of the [2,3]-sigmatropic rearrangement of intermediate allylic ylides. It was demonstrated that simple ruthenium porphyrins are highly effective catalysts for carbenoid reactions with alkyl allyl sulfides and alkyl allyl amines providing

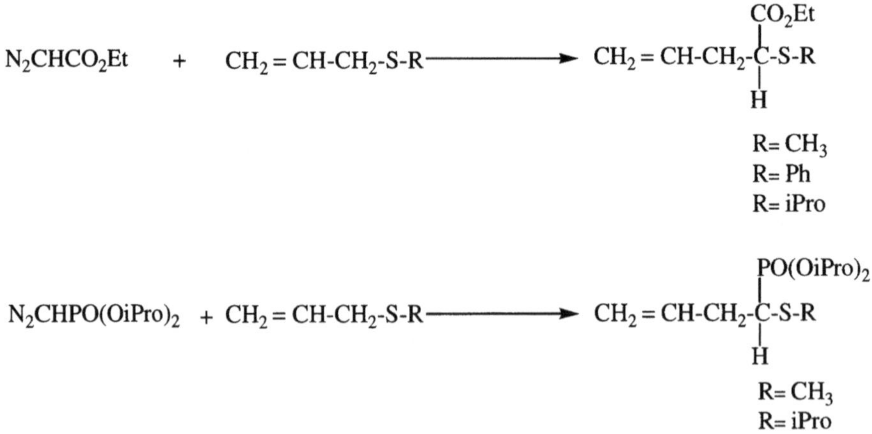

Scheme 20

the formal C – S *or* C – N insertion rather than the more classical cyclopropanation. To fully characterize the catalytic properties of the porphyrin compound, it was also shown that, in the competition process between ylide generation and the insertion of the diazo compound into the heteroatom-hydrogen bond, only the Z – H (Z = N or S) insertion compound is observed [194]. As an application of these 2,3-sigmatropic rearrangements, Che and his group, recently developed a concise and elegant synthesis of racemic Platynecine [195].

5.4 Other Reactions

A novel extension of the catalytic activity of metalloporphyrins with the first use of $(TPP)Fe^{II}$ as a catalyst for the efficient and selective olefination of aldehydes has been recently reported by Woo and coworkers [196]. Olefination of aromatic and aliphatic aldehydes (Eq. 20) was achieved in excellent yield (> 85%) at ambient temperature using ethyl diazoacetate and triphenylphosphine in the presence of a catalytic amount of (TTP)Fe. Ethyl maleate and fumarate were also observed as side products.

$$\mathrm{RCHO + N_2CHCO_2Et + Ph_3P \xrightarrow[-N_2]{(TTP)Fe} RCH = CHCO_2Et + Ph_3PO} \qquad (20)$$

$$\mathrm{R = Ph,}\ \ p\text{-}\mathrm{CH_3C_6H_4},\ \ p\text{-}\mathrm{NO_2C_6H_4},\ \ p\text{-}\mathrm{ClC_6}{}_{H}4,\ \ \mathrm{PhCH_2},\ \ \mathrm{(Ph)_2CH}$$

In the proposed mechanism, the iron porphyrin serves to catalytically convert the diazo reagent and phosphine to the corresponding phosphorane. Then, the phosphorane produces a new olefin and phosphine oxide on reaction with aldehyde [196]. Although other metal complexes can catalyze this reaction [197, 198] the iron system seems to be especially efficient.

Aggarwal et al. [199] reported the preparation of phosphoranes derived from phosphites, generating these previously unknown species, by carbene transfer using decomposition of diazo derivatives catalyzed by iron porphyrins. This new class of ylides led to a high level of *E* selectivity with semi-stabilized ylides in Wittig olefinations.

Recently, ruthenium porphyrins have been used as catalysts for tandem carbonyl ylide formation/cycloaddition transformation [200, 201]. The diastereoselectivity of the reaction with a variety of alkyl- and aryl-substituted α-diazo ketones was found to be highly substrate-dependant. Finally, highly selective intra- and intermolecular coupling reactions of diazocompounds to form cis-alkenes, including organic macrocyclic compounds were developed by Che et al. [202].

6
Perspectives: Bioorganometallic Catalysis with Artificial Heme Proteins

The heme pocket of heme proteins can stabilize the reduced and oxidized states of the iron complex and therefore, heme proteins have a considerable potential for catalytic oxidation of a substrate. If we consider that epoxidation and cyclopropanation are somehow related reactions, carbene transfer reactions catalyzed by heme proteins present another opportunity to use native proteins and even to build artificial proteins with novel properties. Significant results discussed earlier in this review are summarized as follows: (i) carbene adducts of metalloporphyrins have been considered as carbon analogs of the porphyrin-iron-oxo species on the basis of similarity in the electronic spectra [16, 72, 76, 203]; (ii) metalloporphyrins, in particular ruthenium porphyrins, are excellent catalysts for carbene transfer reactions [204, 205]; (iii) evidence for cytochrome P450 carbene complexes is now largely recognized [3]. In view of these results, it is tempting to suggest a combination of heme proteins with carbene precursors, such as diazo derivatives to set up a viable organometallic system for asymmetric synthesis. If native heme proteins can be possible candidates for organometallic reactions, one would also expect ruthenium-reconstituted heme proteins to be good candidates for catalyzing asymmetric cyclopropanations. The ruthenium juxtaposition in the periodic table makes ruthenium an ideal candidate for organometallic reactions with artificial heme proteins. Thus, it should be noted that such hybrid heme proteins have been previously reported, such as ruthenium myoglobin [206, 207], ruthenium hemoglobin [208] and ruthenium peroxidases [209]. Thus, the next stage in this area will be the engineering of new organometallic functions inside the heme proteins. The strategy of heme protein modification can be mainly divided into two approaches [210]: (i) amino acid mutation by site-directed mutagenesis [211]; and (ii) replacement of the native heme with artificially created metalloporphyrins [212, 213]. Some methods have successfully introduced nonbiological functionalities into the proteins which cannot be easily incorporated by genetic or other biological methods. Although many examples have been described an effective system with metalloporphyrins, that utilizes heme proteins as catalysts for carbene transfer, has not yet been forthcoming. However, it is probable that the availability of new artificial metalloproteins will solve this problem in the future.

References

1. Rickes EL, Brink NG, Koniuszy FR, Wood TR, Folkers K (1948) Science 107:396
2. Smith EL, Parker LFJ (1948) Biochem J 43:VIII
3. Mansuy D, Nastainczyk W, Ullrich V (1974) Naunyn-Schmiedeberg's Arch Pharmakol 285:315

4. Mansuy D (1980) Pure Appl Chem 52:681
5. Ortiz de Montellano PR, Reich NO (1986) In: Cytochrome P-450, Structure, Mechanism and Biochemistry. Plenum Press, New York, p 281
6. Wolf CR, Mansuy D, Nastainczyk W, Deutschmann G, Ullrich V (1977) Mol Pharmacol 13:698
7. Mansuy D, Battioni JP, Chottard JC, Ullrich V (1979) J Am Chem Soc 101:3971
8. Groves JT, Avaria-Neisser GE, Fish KM, Imachi M, Kruczkowski RL (1986) J Am Chem Soc 108:3837
9. Mansuy D (1998) Comp Biochem Physiol 121C:5
10. Mansuy D, Battioni P (2000) In: Kadish KM, Smith KM, Guilard R (eds) The Porphyrin Handbook. Academic Press, San Diego, vol 4, p 1
11. Meunier B, de Visser SP, Shaik S (2004) Chem Rev 104:3947
12. Kadish KM, Smith KM, Guilard RE (2000) The Porphyrin Handbook. Academic Press, Boston, (vols 1–10), vol 1
13. Guengerich FP (2001) Chem Res Toxicol 14:611
14. Ortiz de Montellano PRE (1986) Cytochrome P-450: Structure, Mechanism and Biochemistry. Plenum Press, New York
15. Ortiz de Montellano PRE (1995) Cytochrome P-450: Structure, Mechanism and Biochemistry. Plenum Press, New York
16. Mansuy D, Chottard JC, Lange M, Battioni JP (1980) J Mol Catal 7:215
17. Mansuy D (1983) Inorg Chim Acta 79:9
18. Setsune JI, Dolphin D (1987) Can J Chem 65:459
19. Pecknagel RO, Glence EAJ (1973) CRC Crit Rev Toxicol 2:263
20. Ullrich V, Schnabel KH (1973) Drug Metab Disposition 1:176
21. Wolf CR, Harrelson WG, Nastainczyk WM, Philpot RM, Kalyanaraman B, Mason RP (1980) Mol Pharmacol 18:553
22. Nastainczyk W, Ahr H, Ulrich W (1981) Adv Exper Med Biol 136:799
23. Wolf CR, Mansuy D, Nastainczyk W, Ullrich V (1978) Biochem Pharmacol 27:240
24. Hine J, Dowell AM (1954) J Am Chem Soc 76:2688
25. Castro CE, Kray JWC (1966) J Am Chem Soc 88:4447
26. Mansuy D, Battioni JP (1982) J Chem Soc, Chem Commun:638
27. Ahr HJ, King LJ, Nastainczyk W, Ullrich V (1982) Biochem Pharmacol 31:383
28. Ruf HH, Ahr H, Nastainczyk W, Ullrich V, Mansuy D, Battioni JP, Montiel-Montoya R, Trautwein A (1984) Biochemistry 23:5300
29. Castro CE, Bartnicki EW, Belser NO (1978) Biochemistry 17:5582
30. Castro CE, Wade RS (1985) J Org Chem 50:5342
31. Li S, Wackett LP (1993) Biochemistry 32:9355
32. Mansuy D, Lange M, Chottard JC, Guerin P, Morliere P, Brault D, Rougee M (1977) J Chem Soc Chem Commun:648
33. Mansuy D, Lange M, Chottard JC, Bartoli JF, Chevrier B, Weiss R (1978) Angew Chem Int Ed Engl 17:781
34. Ferrara R, Tolando R, King LJ, Manno M (1997) Toxicol Appl Pharmacol 143:420
35. Tolando R, Ferrara R, Eldirdiri NI, King LJ, Manno M (1996) Xenobiotica 26:425
36. Luke BT, Loew GH, McLean AD (1987) J Am Chem Soc 109:1307
37. Luke BT, Loew GH, McLean AD (1988) J Am Chem Soc 110:3396
38. Castro CE, Wade RS, Belser NO (1985) Biochemistry 24:204
39. Krone VE, Thauer RK, Hogenkamp HPC, Steinbach K (1991) Biochemistry 30:2713
40. Dolbier WRJ, Burkholder CR (1990) J Org Chem 55:589
41. Philpot RM, Hogson E (1971) Chem Biol Interact 4:185
42. Franklin MR (1971) Xenobiotica 1:581

43. Mansuy D (1981) Rev Biochem Toxicology 3:283
44. Murray M, Wilkinson CF, Marcus C, Dube CE (1983) Mol Pharmacol 24:129
45. Chiba M, Nishime JA, Wu-Chen I, Vastag KJ, Sahly YS, Kim BM, Dorsey BD, Vacca JP, Lin JH (1998) Biochem Pharmacol 56:223
46. Yu LS, Wilkinson CF, Anders MW (1980) Biochem Pharmacol 29:1113
47. Murray M, Hetnarski K, Wilkinson CF (1985) Xenobiotica 15:369
48. Baek MS, Kim JY, Myung SW, Yim YH, Jeong JH, Kim DH (2001) Drug Metab Dispos 29:381
49. Kumagai Y, Lin LY, Philpot RM, Yamada Y, Oguri K, Yoshimura H, Cho AK (1992) Mol Pharmacol 42:695
50. Stewart FHC (1964) Chem Rev 64:129
51. White EH, Egger N (1984) J Am Chem Soc 106:3701
52. Grab LA, Swanson BA, Ortiz de Montellano PR (1988) Biochemistry 27:4805
53. Artaud I, Gregoire N, Leduc P, Mansuy D (1990) J Am Chem Soc 112:6899
54. Bartnicki EW, Besler NO, Castro CE (1978) Biochemistry 25:5582
55. Wade RS, Castro CE (1973) J Am Chem Soc 95:231
56. Osawa Y, Highet RJ, Murphy CM, Cotter RJ, Pohl LR (1989) J Am Chem Soc 111:4462
57. Tolando R, Cazzaro S, Ferrara R, Rezzadore M, Manno M (1995) Biochem Pharmacol 49:233
58. Chiba M, Jin L, Neway W, Vacca JP, Tata JR, Chapman K, Lin JH (2001) Drug Metab Dispos 29:1
59. Groves JT, Shalyev K, Lee J (2000) In: Kadish KM, Smith KM, Guilard R (eds) Oxometalloporphyrins in Oxidative Catalysis. The Pophyrin Handbook. Academic Press, San Diego 4:17
60. Hamaker CG, Mirafzal GA, Woo LK (2001) Organometallics 20:5171
61. Li Y, Huang JS, Zhou ZY, Che CM, You XZ (2002) J Am Chem Soc 124:13185
62. Fischer EO (1974) Angew Chem Int Ed Engl 86:651
63. Schrock RR (1976) J Am Chem Soc 98:5399
64. Schrock RR (1979) Acc Chem Res 12:98
65. Cardin DJ, Cetinkaya B, Lappert MF (1972) Chem Rev 72:545
66. Herrmann WA (1978) Angew Chem Int Ed Engl 17:800
67. Dotz KH, Fischer H, Hofmann P, Kreissl FR, Schubert U, Weiss K (1983) Transition Metal Carbene Complexes. Verlag Chemie, Weinheim
68. Brookhart M, Studabaker WB (1987) Chem Rev 87:411
69. Demonceau A, Hubert AJ, Noels AF (1991) In: Noels AF, Graziani M, Hubert AJ (eds) Metal Promoted Selectivity in Organic Synthesis. Kluwer Academic Publishers, Dordrecht, p 237
70. Doyle MP (1994) Comprehensive Organometallic Chemistry II 12:421
71. Trnka TM, Grubbs RH (2001) Acc Chem Res 34:18
72. Tatsumi K, Hoffmann R (1981) Inorg Chem 20:3771
73. Brothers PJ, Collman JP (1986) Acc Chem Res 19:209
74. English DR, Hendrickson DN, Suslick KS (1983) Inorg Chem 22:367
75. Lu D, Paeng IR, Nakamoto K (1991) J Coord Rev 83:3
76. Chevrier B, Weiss R, Lange M, Chottard JC, Mansuy D (1981) J Am Chem Soc 103:2899
77. Ikeno T, Iwakura I, Yamada T (2002) J Am Chem Soc 124:15152
78. Iwakura I, Tanaka H, Ikeno T, Yamada T (2004) Chem Lett 33:140
79. Brault D, Rougee M, Momenteau M (1971) J Chim Phys 98:1621
80. Scheidt WR, Reed CA (1981) Chem Rev 81:543

81. Scheidt WR (2000) In: Kadish KM, Smith KM, Guilard R (eds) Systematics of the Stereochemistry of Porphyrins and Metalloporphyrins. The Porphyrin Handbook. Academic Press, San Diego CA 3:49
82. Chottard G, Battioni P, Battioni JP, Lange M, Mansuy D (1981) Inorg Chem 20:1718
83. Mansuy D, Guerin P, Chottard JC (1979) J Organometal Chem 171:195
84. Mansuy D, Lange M, Chottard JC (1978) J Am Chem Soc 100:3213
85. Battioni JP, Mansuy D, Chottard JC (1980) Inorg Chem 19:791
86. Battioni JP, Chottard JC, Mansuy D (1982) Inorg Chem 21:2056
87. Mansuy D, Lecomte JP, Chottard JC, Bartoli JF (1981) Inorg Chem 20:3119
88. Goedken VL, Deakin MR, Bottomley LA (1982) J Chem Soc:607
89. Battioni JP, Dupre D, Mansuy D (1981) J Organometal Chem 214:303
90. Beck W, Knauer W, Robl C (1990) Angew Chem Int Ed Engl 29:318
91. Mansuy D, Lange M, Chottard JC (1979) J Am Chem Soc 101:6437
92. Latos-Grazynski L, Cheng RJ, La Mar GN, Balch AL (1981) J Am Chem Soc 103:4270
93. Balch AL, Cheng RJ, La Mar GN, Latos-Grazynski L (1985) Inorg Chem 24:2651
94. Olmstead MM, Cheng RJ, Balch AL (1982) Inorg Chem 21:4143
95. Mansuy D, Morgenstern-Badarau I, Lange M, Gans P (1982) Inorg Chem 21:1427
96. Ortiz de Montellano PR, Grab LA (1986) J Am Chem Soc 108:5584
97. Artaud I, Gregoire N, Battioni JP, Dupré D, Mansuy D (1988) J Am Chem Soc 110:8714
98. Chevrier B, Weiss R (1976) J Am Chem Soc 98:2985
99. Loew GH, Herman ZS, Rhomer M, Goldblum A, Pudzianowski A (1981) Ann NY Acad Sci 367:192
100. Lange M, Battioni JP, Mansuy D, Lexa D, Saveant JM (1981) J Chem Soc, Chem Commun:888
101. Hanson LK, Eaton WA, Sligar SG, Gunsalus IC, Gouterman M, Connell CR (1976) J Am Chem Soc 98:2672
102. Guerin P, Battioni JP, Chottard JC, Mansuy D (1981) J Organometal Chem 218:201
103. Mansuy D, Battioni JP, Chottard JC (1978) J Am Chem Soc 100:4311
104. Balch AL, Chan YW, La Mar GN, Latos-Grazynski L, Renner MW (1985) Inorg Chem 24:1437
105. Wisnieff TJ, Gold A, Evans JSA (1981) J Am Chem Soc 103:5616
106. Mansuy D, Lange M, Chottard JC, Bartoli JF (1978) Tetrahedron Lett:3027
107. Bruice TC, Furter PG, Ball SS (1981) J Am Chem Soc 103:4578
108. Batten P, Hamilton AL, Johnson AW, Mahendran M, Ward D, King TJ (1977) J Chem Soc Perkin 1:1623
109. Johnson A, Ward D, Batten P, Hamilton AL, Shelton G, Elson CM (1975) J Chem Soc, Perkin Trans 1:2076
110. Ziegler CJ, Suslick KS (1996) J Am Chem Soc 118:5306
111. Ziegler CJ, Suslick KS (1997) J Organometal Chem 528:83
112. Crabtree RH (1994) Organometallic Chemistry of The Transition Metals, 2nd edn. Wiley, New York
113. Rooney AD, McGarvey JJ, Gordon KC (1995) Organometallics 14:107
114. Pourreau DB, Geoffroy GL (1985) Adv Organometal Chem 24:249
115. Battioni JP, Lexa D, Mansuy D, Saveant JM (1983) J Am Chem Soc 105:207
116. Lexa D, Saveant JM, Battioni JP, Lange M, Mansuy D (1981) Angew Chem Int Ed Engl 20:578
117. Guilard R, Tabard A, Van Caemelbecke E, Kadish KM (2000) The Porphyrin Handbook 3:295
118. Setsune JI, Iida T, Kitao T (1988) Tetrahedron Lett 29:5677

119. Buschman J, Angst W, Schwarzenbach RP (1999) Environ Sci Technol 33:1015
120. Chan YW, Renner MW, Balch AL (1983) Organometallics 2:1888
121. Lange M, Mansuy D (1981) Tetrahedron Lett 22:2561
122. Chan YW, Wood FE, Renner MW, Hope H, Balch A (1984) J Am Chem Soc 106:3380
123. Balch AL, Chan YW, Olmstead MM, Renner MW, Wood FE (1988) J Am Chem Soc 110:3897
124. Collman JP, Barnes CE, Swepston PN, Ibers JA (1984) J Am Chem Soc 106:3500
125. Collman JP, Brothers PJ, McElwee-White L, Rose E (1985) J Am Chem Soc 107:6110
126. Collman JP, Brothers PJ, McElwee-White L, Rose E, Wright LJ (1985) J Am Chem Soc 107:4570
127. Collman JP, McElwee-White L, Brothers PJ, Rose E (1986) J Am Chem Soc 108:1332
128. Collman JP, Rose E, Venburg GD (1993) J Chem Soc, Chem Commun:934
129. Camenzind MJ, James BR, Dolphin D (1986) J Chem Soc, Chem Commun:1137
130. Rajapakse N, James BR, Dolphin D (1990) Can J Chem 68:2274
131. Setsune JI, Yoshida ZI, Ogoshi H (1982) J Chem Soc, Perkin Trans 1:983
132. Bonomo L, Stern C, Solari E, Scopelliti R, Floriani C (2001) Angew Chem Int Ed 40:1449
133. Galardon E, Le Maux P, Toupet L, Simonneaux G (1998) Organometallics 17:565
134. Nishiyama H, Aoki K, Itoh H, Iwamura T, Sakata N, Kurihara O, Motoyama Y (1996) Chem Lett:1071
135. Galardon E, Le Maux P, Simonneaux G (1997) J Chem Soc, Chem Commun:927
136. Djukic JP, Smith DA, Young VG, Woo LK (1994) Organometallics 13:3020
137. Boschi T, Licoccia S, Paolesse R, Tagliatesta P, Pelizzi G, Vitali F (1989) Organometallics 8:330
138. Nguyen ST, Johnson LK, Grubbs RH (1992) J Am Chem Soc 114:3974
139. Adams H, Bailey NA, Ridgway C, Taylor BF, Walter SJ, Winter MJ (1990) J Organometal Chem 394:349
140. Che CM, Huang JS, Lee FW, Li Y, Lai TS, Kwong HL, Teng PF, Lee WS, Lo WC, Peng SM, Zhou ZY (2001) J Am Chem Soc 123:4119
141. Li Y, Huang JS, Xu GB, Zu N, Zhou ZY, Che CM, Wong KY (2004) Chem Eur J 10:3486
142. Kawai M, Yuge H, Miyamoto TK (2002) Acta Cryst Sec C 58:m581
143. Harada T, Wada S, Yuge H, Miyamoto TK (2003) Acta Cryst Sec C 59:m37
144. Wada S, Yuge H, Miyamoto TK (2003) Acta Cryst Sec C 59:m369
145. Seyler JW, Safford LK, Fanwick PE, Leidner CR (1992) Inorg Chem 31:1545
146. Ke M, Rettig SJ, James BR, Dolphin D (1987) Chem Soc, Chem Commun:1110
147. Schrock RR (1975) J Am Chem Soc 97:6577
148. Tilset M, Bodner GS, Senn DR, Gladysz JA, Parker VD (1987) J Am Chem Soc 109:7551
149. Woo LK, Smith DA (1992) Organometallics 11:2344
150. Woo LK, Smith DA, Young VG (1991) Organometallics 10:3977
151. Djukic JP, Young VG, Woo LK (1994) Organometallics 13:3995
152. Li Y, Huang JS, Zhou ZY, Che CM (2001) J Am Chem Soc 123:4843
153. Li Y, Huang JS, Zhou ZY, Che CM (2003) J Chem Soc, Chem Commun:1362
154. Sawano K, Yuge H, Miyamoto TK (2005) Inorg Chim Acta 358:1830
155. Nishiyama H, Itoh Y, Matsumoto H, Park SB, Itoh K (1994) J Am Chem Soc 116:2223
156. Nishiyama H, Itoh Y, Suwagara Y, Matsumoto H, Aoki K, Itoh K (1995) Bull Chem Soc Jpn 68:1247
157. Doyle MP, Forbes DC (1998) Chem Rev 98:911
158. Callot HJ, Piechocki C (1980) Tetrahedron Lett 21:3489
159. Callot HJ, Metz F, Piechocki C (1982) Tetrahedron 38:2365

160. Brown KC, Kodadek T (1992) J Am Chem Soc 114:8336
161. Maxwell JL, Brown KC, Bartley DW, Kodadek T (1992) Science 256:1544
162. Maxwell JL, O'Malley S, Brown KC, Kodadek T (1992) Organometallics 11:645
163. O'Malley S, Kodadek T (1992) Organometallics 11:2299
164. Bartley DW, Kodadek T (1993) J Am Chem Soc 115:1656
165. Yang J, Breslow R (2000) Tetrahedron Lett 41:8063
166. Barrett AGM, Braddock DG, Lenoir I, Tone I (2001) J Org Chem 66:8260
167. Doyle MP (1986) Chem Rev 86:919
168. Smith DA, Reynolds DN, Woo LK (1993) J Am Chem Soc 115:2511
169. Hamaker CG, Djukic JP, Smith DA, Woo LK (2001) Organometallics 20:5189
170. Wolf JR, Hamaker CG, Djukic JP, Kodadek T, Woo LK (1995) J Am Chem Soc 117:9194
171. Lo WC, Che CM, Cheng KF, Mak TCW (1997) J Chem Soc, Chem Commun:1205
172. Galardon E, Le Maux P, Simonneaux G (1997) J Chem Soc, Perkin Trans 1:2455
173. Galardon E, Le Maux P, Simonneaux G (2000) Tetrahedron 56:615
174. Zheng SL, Yu WY, Che CM (2002) Org Lett 4:889
175. Le Maux P, Bahri H, Simonneaux G (1991) J Chem Soc Chem Commun:1350
176. Halterman RL, Jan ST (1991) J Org Chem 56:5253
177. Ini S, Kapon M, Cohen S, Gross Z (1996) Tetrahedron: Asymmetry 7:659
178. Lai TS, Kwong HL, Zhang R, Che CM (1998) J Chem Soc, Dalton Trans:3559
179. Gross Z, Ini S, Kapon M, Cohen S (1996) Tetrahedron Lett 37:7325
180. Berkessel A, Frauenkron M (1997) J Chem Soc, Perkin Trans:2265
181. Frauenkron M, Berkessel A (1997) Tetrahedron Lett 38:7175
182. Gross Z, Galili N, Simkhovich L (1999) Tetrahedron Lett 40:1571
183. Zhang JL, Chan PWH, Che CM (2003) Tetrahedron Lett 44:8733
184. Ferrand Y, Poriel C, Le Maux P, Rault-Berthelot J, Simonneaux G (2005) Tetrahedron Asymmetry 16:1463
185. Ferrand Y, Le Maux P, Simonneaux G (2005) Tetrahedron Asymmetry 16:3829
186. Zhang JL, Liu YL, Che CM (2002) Chem Commun: 2906
187. Poriel C, Ferrand Y, Le Maux P, Paul C, Rault-Berthelot J, Simonneaux G (2003) Chem Commun:2308
188. Paul-Roth C, Rault-Berthelot J, Simonneaux G (2004) Tetrahedron 60:12169
189. Callot H, Metz F (1982) Tetrahedron Lett 23:4321
190. Hayashi T, Kato T, Kaneko T, Asai T, Ogoshi H (1994) J Organometal Chem 473:323
191. Galardon E, Roué S, Le Maux P, Simonneaux G (1998) Tetrahedron Lett 39:2333
192. Zheng SL, Yu WL, Xu MX, Che CM (2003) Tetrahedron Lett 44:1445
193. Cheung WH, Zheng SL, Yu WY, Zhou GC, Che CM (2003) Org Lett 5:2535
194. Simonneaux G, Galardon E, Paul-Roth C, Gulea M, Masson S (2001) J Organometal Chem 617–618:360
195. Zhou CY, Yu WY, Chan PWH, Che CM (2004) J Org Chem 69:7072
196. Mirafzal GA, Cheng G, Woo LK (2002) J Am Chem Soc 124:176
197. Fujimura O, Honma T (1998) Tetrahedron Lett 39:625
198. Shen Y, Zhang Y, Zhou Y (1998) J Chem Soc, Chem Commun:2195
199. Aggarwal VK, Fulton JR, Sheldon CG, de Vicente J (2003) J Am Chem Soc 125:6034
200. Zhou CY, Yu WY, Che CM (2002) Org Lett 4:3235
201. Zhou CY, Chan PWH, Yu WY, Che CM (2003) Synthesis 9:1403
202. Li GY, Che CM (2004) Org Lett 6:1621
203. Tatsumi K, Hoffmann R, Whangbo MH (1980) J Chem Soc, Chem Commun:509
204. Simonneaux G, Le Maux P (2002) Coord Chem Rev 228:43
205. Che CM, Huang JS (2002) Coord Chem Rev 231:151

206. Paulson DR, Addison AW, Dolphin D, James BR (1979) J Biol Chem 254:7002
207. Srivastava TS (1981) Inorg Chim Acta 55:161
208. Ishimori K, Morishima I (1988) Biochemistry 27:4060
209. Morishima I, Shiro Y, Nakajima K (1986) Biochemistry 25:3576
210. Lu Y, Berry SM, Pfister TD (2001) Chem Rev 101:3047
211. Ozaki S, Matsui T, Roach MP, Watanabe Y (2000) Coord Chem Rev 198:39
212. Hayashi T, Hisaeda Y (2002) Acc Chem Res 35:35
213. Yamamoto Y, Nagao S, Hirai Y, Inose T, Terui N, Mita H, Suzuki A (2004) J Biol Inorg Chem 9:152

Top Organomet Chem (2006) 17: 123–142
DOI 10.1007/3418_002

Published online: 20 January 2006

Organometallic Receptors for Biologically Interesting Molecules

Kay Severin

Institut des Sciences et Ingénierie Chimiques,
École Polytechnique Fédérale de Lausanne (EPFL), 1015 Lausanne, Switzerland
kay.severin@epfl.ch

Abstract Trinuclear organometallic macrocycles are able to complex amino acids and peptides in aqueous solution. The complexation, which is driven by hydrophobic and $\pi-\pi$ interactions, induces a large shift of the ^{1}H NMR signals of the guest. The organometallic host can thus be employed as an NMR shift reagent. Structurally related trinuclear macrocycles selectively bind lithium and fluoride ions in protic as well as non-protic solvents. The binding event can be followed electrochemically or can be transduced via a subsequent chemical reaction into a change of color, which offers the possibility to construct chemosensors for these ions. The organometallic complex $[Cp^*RhCl_2]_2$ can be combined with the dye azophloxine to build an indicator displacement assay for the sequence-specific detection of histidine- and methionine-containing peptides. The assay allows one to detect peptides with His/Met residues close to the *N*-terminus down to a concentration of 0.3 μM in water at neutral pH.

Keywords Amino Acid · Chemosensor · Lithium · Metallamacrocycle · Peptide · Receptor

1 Introduction

Organometallic complexes are increasingly being used for analytical purposes in medicinal and biological chemistry. So far, most efforts have focused on the utilization of organometallic complexes as markers [1], but recent results show that they can also be employed to construct specific receptors for biologically interesting molecules. The latter topic will be discussed in this article. In the first part, the host-guest chemistry of trinuclear, metallamacrocyclic complexes will be described. These complexes can easily be obtained by self-assembly using commercially available half-sandwich complexes of ruthenium, rhodium or iridium. Depending on the ligands employed, they can either bind amino acids or peptides or selectively complex lithium and sodium halides. Potential applications include the utilization as an NMR shift reagent or as chemosensors. In the second part, a recently developed colorimetric assay for histidine- and methionine-containing peptides will be described. This assay is based on the competitive interaction of peptides and a dye with a Cp*Rh(III) complex. It allows one to detect His/Met peptides in a sequence-specific fashion down to a concentration of 0.3 μM. Future developments in this area will be discussed in the final part of this overview.

2 Trinuclear, Metallamacrocyclic Hosts

2.1 Syntheses and Structures of the Hosts

Trinuclear macrocycles containing organometallic half-sandwich complexes of ruthenium, rhodium and iridium can be obtained with various bridging ligands (Scheme 1). In 1992, Fish and coworkers reported that the aqua complex $[Cp^*Rh(OH_2)_3](OTf)_2$ reacts with 9-methyladenine (**L1**) to give a cationic macrocycle in which the metals are connected by the *N*1, *N*6, and *N*7 atoms of the deprotonated nucleobase [2]. Similar structures were obtained with adenosine [2–4], the phosphate methyl ester of adenosine 5′-monophosphate [5], 9-ethylhypoxanthin [6] and 2′-deoxyadenosine (**L2**) [4, 7–9]. The latter complex displays a particularly interesting host-guest chemistry, which will be described in detail in Sect. 2.2. The adenine-based complexes are formed at pH 6–9 and are stable in solution for at least two weeks. With 9-ethylhypoxanthin, on the other hand, trimer-formation is only observed at a pH of ∼6 [6]. Sheldrick [10–12] and Yamanari [3, 4] have shown that the Cp*Rh fragment can be substituted with other halfsandwich complexes such as Cp*Ir, (C_6H_6)Ru and (cymene)Ru.

Neutral metallamacrocycles, which are soluble in organic solvents, were obtained with 2,3-dihydroxypyridine (**L3**) as the bridging ligand. Here, the chloro-bridged complexes $[(arene)RuCl_2]_2$ [13–16], $[Cp^*RhCl_2]_2$ [16, 17] and $[Cp^*IrCl_2]_2$ [16–18] instead of cationic aqua complexes were used for the synthesis. Typically, the dimeric metal complexes were allowed to react with two equivalents of **L3** in methanol in the presence of Cs_2CO_3. Extraction with dichloromethane gave the macrocyclic products in good yields. Complexes

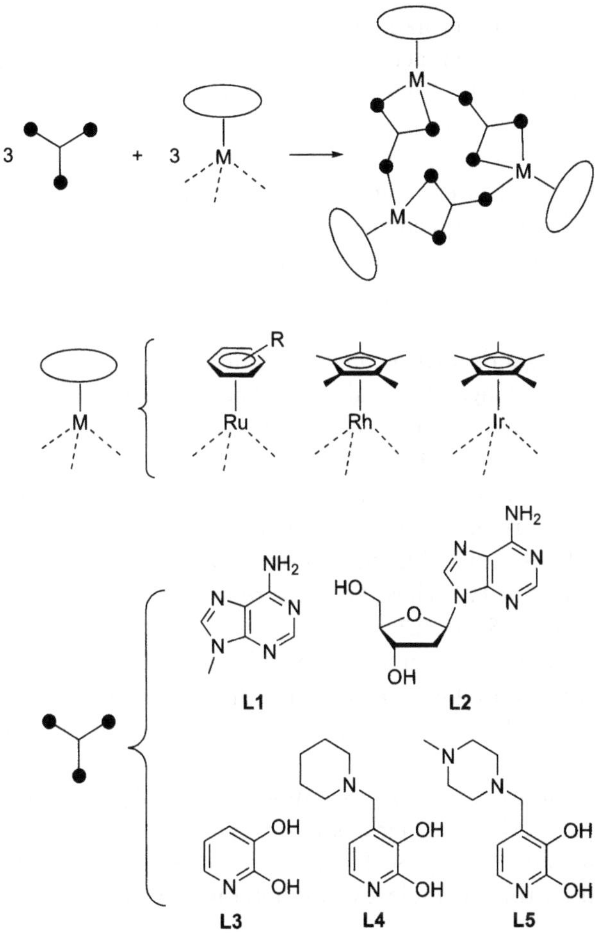

Scheme 1 Trinuclear macrocycles containing organometallic halfsandwich complexes of ruthenium, rhodium and iridium can be obtained with the bridging ligands **L1–L5**

build with the piperidine- and the *N*-methyl piperazine-substituted ligands **L4** [19] and **L5** [20] show a unique behavior: they are soluble in common organic solvents such as chloroform but also in buffered aqueous solution at pH 7–8 because under the latter conditions, the amine side chains are protonated. As a consequence, complexes with the ligands **L4** and **L5** can be obtained simply by dissolving the complex $[(\pi\text{-ligand})MCl_2]_2$ together with the respective ligand in phosphate buffer [19, 20]. The macrocycles are then formed by self-assembly as soon as the starting materials have dissolved.

The trimeric complexes display common spectroscopic and structural characteristics. The high symmetry is reflected by the NMR spectra of the complexes: with the achiral ligands **L1** and **L3–L5**, only one set of signals is observed for the bridging ligands as well as for the π-ligands. For the chiral ligand **L2**, on the other hand, two diastereoisomers and thus a double set of signals are found. The macrocyclization typically results in large shifts for the ^{1}H NMR signals of **L1**–**L5** when compared to the non-coordinated ligands. The metal centers in these complexes are chiral and epimerization is slow compared to the NMR time scale. Apart from spectroscopic investigations, the macrocycles were comprehensively studied by single crystal X-ray analysis. The three metal centers have always the same absolute configuration and are ~ 5.4 Å apart from each other. As a representative example, the molecular structure of a Cp*Rh complex with the ligand **L3** is shown below (Fig. 1).

The bridging ligands form a concave, dome-like structure, which is capped by the three Cp* π-ligands. A similar concave geometry is found for all complexes of this type, regardless of the metal fragment employed. For complexes with **L2**, the hydrophobic cavity formed by the ligands can be used to bind aromatic guests as detailed in Sect. 2.2. The view along the pseudo C_3-symmetry axis reveals that the 12-membered macrocycle contains three O-atoms, which are situated in close proximity to each other (NH groups for complexes with **L1** or **L2**). These O-atoms represent a binding site for lithium and sodium halides (Sects. 2.3 and 2.4).

It should be noted that trinuclear metallamacrocycles with half-sandwich complexes cannot only be obtained with the ligands mentioned above but also with amino acids [21–25] and 3,4-dihydroxy-2-methylpyridine [26]. Furthermore, larger macrocycles have been synthesized with the ligands adenine [10, 11], 6-purinethione [27], 2-amino-6-purinethione [27], 4-imidazolecarboxylic acid [28] (tetrameric structures) and 6-purinethione riboside (hexameric structure) [29]. So far, no special host-guest interactions have been reported for these complexes and therefore they will not be described in more detail.

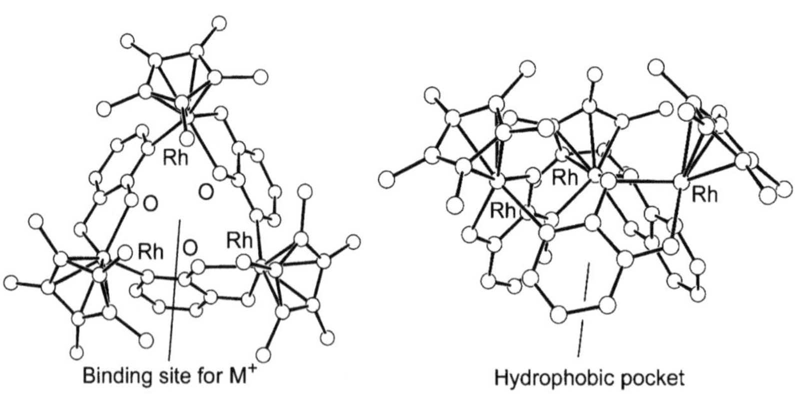

Fig. 1 Molecular structurr of a Cp*Rh complex with the bridging ligand **L3** in the crystal. *Left*: view along the pseudo C_3-symmetry axis highlighting the binding site for alkali metal ions; *Right*: view from the side highlighting the hydrophobic pocket

2.2 Molecular Recognition of Amino Acids and Peptides

The host-guest chemistry of several trimeric Cp*Rh complexes with adenine-based ligands was investigated by Fish and coworkers [7]. It turned out that complex **1** containing the 2′-deoxyadenosine ligand **L2** showed the highest binding constants. The following discussion will thus focus on this receptor. The complexation of several amino acids in buffered aqueous solution (pH 7) was investigated by NMR spectroscopy. The relative binding affinity of the host **1** for different amino acids was determined by the magnitude of the complexation-induced ^{1}H NMR chemical shifts (CICS) with a host/guest ratio of 1.0 : 1.2 ([**1**] = 10–20 mM). Large CICS values were observed for the aromatic protons of the amino acids *L*-phenylalanine (**2**) and *L*-tryptophan (**3**) with no apparent diastereoselectivity in the recognition process. A weak binding was found for the hydrophobic amino acids *L*-leucine and *L*-isoleucine whereas for *L*-valine, *L*-glutamic acid, *L*-histidine, *L*-alanine and *L*-proline, no interaction was observed at all. This data points to $\pi - \pi$ and/or hydrophobic interaction as the main driving force for complexation. Evidence for the assumption that the aromatic amino acids are bound in the cavity and not near the Cp* π-ligands was provided by the fact that substitution at the sugar group lead to significant changes in the binding affinity.

Several other aromatic carboxylic acids were investigated as potential guests and their binding constants were determined by NMR. 2-Aminobenzoic acid (**4**) was bound even better than *L*-tryptophane but for its regioisomer **5**, the host-guest interaction was dramatically decreased. The presence of an amino group is not a prerequisite for binding as evidenced by the values

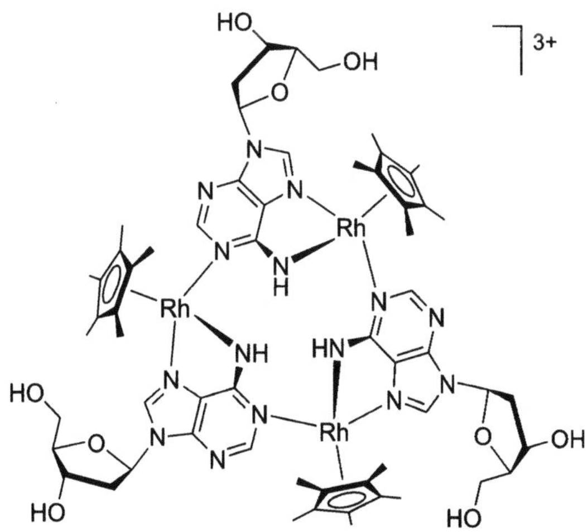

1

for the carboxylic acids **6** and **7**. Sterically rigid aliphatic acids such as **8** and **9** were also able to bind to the host **1**. This suggests that $\pi - \pi$ and hydrophobic interactions are of similar importance in aqueous solution.

Interestingly, the binding of *L*-tryptophane with the host **1** could also be detected by electrospray ionization mass spectrometry [30]. When a mixture of host **1** with guest **2** in an aqueous solution (10 mM NH_4OAc, pH 7.0) was investigated by ESI-MS, ions for **1** and for the host-guest complex [**1** · **2**] were observed at m/z 488 (100%) and m/z 556 (35%), respectively. By further increasing the orifice potential from 45 to 80 V, it was possible to follow the dissociation of the m/z 556 ion for the host-guest complex in the gas phase. Contrary to what was found in solution, an interaction between the aliphatic acid **8** and host **1** could not be observed by ESIMS.

In continuation of their studies, Fish and coworkers reported that host **1** can be used as an aqueous ^{1}H NMR shift reagent [31]. Since the cavity of the receptor is rather shallow – molecular modeling suggests a depth of ~ 4 Å – aromatic guest molecules are only partially surrounded by the host. As a consequence, the CICS values for the aromatic protons differ substantially. This can be used to identify certain protons. The ^{1}H NMR spectrum of 1-naphtoic acid, for example, displays several overlapping signals in the aromatic region (500 MHz, D_2O, pH 9.4). Upon addition of one equivalent of the receptor **1**, some of the signals are shifted due to complexation and all seven aromatic protons show baseline-separated signals. This effect can also be seen for more complicated biomolecules such as the dipeptide *L*-Trp-*L*-Phe (**10**) and the tetrapeptide *L*-Trp-*L*-Met-*L*-Asp-*L*-Phe (**11**). For several of the protons of the aromatic side chains, strong CICS can be observed upon addition of the ^{1}H NMR shift reagent **1**. These effects may help to elucidate the structures of such peptides.

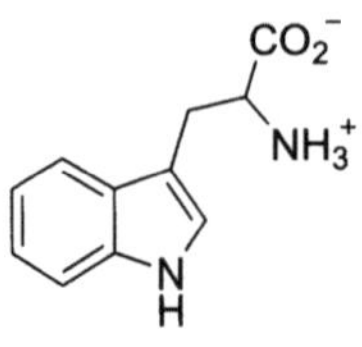

2
$K = 607\ M^{-1}$

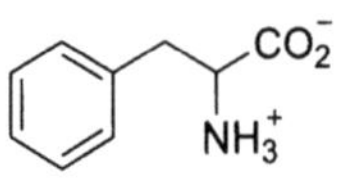

3
$K < 10\ M^{-1}$

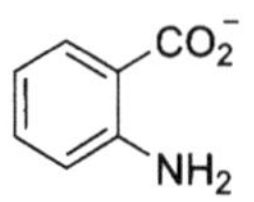

4
$K = 810\ M^{-1}$

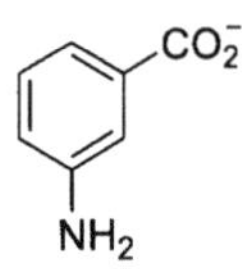

5
$K = 15\ M^{-1}$

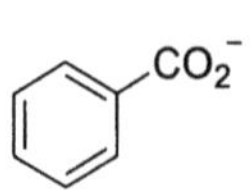

6
$K = 710\ M^{-1}$

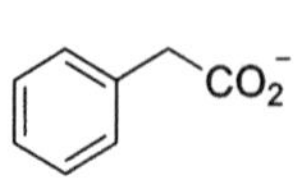

7
$K = 710\ M^{-1}$

8
$K = 760\ M^{-1}$

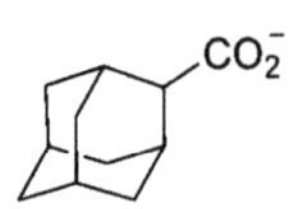

9
$K = 15\ M^{-1}$

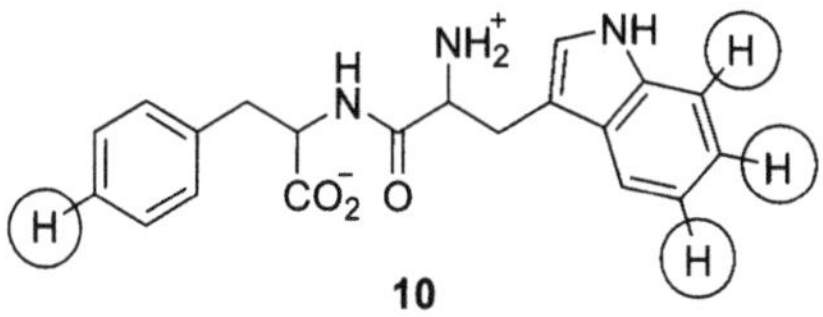

10

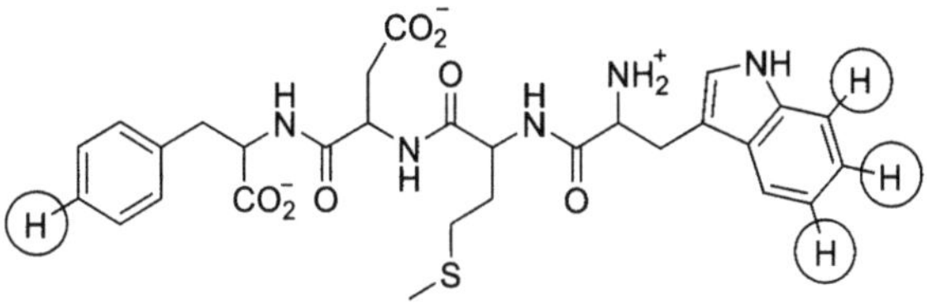

11

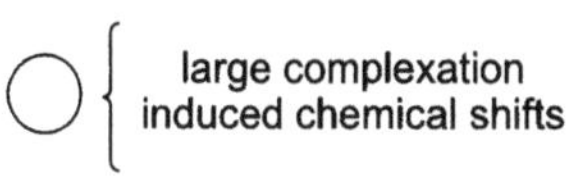

2.3
Selective Recognition and Sensing of Lithium Ions

Metallacrown complexes are analogues of crown ethers, in which metal atoms constitute an integral part of the macrocyclic framework. Complexes of this kind were first reported in 1989 by Pecoraro and coworkers [32, 33]. Today, structurally diverse metallacrown complexes with ring sizes up to 30 atoms are known [34, 35]. Generally, they are obtained in self-assembly reactions using transition metal salts and suited multidentate ligands. Similar to their organic counterparts, metallacrown complexes are able to bind cationic guests such as alkali metal ions.

Trinuclear macrocycles, which contain the dihydroxypyridine ligands **L3**–**L5**, represent analogues of 12-crown-3. When investigated for their ability to bind alkali metal ions, it was found that the complexes display a very high affinity for lithium and sodium salts [13, 17] (Eq. 1). Simple salts such as LiCl and NaCl are bound extremely strong: competition experiments have shown that in chloroform, the binding constants are comparable to those of cryptands [17].

The selectivity of the metallacrown complexes was found to depend on the nature of the π-ligand and the solvent. In organic solvents, the (C_6H_6)Ru and $(p\text{-MeC}_6H_4{}^i\text{Pr})$Ru complexes bind both Li^+ and Na^+ salts whereas the $(C_6H_3Et_3)$Ru, (C_6Me_6)Ru, Cp*Rh and Cp*Ir complexes are specific for Li^+ salts [17]. None of the receptors are able to bind K^+ salts. This pronounced selectivity for small cations can be explained by the steric requirements of the π-ligands. A view along the pseudo C_3 symmetry axis of the Li^+ and Na^+ adducts of a receptor built with (C_6H_6)Ru complexes and **L3** illustrates the close encapsulation of the cations by the benzene ligands of the host (Fig. 2).

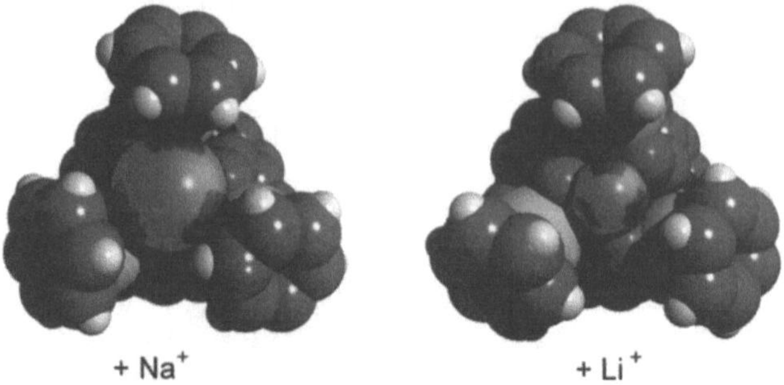

Fig. 2 Space filling representation of a receptor comprised of (benzene)Ru complexes and the ligand **L3** with a sodium ion (*left*) or a lithium ion (*right*) in the binding pocket. Larger cations do not bind to the receptor because they are efficiently blocked by the arene π-ligands

Given the rigidity of the receptor, it is evident that larger cations such as K^+ are not able to bind to the three adjacent O-atoms, which constitute the binding site.

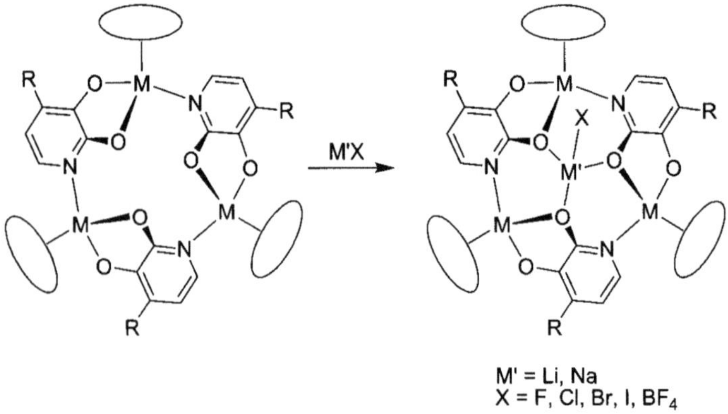

(1)

Since the mid-1960s, lithium salts are among the most frequently used drugs for patients suffering from bipolar disorder [36–38]. Recent studies suggest that lithium could also be used for the treatment of Alzheimer's disease [39, 40] and it was speculated whether lithium could become the "aspirin of the brain" [41]. Apart from applications in the field of psychiatry and neurology, lithium salts were shown to inhibit the replication of certain viruses, have been employed to treat skin diseases and can affect the immune response [38]. Given the pharmacological relevance, it is not surprising that considerable efforts have been devoted towards the synthesis of specific receptors [42]. So far, most investigations have focused on organic ionophores but it was shown that the organometallic crown complexes discussed above are an interesting alternative for the selective sequestering and sensing of lithium ions.

In organic solvents, a receptor based on $(C_6H_5CO_2Et)Ru$ complexes and the ligand **L3** (**12**) showed the most promising characteristics of all complexes investigated [14]. Although this macrocycle was able to bind sodium ions, it showed an outstanding affinity and selectivity for Li^+ salts. This was demonstrated by the following experiment: If an aqueous solution containing NaCl, KCl, CsCl, $CaCl_2$ and $MgCl_2$ (1.00 M each) together with a small amount of LiCl (0.05 M) was shaken with a chloroform solution of complex **12**, the quantitative and exclusive extraction of LiCl was observed (Scheme 2). This is remarkable because: 1) the extraction of LiCl from water is in principle a very difficult thing to accomplish due to the high solvation energy and 2) the very high selectivity for Li^+.

An interesting aspect of the receptor **12** is that the presence of lithium ions can be detected electrochemically. For the free complex **12**, three irreversible oxidations were observed at 683, 963 and 1150 mV (CH_3CN/CH_2Cl_2, against Ag/AgCl). In the presence of LiCl, the first peak potential was shifted by more than 350 mV toward anodic potential [14]. This offers the possibility to use receptor **12** as the recognition unit of an amperometric chemosensor for lithium ions.

Since water itself is a good ligand for lithium ions, the binding constants of synthetic hosts are significantly lower in water when compared to or-

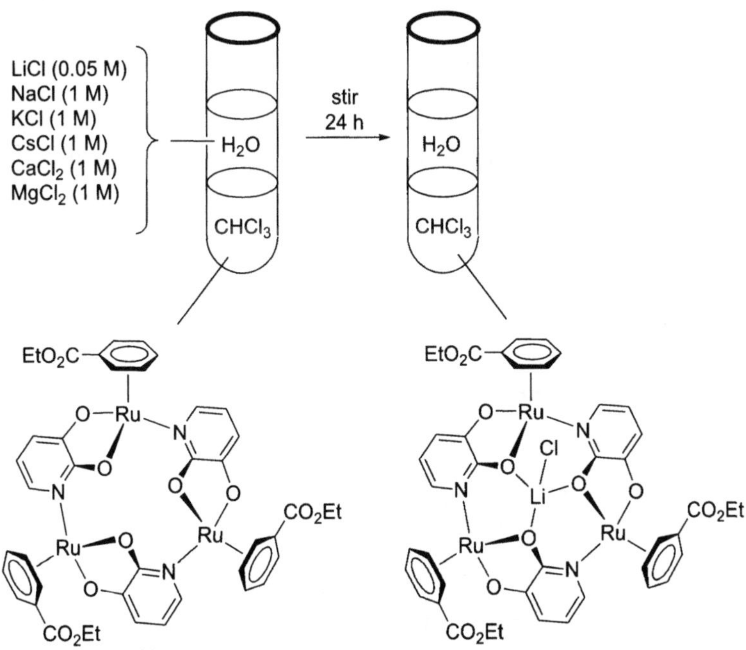

12

Scheme 2 Selective extraction of LiCl from an aqueous solution containing a large excess of alkali and earth alkaline metal salts using the receptor **12**

ganic solvents. The Li^+ specific ionophore 12-crown-4, for example, displays a binding constant of $\log K = 4.25$ in acetonitrile whereas in water, no interaction with Li^+ could be detected at all ($\log K \sim 0$) [43]. Cryptand (2,1,1) is amongst the few receptors, which are able to bind Li^+ in water with high affinity ($\log K = 5.5$) [44]. This receptor also displays a good selectivity for Li^+ over Na^+ (200 : 1). For potential applications, however, aza-macrobicyclic ionophores of this kind show a severe disadvantage: the synthesis generally requires several steps and proceeds with modest overall yields.

As mentioned in Sect. 2.1, water-soluble 12-metallacrown-3 complexes could be obtained with the amino-substituted ligands **L4** [19] and **L5** [20]. The synthesis of these receptors was remarkably simple: they were formed in nearly quantitative yield if the ligands were dissolved together with the corresponding half-sandwich complex $[(\pi\text{-ligand})MCl_2]_2$ in water containing phosphate buffer (pH 7–8). Macrocycles were also obtained if 1–2 equivalents of CsOH were added to an aqueous solution of $[(\pi\text{-ligand})MCl_2]_2$ and **L4**/**L5** [20]. The binding constant for the complexation of Li^+ in water was found to depend strongly on the nature of the (π-ligand)M fragment and on the pH. Among the various half-sandwich complexes that were tested, the ruthenium complex (p-$MeC_6H_4{}^iPr$)Ru gave the best results. In combination with ligand **L4**, a receptor **13** with a binding constant of $K(LiCl) = 2 \times 10^3\ M^{-1}$ was obtained in phosphate buffer at pH 7 [19]. Using the N-methyl piperazine substituted ligand **L5** and CsOH instead of phosphate buffer (receptor **14**), it was possible to increase the affinity by more than one order of magnitude to $K(LiCl) = 6 \times 10^4\ M^{-1}$ [20]. This value is sufficient to achieve a nearly quantitative complexation of Li^+ at the pharmacologically relevant concentration of $\sim$ 1 mM. Sodium salts do not interfere with complexation because the binding constants are more than three orders of magnitude lower than what is found for lithium salts.

In order to use receptor **14** as a chemosensor, a unique way to transduce the binding of lithium ions into a signal was devised [20]. When $FeCl_3$ was added to an aqueous solution of **14**, the receptor immediately decomposed

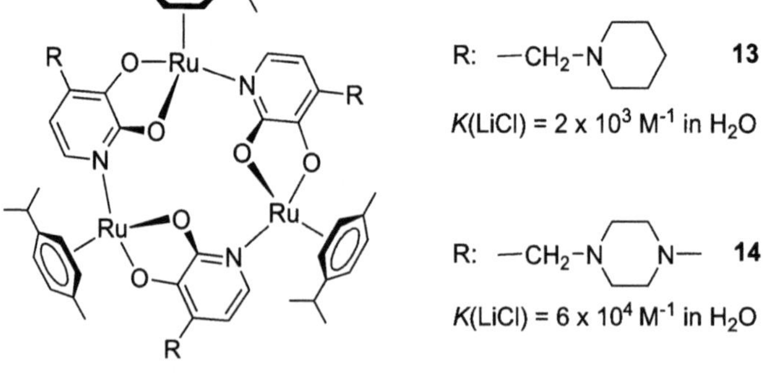

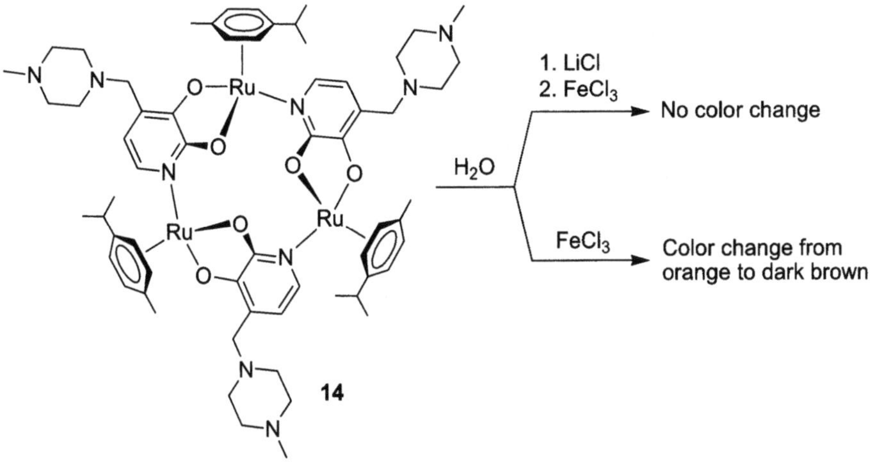

Scheme 3 In the presence of lithium ions, the decomposition of the receptor **14** by $FeCl_3$ is kinetically inhibited. Since the reaction with $FeCl_3$ results in a change of color (*orange to dark brown*), it can be used as a colorimetric test for the detection of lithium ions

giving a dark brown solution from which a brown powder slowly precipitated. In the presence of lithium ions, this reaction was kinetically inhibited and addition of $FeCl_3$ lead to no immediate color change. This difference in reactivity can be used for the "naked eye" detection of low millimolar concentrations of lithium ions in water.

2.4 Selective Recognition and Sensing of Fluoride Ions

Extensive spectroscopic and crystallographic investigations have revealed that in organic solvents, lithium halides are bound to the 12-metallacrown-3 complexes as ion pairs. Based on this observation, a specific chemosensor for the pharmacologically [45] as well as toxicologically [46] interesting fluoride ion was developed [18]. The basic idea is shown in Scheme 4. A lithium ion is coordinated inside a 12-metallacrown-3 complex. The accessibility of the Li^+ center is controlled by the steric requirements of the π-ligand. If bulky ligands are employed, only the small fluoride anion is able to enter the rigid cavity whereas larger anions are efficiently blocked. Since the radius of the fluoride ion is shorter than that of most other anions, a specific receptor is obtained. The selective formation of LiF ion-pairs is further enhanced by the intrinsic affinity of the hard Lewis acid Li^+ to the hard Lewis base F^-.

This concept was realized with the $LiBF_4$ adduct of receptor **15**, comprised of Cp*Ir complexes and the ligand **L3** [18]. NMR data showed that in solution ($CDCl_3/CD_3CN$, 2 : 1), the weekly bound BF_4^- ion was not coordinated to the lithium ion. If Bu_4NF was added to this solution, signals of the ion-paired complex [**15** · LiF] could be observed by ^{19}F and ^{7}Li NMR spectroscopy.

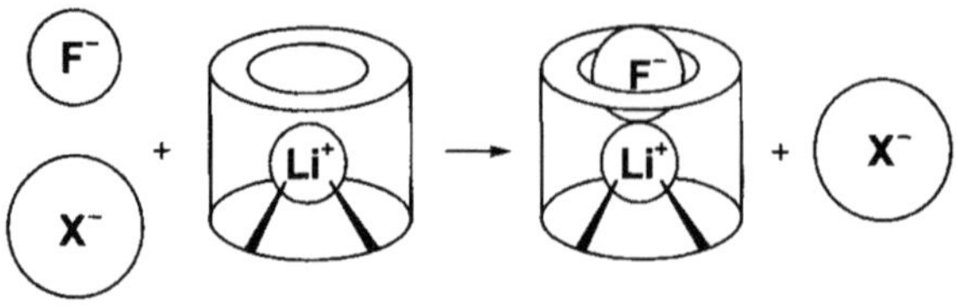

Scheme 4 Schematic representation of a specific fluoride receptor based on a Li^+ containing metallamacrocycle

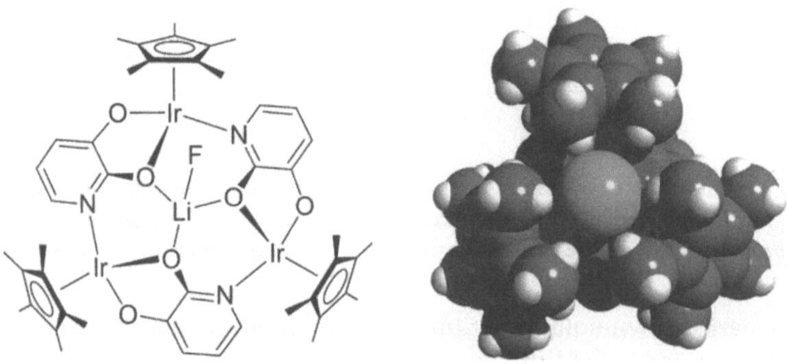

Fig. 3 ChemDraw representation (*left*) or space filling representation (*right*) of receptor **15**, comprised of Cp*Ir complexes and the bridging ligand **L3**, with LiF in the binding pocket. The fluoride anion is closely encapsulated by the Cp* π-ligands

The LiF complex was also obtained in the presence of a large excess of other anions X ($X^- = Cl^-$, Br^-, I^-, NO_3^-) indicating a fluoride-X^- selectivity of > 1000 : 1. The structure of the complex [**15** · LiF] in the crystal gives an explanation for this pronounced selectivity (Fig. 3). As a result of the high intrinsic affinity of Li^+ for F^-, a very short bond of Li – F = 1.755(11) Å is observed. The three Cp* ligands closely surround the fluoride ion. This tight encapsulation of the fluoride ion is expected to contribute to the overall stability of the host-guest complex (CH ··· F contacts) and prevents the coordination of larger anions.

Complex [**15** · $LiBF_4$] was shown to act as a redox-responsive receptor: in the presence of F^-, the complex was significantly easier to oxidize ($\Delta E =$ – 203 mV) whereas only small changes were observed upon addition of Cl^-, Br^-, NO_3^-, HSO_4^- or ClO_4^- salts ($\Delta E < 24$ mV) [18]. Similar results were obtained in solutions containing methanol. The complex [**15** · $LiBF_4$] can, therefore, be used as a highly selective chemosensor, which allows the detection of fluoride anions by electrochemical means, even in protic solvents.

3
Indicator Displacement Assays with an Organometallic Complex

3.1
General Principle

In traditional chemosensors, a receptor unit is covalently connected with a reporter unit. A new approach is the utilization of a synthetic receptor, which is bound via non-covalent interactions to an indicator. This ensemble is able to function as a sensor, given that the displacement of the indicator by an analyte results in a change of its optical properties (Scheme 5) [47, 48]. An indicator-displacement assay (IDA) of this kind has been used to detect citrate [49], tartrate [50], gallic acid [51], phosphate [52], carbonate [53], and amino acids [54–56], among others. Compared to traditional chemosensors, IDAs have two major advantages: 1) they are very flexible because the physical properties of the indicator (color, affinity for the host, solubility) as well as the indicator–receptor ratio can be varied according to specific needs [57]; 2) they are easy to generate because the signaling unit is attached by non-covalent interactions. The indicator can be attached to the receptor via electrostatic interactions, hydrogen bonding and/or metal ligand interactions. The latter mode of attachment is particularly interesting for sensing in water since high association constants can be achieved. So far, "classical" coordination compounds of the 3*d* transition metal ions were used almost exclusively for this purpose but the results summarized below suggest that organometallic complexes are a potentially very attractive alternative.

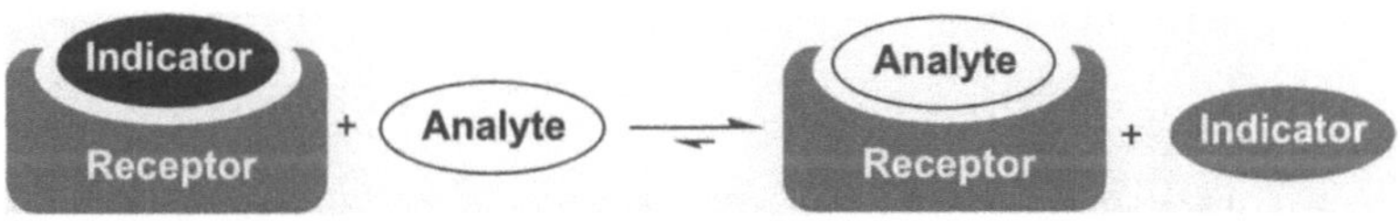

Scheme 5 Basic concept of an indicator displacement assay (IDA): the optical properties of an indicator (color, fluorescence) change upon replacement by an analyte

3.2
A Chemosensor for Histidine- and Methionine-Containing Peptides

The coordination chemistry of organometallic half-sandwich complexes with amino acids and peptides has been investigated intensively [58]. Peptides are known to preferentially bind to $Cp^{*}Rh^{III}$-, $Cp^{*}Ir^{III}$- and (arene)Ru^{II}-fragments via the terminal amino group and deprotonated amide bond(s). For histidine and methionine, an additional interaction between the *N*- or *S*-donor group of the side-chain and the metal is generally observed. These findings were the basis for the development of an organometallic IDA for histidine- and methionine-containing peptides [59].

As the receptor unit, the Cp*Rh complex **16** was chosen because it is soluble in water and because the exchange kinetics for the three facial coordination sites opposite to the π-ligand are fast [60]. As the indicator, the dye azophloxine (**17**) was found to be well suited because it undergoes a strong color change upon complexation to Cp*Rh. The UV/visible absorption spectrum of **17** (25 μm) after addition of increasing amounts of **16** is shown in Fig. 4 (H_2O, 100 mM phosphate buffer, pH 7.0). The complexation reaction results in a pronounced decrease of the absorption in the region of 500 nm with a new local maximum being found at $\lambda = 549$ nm. A single complex is formed as indicated by clear isosbestic points at $\lambda = 554$ and 443 nm. At room temperature, the reaction between **16** (12.5 μM) and **17** (25 μM) proceeds with a half-life of $t_{1/2} = 2.5$ min, whereas at 50 °C the reaction is complete within 5 min. The interaction between the indicator and the rhodium complex was found to be very strong: fitting of the titration data to a 1 : 1 binding algorithm yielded a binding constant of $3.2 \pm 1.0 \times 10^7$ M^{-1}.

In order to determine whether a mixture of **16** and **17** can be employed to detect histidine- or methionine-containing peptides, competition experiments with **16**, **17** and the dipeptides His-Ala or Val-Phe were performed [59]. When a mixture of **16** and **17** ([Rh] = [**17**] = 50 μM) in buffered aqueous so-

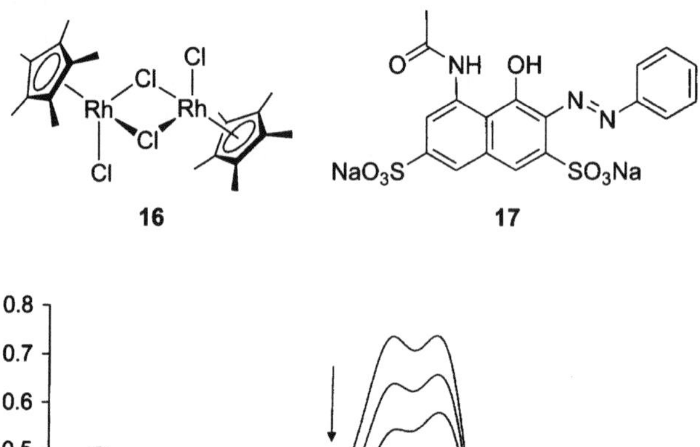

16 **17**

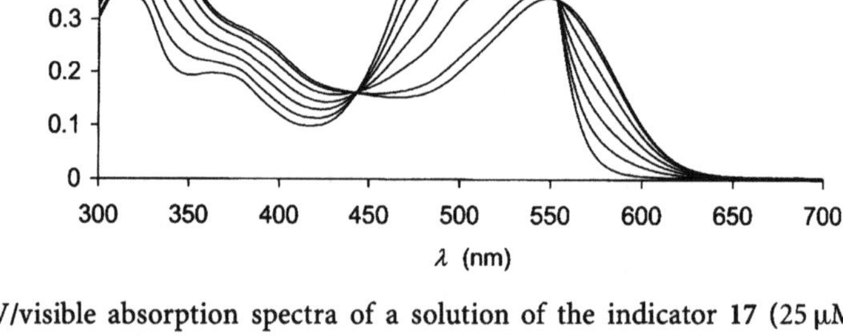

Fig. 4 UV/visible absorption spectra of a solution of the indicator **17** (25 μM) in H_2O (100 mM phosphate buffer, pH 7.0) upon addition of complex **16** (final Rh concentration: 0, 4.8, 9.6, 14.4, 19.2, 24.0 and 26.4 μM). The spectra were recorded after equilibration

lution was mixed with one equivalent of His-Ala, the original red color of the non-coordinated indicator **17** re-appeared. This showed that the stability of the complex between Cp*Rh and His-Ala is considerably higher than that of the adduct between Cp*Rh and the indicator **17**. When a similar experiment was performed with Val-Phe instead of His-Ala, on the other hand, the purple color of the complex between the receptor **16** and the indicator **17** was observed (Scheme 6). The affinity of Val-Phe to the Cp*Rh complex is thus not sufficient to replace the indicator **17** to a significant extent. This shows that an IDA with **16** and **17** can be used to differentiate His-Ala and Val-Phe. For detection by the "naked-eye", the required minimum peptide concentration is approximately 20 μM.

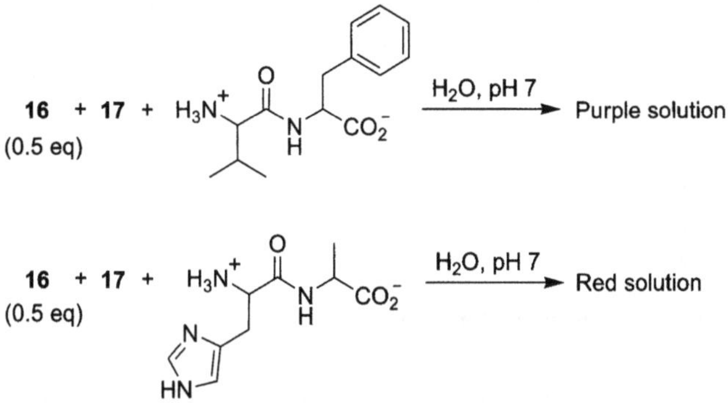

Scheme 6 Pronounced color changes are observed upon equilibration of a solution of the receptor **16**, the indicator **17** and His-Ala or Val-Phe, respectively

UV/visible spectroscopy was used in order to obtain more information about the relative binding constants $K_r = K_{peptide}/K_{indicator}$ of the Cp*Rh complex **16** for various peptides [59]. The results are summarized in Table 1. For peptides containing either His or Met at – or adjacent to – the *N*-terminus, the nearly complete replacement of the indicator upon addition of the peptide was observed ($K_r \geq 740$) (entries 1–6). It was suggested that for those peptides, a simultaneous coordination of the amino and the respective side-chain occurs. Such a mode of coordination is less likely for tri- or longer peptides with His/Met residues in the middle or at the C-terminus and consequently lower relative binding constants were observed (entries 7–10).

The interaction of the thioether side-chain of methionine with the Cp*Rh complex appears to be less strong than the interaction of the imidazole side chain of histidine as suggested by the values for Gly-Gly-His and Gly-Gly-Met (entry 7 and 10). Several peptides without His/Met residues were investigated, all of which showed a very low relative binding constant (entries 11–15). It is interesting to note that the substitution of a single His residue with an Ala residue resulted in a drop of affinity of more than five orders of magnitude (entry 1 and 12).

Table 1 Relative binding constants $K_r = K_{peptide}/K_{indicator}$ of the Cp*Rh complex **16** for various peptides

Entry	Peptide	K_r [a]
1	His-Ala	$> 1 \times 10^3$
2	His-Gly-Gly	$> 1 \times 10^3$
3	Gly-His-Gly	$> 1 \times 10^3$
4	Leu-His-Leu	$> 1 \times 10^3$
5	Gly-Met-Gly	$> 1 \times 10^3$
6	Met-Leu-Phe	7.4×10^2 $(\pm 1.5 \times 10^2)$
7	Gly-Gly-His	1.6×10^1 $(\pm 0.2 \times 10^1)$
8	Tyr-Gly-Gly-Phe-Met-Arg-Phe	6.9 (± 0.2)
9	Ala-Ser-His-Leu-Gly-Leu-Ala-Arg	9.1×10^{-1} $(\pm 0.2 \times 10^{-1})$
10	Gly-Gly-Met	1.7×10^{-1} $(\pm 0.2 \times 10^{-1})$
11	Val-Gly-Gly	7.4×10^{-2} $(\pm 0.2 \times 10^{-2})$
12	Ala-Ala	2.2×10^{-3} $(\pm 0.3 \times 10^{-3})$
13	Pro-Glu	$< 1 \times 10^{-3}$
14	Val-Phe	$< 1 \times 10^{-4}$
15	Lys-Tyr	$< 1 \times 10^{-4}$

[a] Conditions: 100 mM phosphate buffer, pH 7.0

Based on the high selectivity of the organometallic receptor it was possible to carry out IDAs for the quantitative detection of peptides containing His/Met residues close to the *N*-terminus. This was demonstrated by the following experiments [59]: The receptor **16** (final Rh conc.: 12 μM) and the indicator **17** (final conc.: 50 μM) were added successively to a solution containing a variable amount of His-Ala (0–10 μM). The resulting mixture was tempered for 10 min at 50 °C and then the absorption at 580 nm was determined. A linear calibration curve was observed with a detection limit of ~ 0.3 μM (Fig. 5). If Val-Phe was used instead of His-Ala, no significant change in absorption was observed. This pronounced selectivity allowed to sense low micromolar concentrations of His-Ala in the presence of a 100-fold excess of Val-Phe.

A major advantage of IDAs over traditional chemosensors is the fact that the receptor–indicator ratio can be adjusted for specific sensing problems [57]. For the IDA based on the organometallic receptor **16**, for example, it was demonstrated that it is possible to quantify *and* distinguish the tripeptides His-Gly-Gly and Gly-Gly-His [59]. Using a peptide concentration of 5 μM, a first IDA was performed using an excess of the receptor and the indicator ([**16**] = 25 μM; [**17**] = 60 μM). Under these conditions, the replacement of the indicator by both peptides was almost quantitative, which allowed us to determine their concentration without knowing their identity. In a second experiment, the concentration of the receptor was reduced from 25 μM

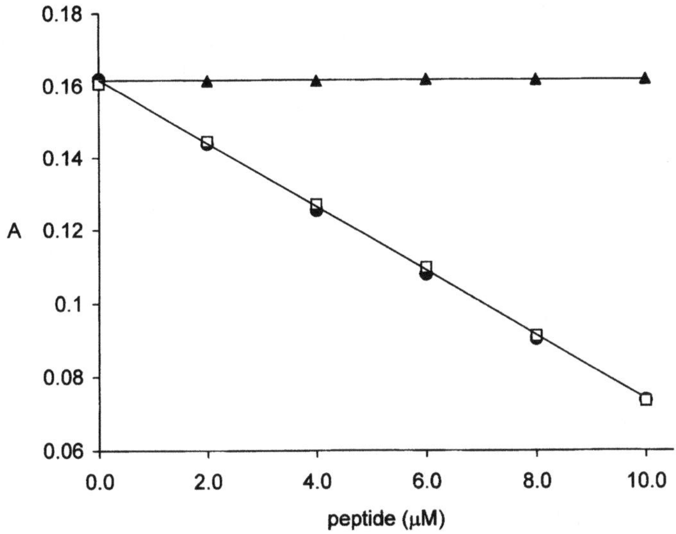

Fig. 5 Absorbance at 580 nm for solutions containing receptor **16** (6 μM), the indicator **17** (50 μM) and different amounts of His-Ala (•), Val-Phe (▲) or His-Ala in the presence of a 100-fold excess of Val-Phe (□)

Table 2 Difference in absorbance at 504 nm for solutions containing the dipeptides His-Gly-Gly and Gly-Gly-His after addition of various amounts of receptor **16** and indicator **17**

Peptide	[Peptide] [a]	[**16**] [a]	[**17**] [a]	ΔA
His-Gly-Gly	5.0 μm	25 μm	60 μm	0.108
Gly-Gly-His	5.0 μm	25 μm	60 μm	0.103
His-Gly-Gly	5.0 μm	2.5 μm	60 μm	0.108
Gly-Gly-His	5.0 μm	2.5 μm	60 μm	0.057

[a] Final concentrations are given

to 2.5 μM. Due to the excess of the indicator (60 μM), only the high affinity analyte His-Gly-Gly resulted in a complete replacement of the indicator whereas for the lower affinity analyte Gly-Gly-His, the UV/vis response was much smaller. Based on this data it was possible to identify the two peptides. The experimental values are given in Table 2.

4 Summary and Outlook

In the first part of this overview, macrocyclic receptors containing organometallic half-sandwich complexes are described. They can easily be obtained by self-assembly using commercially available metal complexes. The host-guest chemistry and the solubility of these receptors strongly depends on the ligands, which are employed to assemble the complexes. With the adenine-based ligands **L1** and **L2**, cationic macrocycles are obtained. They can be used to selectively bind amino-acids and peptides with aromatic side chains (Phe and Trp) in aqueous solution at neutral pH. Since the binding event is accompanied with large ^{1}H NMR chemical shifts, receptors of this kind may find applications as NMR shift reagents. Using the dihydroxypyridine ligands **L3–L5**, it is possible to generate metallamacrocycles with high affinity for lithium salts. The complexation of LiX can be followed electrochemically or can be transduced via a subsequent chemical reaction into a change of color, which offers the possibility to construct chemosensors for lithium and fluoride ions. It appears likely that the synthetic concept to combine multidentate ligands with organometallic complexes can be expanded. Given the size and the rigidity of the structures that are accessible in this fashion and given the interesting host-guest chemistry that has already been reported, an increasing interest in macrocyclic receptors containing organometallic complexes as building blocks can be expected for the future.

As described in the second part, it is possible to use an organometallic 4*d* transition metal complex to build an IDA for the sequence-specific detection of histidine- and methionine-containing peptides. The following characteristics make this new assay especially appealing: 1) it can be performed in water at neutral pH and can be completed within a few minutes; 2) the receptor **16** and the indicator **17** are both commercially available; 3) it allows to detect peptides with His/Met residues close to the *N*-terminus at concentrations as low as 0.3 μM; 4) it is very selective, which allows to perform the analysis in the presence of 100 mM phosphate buffer and in the presence of a large excess of competing peptides. Although this is the first IDA, which utilizes an organometallic receptor, it is conceivable that related systems can be constructed for other analytes (e.g., amino acids) [60] or with different indicators (e.g., fluorescence dyes).

References

1. Metzler-Nolte N (2001) Angew Chem Int Ed 40:1040
2. Simth DP, Baralt E, Morales B, Olmstead MM, Maestre MF, Fish RH (1992) J Am Chem Soc 114:10647
3. Yamanari K, Ito R, Yamamoto S, Fuyuhiro A (2001) Chem Commun 1414
4. Yamanari K, Ito R, Yamamoto S, Konno T, Fuyuhiro A, Kobayashi M, Arakawa R (2003) Dalton Trans 380
5. Smith DP, Kohen E, Maestre MF, Fish RH (1993) Inorg Chem 32:4119
6. Chen H, Olmstead MM, Smith DP, Maestre MF, Fish RH (1995) Angew Chem Int Ed Engl 24:1514
7. Chen H, Ogo S, Fish RH (1996) J Am Chem Soc 118:4993

8. Fish RH (1999) Coord Chem Rev 185–186:569
9. Fish RH, Jaouen G (2003) Organometallics 22:2166
10. Annen P, Schildberg S, Sheldrick WS (2000) Inorg Chim Acta 307:115
11. Korn S, Sheldrick WS (1997) Inorg Chim Acta 254:85
12. Korn S, Sheldrick WS (1997) J Chem Soc, Dalton Trans 2191
13. Piotrowski H, Polborn K, Hilt G, Severin K (2001) J Am Chem Soc 123:2699
14. Piotrowski H, Severin K (2002) Proc Natl Acad Sci USA 99:4997
15. Grote Z, Scopelliti R, Severin K (2003) Angew Chem Int Ed 42:3821
16. Severin K (2003) Coord Chem Rev 245:3
17. Piotrowski H, Hilt G, Schulz A, Mayer P, Polborn K, Severin K (2001) Chem Eur J 7:3196
18. Lehaire ML, Scopelitti R, Piotrowski H, Severin K (2002) Angew Chem Int Ed 41:1419
19. Grote Z, Lehaire ML, Scopelitti R, Severin K (2003) J Am Chem Soc 125:13638
20. Grote Z, Scopelitti R, Severin K (2004) J Am Chem Soc 126:16959
21. Krämer R, Polborn K, Robl C, Beck W (1992) Inorg Chim Acta 198–200:415
22. Ogo S, Chen H, Olmstead MM, Fish RH (1996) Organometallics 15:2009
23. Sünkel K, Hoffmüller W, Beck W (1998) Z Naturforsch 53b:1365
24. Carmona D, Lahoz FJ, Atencio R, Oro LA, Lamata MP, Viguri F, José ES, Vega C, Reyes J, Joó F, Kathó Á (1999) Chem Eur J 5:1544
25. Kathó Á, Carmona D, Viguri F, Remacha CD, Kovács J, Joó F, Oro LA (2000) J Organomet Chem 593–594:299
26. Habereder T, Warchhold M, Nöth H, Severin K (1999) Angew Chem Int Ed 38:3225
27. Yamanari K, Ito R, Yamamoto S, Konno T, Fuyuhiro A, Fujioka K, Arakawa R (2002) Inorg Chem 41:6824
28. Lehaire ML, Scopelliti R, Herdeis L, Polborn K, Mayer P, Severin K (2004) Inorg Chem 43:1609
29. Yamanari K, Yamamoto S, Ito R, Kushi Y, Fuyuhiro A, Kubota N, Fukuo T, Arakawa R (2001) Angew Chem Int Ed 40:2268
30. Bakhtiar R, Chen H, Ogo S, Fish RH (1997) Chem Commun 2135
31. Ogo S, Nakamura S, Chen H, Isobe K, Watanabe Y, Fish RH (1998) J Org Chem 63:7151
32. Lah MS, Pecoraro VL (1989) J Am Chem Soc 111:7258
33. Lah MS, Kirk ML, Hatfield W, Pecoraro VL (1989) J Chem Soc, Chem Commun 1606
34. Pecoraro VL, Stemmler AJ, Gibney BR, Bodwin JJ, Wang H, Kampf JW, Barwinski A (1997) Prog Inorg Chem 45:83
35. Bodwin JJ, Cutland AD, Malkani RG, Pecoraro VL (1989) Coord Chem Rev 216–217:489
36. McIntyre RS, Mancini DA, Parikh S, Kennedy SH (2001) Can J Psychiatry 46:322
37. Birch NJ (1999) Chem Rev 99:2659
38. Dalay I (1997) Lancet 349:1157
39. Phiel CJ, Wilson CA, Lee VMY, Klein PS (2003) Nature 423:435
40. De Strooper B, Woodgett J (2003) Nature 423:392
41. Pilcher HR (2003) Nature 425:118
42. Bartsch RA, Ramesh V, Bach RO, Shono T, Kimura K. In: Sapse AM, von Ragué Schleyer P (eds) (1995) Lithium Chemistry. Wiley, New York, p 393
43. Smetana AJ, Popov AI (1980) J Solut Chem 9:183
44. Cox BG, Garcia-Rosas J, Schneider H (1981) J Am Chem Soc 103:1384
45. Rubin MR, Bilezikian JP (2003) Endcrinol Metab Clin North Am 32:285
46. Aaseth J, Shimshi M, Gabrilove JL, Birketvedt GS (2004) J Trace Elem Exp Med 17:83
47. Wiskur SL, Aït-Haddou H, Lavigne JJ, Anslyn EV (2001) Acc Chem Res 34:963

48. Fabbrizzi L, Licchelli M, Taglietti A (2003) Dalton Trans 3471
49. Metzger A, Anslyn EV (1998) Angew Chem Int Ed Engl 37:649
50. Lavigne JJ, Anslyn EV (1999) Angew Chem Int Ed 38:3666
51. Wiskur SL, Anslyn EV (2001) J Am Chem Soc 123:10109
52. Han MS, Kim DH (2002) Angew Chem Int Ed 41:3809
53. Fabbrizzi L, Leone A, Taglietti A (2001) Angew Chem Int Ed 40:3066
54. Hortalá MA, Fabbrizzi L, Marcotte N, Stomeo F, Taglietti A (2003) J Am Chem Soc 125:20
55. Aït-Haddou H, Wiskur SL, Lynch VM, Anslyn EV (2001) J Am Chem Soc 123:11296
56. Klein G, Reymond JL (2001) Angew Chem Int Ed 40:1771
57. Piątek AM, Bomble YJ, Wiskur SL, Anslyn EV (2004) J Am Chem Soc 126:6072
58. Severin K, Bergs R, Beck W (1998) Angew Chem Int Ed Engl 37:1635
59. Buryak A, Severin K (2004) Angew Chem Int Ed 43:4771
60. Buryak A, Severin K (2005) J Am Chem Soc 127:3700

Top Organomet Chem (2006) 17: 143–175
DOI 10.1007/3418_005

Published online: 30 March 2006

Ferrocene–Peptide Bioconjugates

Toshiyuki Moriuchi · Toshikazu Hirao (✉)

Department of Applied Chemistry, Graduate School of Engineering, Osaka University, Yamada-oka, Suita, 565-0871 Osaka, Japan
hirao@chem.eng.osaka-u.ac.jp

Abstract This chapter sketches an outline of ferrocene–peptide bioconjugates. A variety of ferrocene–peptide bioconjugates have been designed to induce highly ordered structures of peptides and develop new biomaterials. The ferrocene serves as a reliable organometallic scaffold for the construction of an ordered structure via intramolecular hydrogen bonding, wherein the attached peptide strands are regulated in appropriate dimensions. Another noteworthy feature of ferrocene–peptide bioconjugates is their strong tendency to self-assemble through the contribution of available hydrogen bonding sites in the solid state. Furthermore, the ferrocene moiety of ferrocene–peptide bioconjugates is able to act as a redox-switching center and an electrophore in the receptors and biomaterials.

Keywords Ferrocene · Amino acid · Peptide · Hydrogen bond · Chirality-organized structure

Abbreviations

Ac	Acetyl
Bn	Benzyl
Boc	*tert*-Butoxycarbonyl
i-Bu	Isobutyl
CAN	Ceric ammonium nitrate
CD	Circular dichroism
COSY	Correlation spectroscopy

Cp	Cyclopentadienyl
Cp*	Pentamethyl cyclopentadienyl
DNA	Deoxyribonucleic acid
ET	Electron transfer
Et	Ethyl
FT-IR	Fourier-transform infrared spectrometry
L	Liter(s)
Me	Methyl
mol	Mole(s)
NMR	Nuclear magnetic resonance
NOE	Nuclear Overhauser effect
NOESY	Nuclear Overhauser and exchange spectroscopy
Ph	Phenyl
PNA	Peptide nucleic acid
Pr	Propyl
i-Pr	Isopropyl
SAM	Self-assembled monolayer

1 Introduction

Recently the research field of bioorganometallic chemistry, which is a hybrid area between biochemistry and organometallic chemistry, has drawn much attention. Conjugation of organometallic compounds with biomolecules such as DNA, amino acids, and peptides is envisioned to provide novel systems depending on the properties of both types of molecules. In these bioconjugates, the organometallic moiety can serve as a molecular scaffold, a sensitive probe, a chromophore, a biological marker, a redox-active site, a catalytically active site, etc. Considerable effort has been devoted to designing bioconjugates composed of organometallic compounds and biomolecules [1–3].

Architectural control of molecular self-organization is of importance for the development of functional materials [4–6]. Regulation of hydrogen bonding [7] is a key factor in the design of various molecular assemblies by virtue of its directionality and specificity [8–10]. The reversibility and tunability of hydrogen bonding is also of fundamental importance in the chemical and/or physical properties of molecular assemblies. The utilization of the self-assembling properties of short peptides, which possess chiral centers and hydrogen bonding sites, is considered to be a relevant approach to highly ordered molecular assemblies. Hydrogen bonding regulates the three-dimensional structure and function of biological systems. Highly ordered molecular assemblies are constructed in proteins to fulfill unique functions as observed in enzymes, receptors, etc. Secondary structures such as α-helices, β-sheets, and β-turns play an important role in protein folding, which is mostly stabilized by hydrogen bonding and hydrophobic interaction of side chains [11, 12]. Highly specific patterns of complementary intra- and intermolecular hydrogen bonds are created in such secondary structures. On the other hand, ferrocene, which is one of the most stable organometallic compounds and a most useful one among metallocenes, has attracted much attention in its application to materials due to a reversible redox couple and two rotatory coplanar cyclopentadienyl (Cp) rings [13]. The inter-ring spacing of ferrocene is appropriate for hydrogen bonding of the attached peptide strands. The utilization of a ferrocene unit as an organometallic scaffold is considered to be one strategy for studying the hydrogen bonding ability of various peptide strands. Ferrocenylalanine, which is the first example of a ferrocene–amino acid bioconjugate, was synthesized in 1957 [14–16]. After its discovery, a variety of ferrocene–amino acid or ferrocene–peptide bioconjugates were prepared [17–31]. This chapter sketches an outline of ferrocene–peptide bioconjugates. Especially, focus will be on the hydrogen bonding properties of ferrocene–peptide bioconjugates in creating highly ordered molecular structures, including their applications.

2
Design of Ferrocene–Peptide Bioconjugates

Although β-sheets are the key structural elements in a three-dimensional structure that fulfill the biological activity of proteins, the structure and stability of β-sheets are less understood compared to those of α-helices. A series of 12-membered hydrogen-bonded rings are formed in parallel β-sheets, while an alternating series of 10- and 14-membered hydrogen-bonded rings are organized in antiparallel β-sheets. It is difficult to predict the pattern of protein folding from the sequence of amino acids. Considerable efforts have focused on designing secondary structure mimics composed of short peptides to gain fundamental insight into the factors affecting the protein structure and stability, and to facilitate the rational design of pharmacologically useful compounds. Generally, the preparation of chemical models of β-sheets is difficult due to the complexity of their folding and their propensity for self-association. The utilization of molecular scaffolds is a potential strategy for organization of peptide structures, which allows control of intramolecular interaction of peptide or peptidomimetic strands. Therefore, various molecular scaffolds such as rigid aromatic [32–35], epindolidione [36–40], dibenzofuran [41–46], oligourea [47–56], and *endo-cis*-(2*S*,3*R*)-norbornene [57–59] scaffolds have been employed to create the β-sheet-like structure of attached peptide chains and serve as a substitute for the β-turn in the chemical models of protein secondary structures.

In addition to organic molecular scaffolds, ferrocenes are recognized as an organometallic scaffold with a central reverse-turn unit based on the inter-ring spacing of ferrocene of about 3.3 Å, which is suitable for hydrogen bonding of the attached peptide strands as observed in β-sheets. Another advantage in the use of ferrocene as a scaffold depends on the electrochemical reversibility of the redox couple, which permits its usage as a redox-switching center and an electrophore. Furthermore, the architectural control of molecular assemblies utilizing chiral centers and hydrogen bonding sites of peptide chains is considered to be a useful approach to artificial highly ordered supramolecular systems. From these points of view, a variety of ferrocene–peptide bioconjugates have been designed to obtain a peptidomimetic basis for protein folding and to construct highly ordered molecular assemblies.

2.1 Hydrogen Bonding Properties of 1,1′-Disubstituted Ferrocene–Peptide Bioconjugates

The ability of ferrocenes to act as a molecular scaffold for an ordered conformation through intramolecular hydrogen bonds has been demonstrated in a preliminary study using a valine unit [60]. Two identical intramolecular interchain hydrogen bonds are formed between the valine carbonyl and the NH of another valine unit in $CDCl_3$ to give a ten-membered hydrogen-bonded ring in the case of the ferrocene **1** bearing the podand amino acid chains (–L–Val–OMe), which resembles the hydrogen bonding pattern observed in an antiparallel β-sheet. The ordered conformation of **1** is confirmed by single-crystal X-ray structure determination [61]. The same ordered conformation is also formed in the ferrocene **2** bearing the podand amino acid chains ($-\mathrm{Gly}-\mathrm{NH_2}$), in which the $\mathrm{N}\cdots\mathrm{O}$ distance of **2** is shorter than that of **1** due to the better hydrogen acceptor ability of amide carbonyl oxygen compared to ester carbonyl oxygen [62]. The adjacent molecules of **2** are connected by intermolecular hydrogen bonding between the terminal NH and CO adjacent to the ferrocene unit of another molecule to form a two-dimensional sheet. These sheets are then cross-linked by hydrogen bonds to the included water molecules which are intercalated between the sheets. On the other hand, the ferrocene **3** bearing the podand amino acid chains (–L–Phe–OMe) is characterized by only one intramolecular interchain hydro-

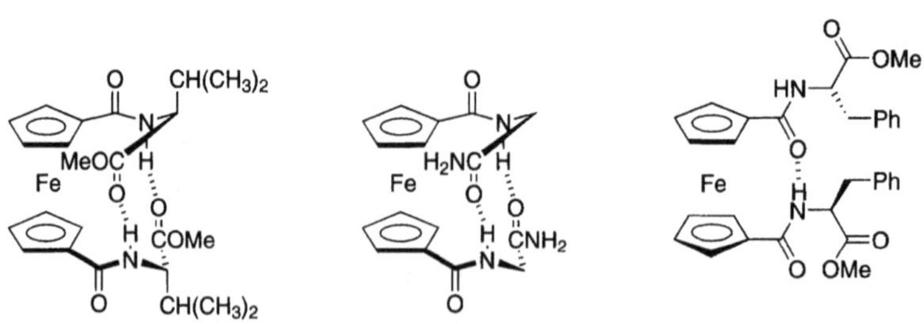

1 **2** **3**

gen bond between CO adjacent to the ferrocenyl unit and the NH of another strand [63]. The molecules of **3** arrange in a helical fashion with 10.37 Å pitch, in which the molecules are linked together via intermolecular hydrogen bonds.

Conformational enantiomers based on the torsional twist about the Cp(centroid)–Fe–Cp(centroid) axis are possible in the case of the 1,1′-disubstituted ferrocene, as shown in Fig. 1 [13, 64]. Conformational enantiomers can interconvert with ease due to the low barrier of Cp ring twisting. The introduction of peptide chains into a ferrocene scaffold is envisaged to induce conformational enantiomerization by restriction of the torsional twist through intramolecular interchain hydrogen bonding.

The X-ray crystal structure of **4a** bearing the podand L-dipeptide chains (-L-Ala-L-Pro-OEt) reveals the formation of two C_2-symmetrical intramolecular interchain hydrogen bonds between CO (Ala) and NH (Ala of another strand) of each podand dipeptide chain to give a ten-membered hydrogen-bonded ring [65–67]. An advantage in the use of L-alanyl-L-proline

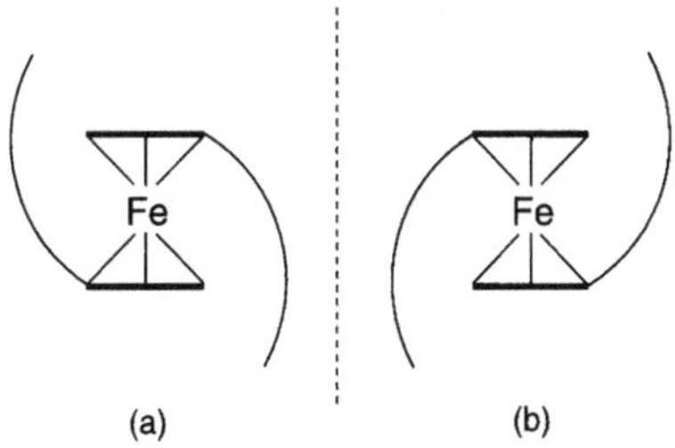

Fig. 1 Enantiomorphous conformations of the 1,1′-disubstituted ferrocene. The enantiomorphs are related by the mirror plane

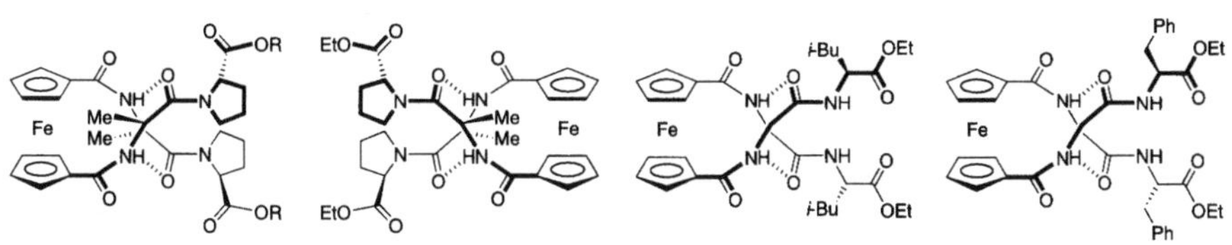

4a: R = Et
4b: R = Me
4c: R = Pr
4d: R = Bn

5 **6** **7**

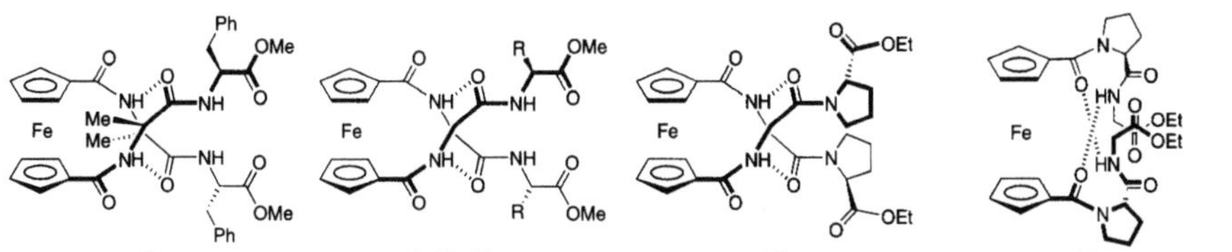

8

9a: R = H
9b: R = Me
9c: R = Bn

10 **11**

as a dipeptide chain unit depends on the hydrogen bonding site and a sterically constrained proline as a well-known turn inducer in proteins.

The molecular structures of **4a** and **5** composed of the enantiomeric dipeptide chains (–D–Ala–D–Pro–OEt) are in a good mirror image relationship as shown in Fig. 2, indicating that they are conformational enantiomers (Fig. 1). As a result, the introduction of the chiral dipeptide chains into ferrocene induces the chirality organization by restriction of the torsional twist through the intramolecular interchain hydrogen bonds [67]. Furthermore, a mutually opposite helically ordered molecular arrangement with one turn of pitch height 14.70 and 14.95 Å and a separation (Fe – Fe) of 8.35 and 8.16 Å between the closest ferrocene units, respectively, is observed in the crystal packings of **4a** and **5** (Fig. 3a and b), in which the ferrocene moieties arrange in a herringbone motif to form the columns of the proline and ethyl ester moieties (Fig. 3c and d). Protein folding is driven by not only the hydrogen bonding but also the hydrophobic effect [11, 12]. The podand dipeptide chains (– Ala – Pro – OEt) are considered to induce such molecular aggregation through a hydrogen bonding site (Ala) and a sterically constrained hydrophobic moiety (Pro). X-ray crystallographic analyses of **4b–d** bearing the methoxycarbonyl, propoxycarbonyl, and benzyloxycarbonyl groups, respectively, show a chirality-organized structure similar to **4a** based on two identical intramolecular interchain hydrogen bonds between C = O (Ala) and N – H (Ala of another strand). Conversely, a helical molecular arrangement is not induced in the crystal packing of **4c** and **4d**, although **4b** exhibits the same helical molecular arrangement as that observed in **4a** [66]. The crystal packing is influenced by the difference in the ester moieties, whereas the molecular conformation is not changed.

Circular dichroism (CD) spectrometry is a useful tool for determining an ordered structure in solution. The ferrocene **4a** exhibits an induced CD

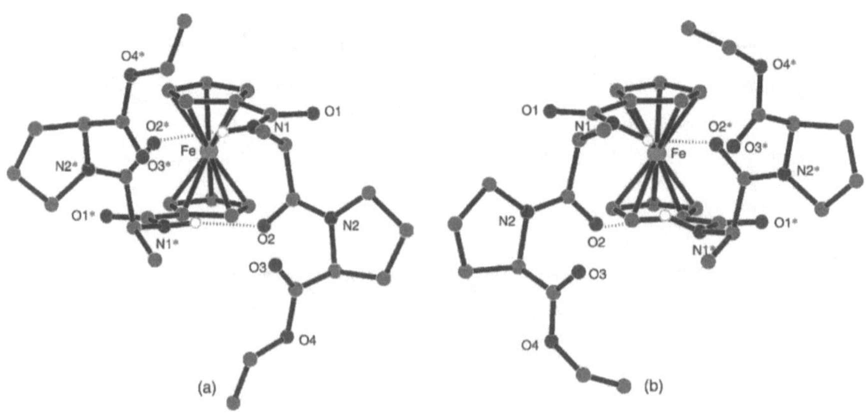

Fig. 2 Molecular structures of **a 4a** and **b 5**

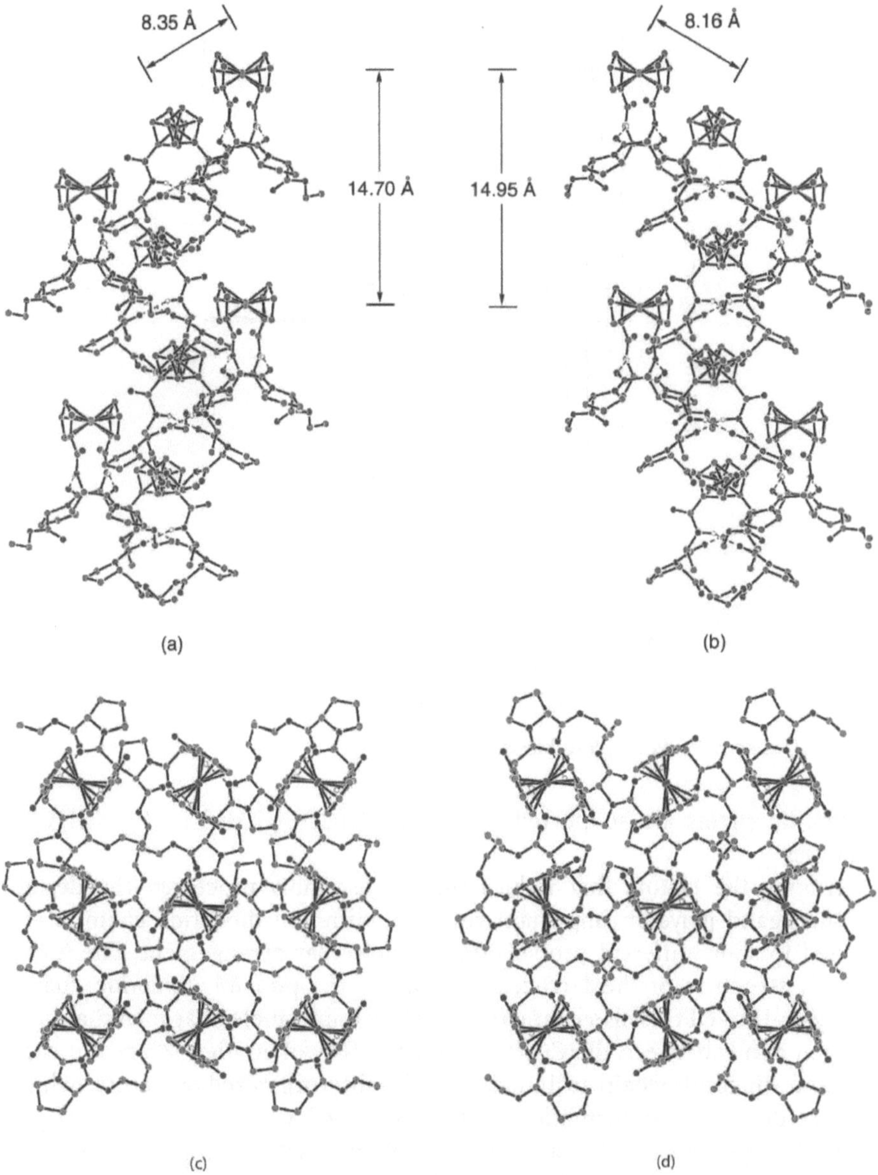

Fig. 3 Top view of a portion of a layer containing the helical arrangement of crystal packing of **a** **4a** and **b** **5**, and side view of crystal packing of **c** **4a** and **d** **5**

around the absorbance of the ferrocenyl function in MeCN (Fig. 4). The mirror image of the signals is obtained in the CD spectrum of **5**, indicating that a chiral molecular arrangement based on an ordered structure via intramolecular interchain hydrogen bonds is formed even in solution [67]. Furthermore, two identical intramolecular hydrogen bonds between the podand

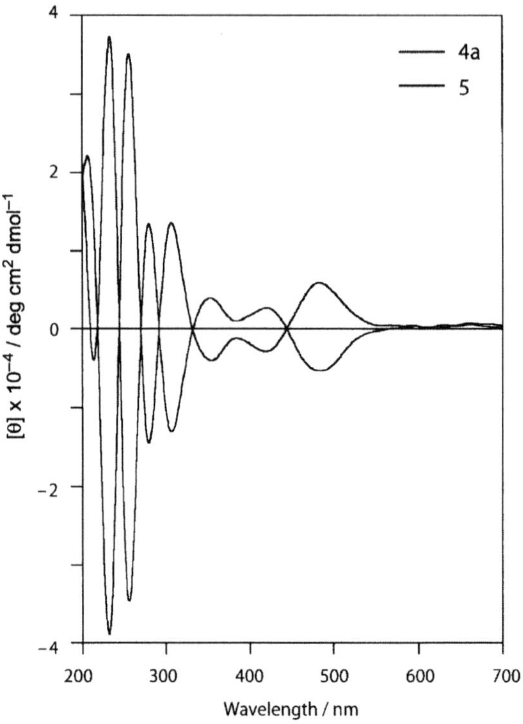

Fig. 4 CD spectra of **4a** and **5** in MeCN (1.0×10^{-4} M)

dipeptide chains are supported by ^{1}H NMR ($CDCl_3$) and FT-IR (CH_2Cl_2) analyses.

Among the factors that induce the self-assembly processes, the design of the ligand plays an important role. The different dipeptide chain is expected to show different propensities of molecular arrangement. The X-ray crystal structure of the ferrocene **6** bearing the podand dipeptide chains (-Gly-L-Leu-OEt) is also characterized by two intramolecular interchain hydrogen bonds between the CO (Gly) and NH (Gly of another strand) of each podand dipeptide chain to induce the chirality-organized structure, which adopts the same conformation as that observed with **4a**. The NH of the Leu residue in this conformation is available for participating in intermolecular hydrogen bonding with the CO adjacent to the ferrocene unit as shown in Fig. 5, thus creating highly organized self-assembly in the crystal packing, wherein each molecule is connected to four neighboring molecules [67].

The kind and grouping of amino acid side chains are known to determine protein secondary structures. The X-ray crystal structure of the ferrocene **7** bearing the podand dipeptide chains (-Gly-L-Phe-OEt) shows the same chirality-organized structure based on two C_2-symmetrical intramolecular interchain hydrogen bonds between CO (Gly) and NH (Gly of another strand) of each podand dipeptide chain. However, the ferrocene **7** exhibits

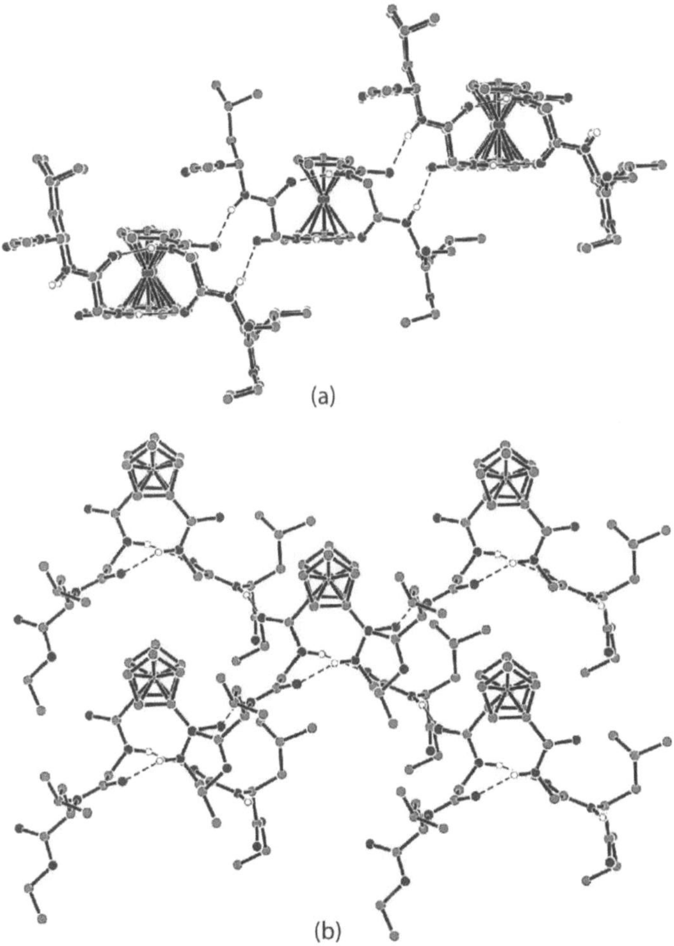

Fig. 5 **a** Top view and **b** side view of a hydrogen-bonded chain assembly of **6**

a different self-assembly in the crystal (Fig. 6). Each molecule is bonded to two neighboring molecules, wherein each podand dipeptide chain forms a 14-membered intermolecularly hydrogen-bonded ring with the podand dipeptide chain of the neighboring molecule through two pairs of symmetrical intermolecular hydrogen bonds [67]. A similar organization is also observed in the case of the ferrocene **8** bearing the podand dipeptide chains (–L–Ala–L–Phe–OMe), although the unit cell of **8** contains two crystallographically independent molecules [63]. From ^{1}H NMR ($CDCl_3$) and FT-IR (CH_2Cl_2) analyses, the ferrocenes **6–8** are likely to form an ordered structure via intramolecular hydrogen bonding, as observed in the crystal structure, although the N – H of –L–Leu– and –L–Phe– is not thought to participate in the hydrogen bonding in solution [63, 67]. An ordered structure via intramolec-

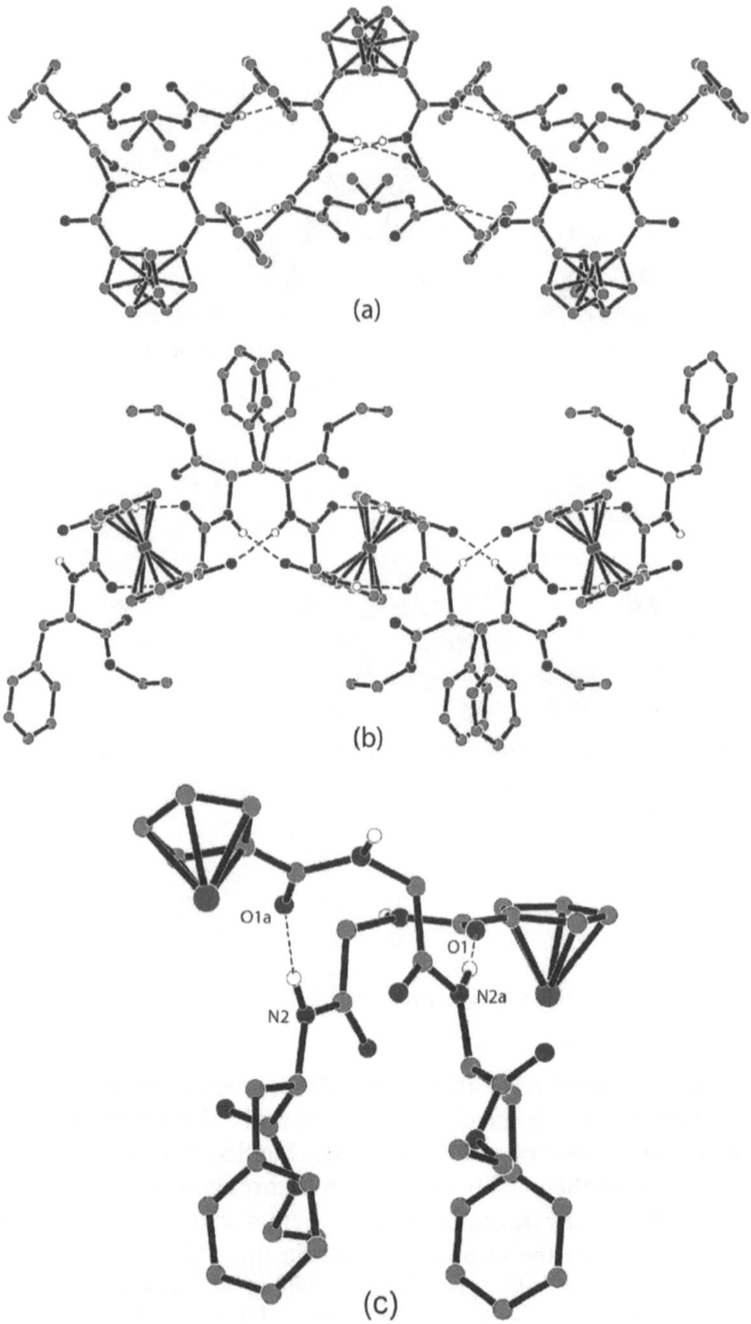

Fig. 6 **a** Top view and **b** side view of a self-assembly of 7 via the formation of a 14-membered intermolecularly hydrogen-bonded ring. **c** A 14-membered intermolecularly hydrogen-bonded ring of 7. Only half of the molecule is shown for clarity

ular hydrogen bonds is also formed in the case of the ferrocenes **9a–c** [68]. The ferrocene **10** composed of the dipeptide chains (–Gly–L–Pro–OEt) forms a ten-membered hydrogen-bonded ring to induce the chirality-organized structure as observed with **4a**, although a 14-membered hydrogen-bonded ring is assumed to be formed in the case of the ferrocene **11** composed of the dipeptide chains (–L–Pro–Gly–OEt) [69]. The alternation of the sequence of amino acids is considered to control the intramolecular conformational regulation.

The single-crystal X-ray structure determination of the ferrocene **12** bearing the podand dipeptide chains (–L–Ala–D–Pro–NHPy) reveals intramolecular antiparallel *β*-sheet-like hydrogen bonds between the CO (Ala) and NH (Ala of another strand) of each dipeptide chain to induce the chirality-organized structure shown in Fig. 7 [70]. The chiral molecular conformation of the ferrocene moiety of **12** is the same as that observed with **4a**, indicating that the chiral molecular conformation of the ferrocene moiety appears to be controlled by the configuration of the alanyl *α*-carbon atom. Another remarkable structural feature is that the NH adjacent to the pyridyl moiety participates in the intramolecular hydrogen bonding with the CO adjacent to the ferrocene unit to nucleate a type II *β*-turn-like structure in each podand dipeptide chain. The ferrocene serves as a reliable organometallic scaffold in a central reverse-turn unit for the formation of an antiparallel-like *β*-sheet-like and a type II *β*-turn-like structure simultaneously, thus constructing a chirality-organized structure. The molecular structures of **13** composed of the dipeptide chains (–D–Ala–L–Pro–NHPy) and **12** are in a good mirror image relationship, indicating that they are conformational enantiomers (Fig. 7). The ^{1}H NMR ($CDCl_3$) including NOE ($CDCl_3$) studies indicate the preservation of a type II *β*-turn-like structure in solution.

The X-ray crystal structure of the ferrocene **14** bearing the podand amino acid chains (–L–Pro–OMe) reveals a 1,3′ conformation of the two amino acid chains, which minimizes steric interactions of the amino acid chains [71]. The growing oligoproline chain of **15** adopts a stable polyproline-II helix in solu-

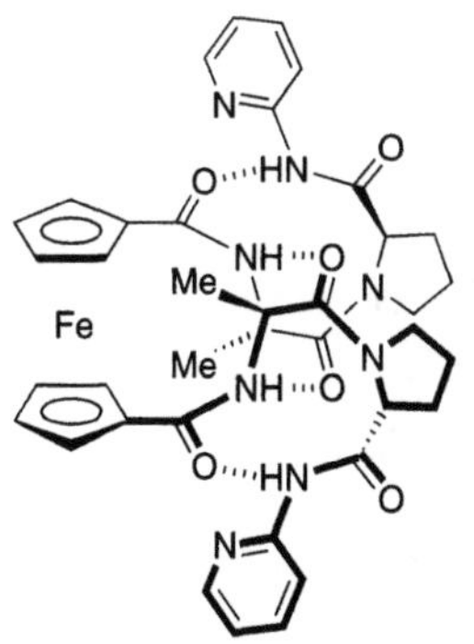

12

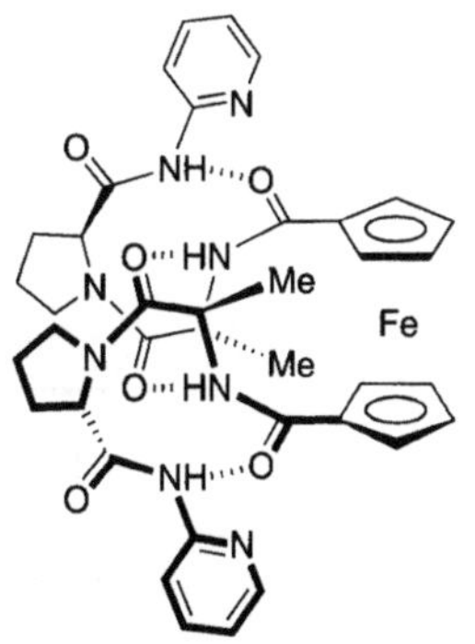

13

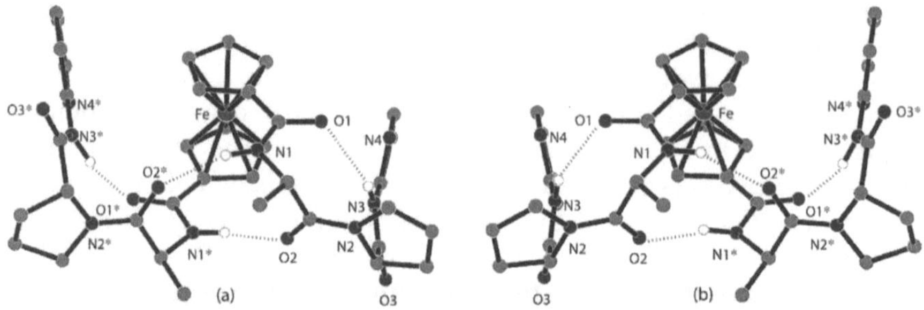

Fig. 7 Molecular structures of **a** 12 and **b** 13

tion. Furthermore, the ferrocene moiety becomes more easily oxidized with growing peptide length.

Temperature-dependent NMR, COSY, and NOESY experiments of the constrained Leu(5)-enkephalin amide analogue **16** containing a ferrocene substructure indicate an intramolecular hydrogen bond between Fc(1)CO and Fc(4)NH, creating a *β*-turn structure [72]. The rotational barrier

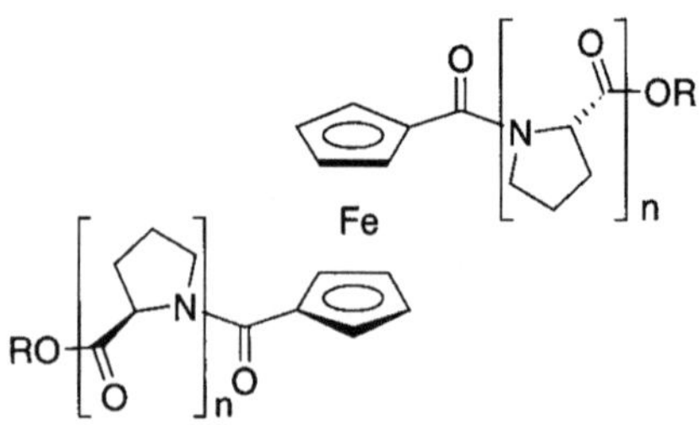

14: R = Me, n = 1
15a: R = Bn, n = 1
15b: R = Bn, n = 2
15c: R = Bn, n = 3
15d: R = Bn, n = 4

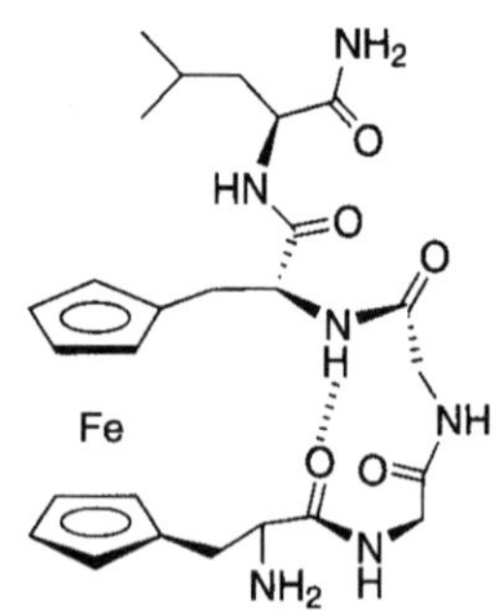

16

(16.8 kcal mol^{-1}) of the *C*-terminal amide bond of **16** is essentially the same as that for Leu(5)-enkephalin.

2.2 Hydrogen Bonding Properties of Monosubstituted Ferrocene–Peptide Bioconjugates

Generally, an organized structure based on intramolecular interchain hydrogen bonds is not formed in the case of the ferrocenes bearing only one peptide chain, and a network of intermolecular hydrogen bonds is created in the solid state. The ferrocene **17** bearing only one amino acid chain (–L–Glu–(OBn)$_2$) shows a linear chain-like structure of molecules linking through intermolecular hydrogen bonding in a crystal structure [73]. The individual molecules of the ferrocene **18** bearing only one amino acid chain (–L–Cys(SBn)–OMe) are also linked by intermolecular hydrogen bonding to form endless chains, as observed in **17** [73]. The X-ray crystal structure of the ferrocene **19a** bearing only one amino acid chain (-Gly–OMe) indicates an intramolecular interaction between CO (Gly) and CH (Cp) together with intermolecular hydrogen bonds [74]. The hydrogen bonding interactions of the CO adjacent to the ferrocene unit with the OHs of the α-carboxyl group and the Asp acid side chain of two adjacent molecules, which form a bifurcated hydrogen bond, are observed in the crystal structure of the ferrocene **20** bearing only one amino acid chain (–L–Asp–OH) [75]. The crystal structure of **20** accommodates 1/3 molecule of water, which enables hydrogen bonding with NH (Asp) and COs (Asp). The CH···O interaction between CH (Cp) and the water molecule is also indicated. These hydrogen bonds create

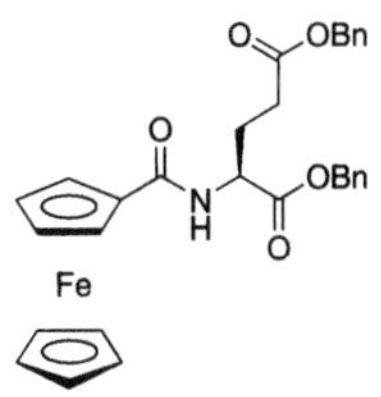

17

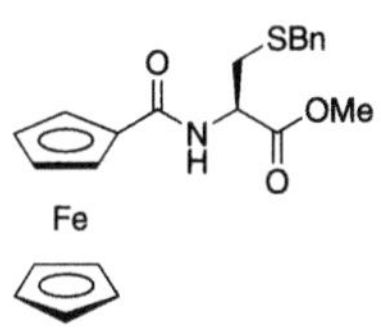

18

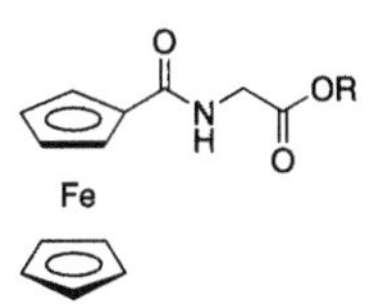

19a: R = Me
19b: R = Et
19c: R = Bn

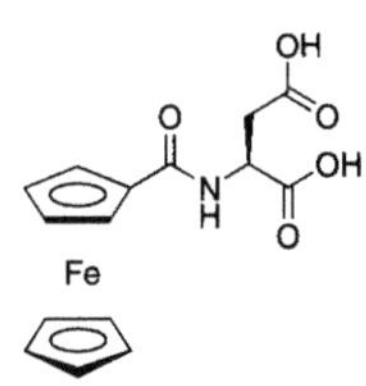

20

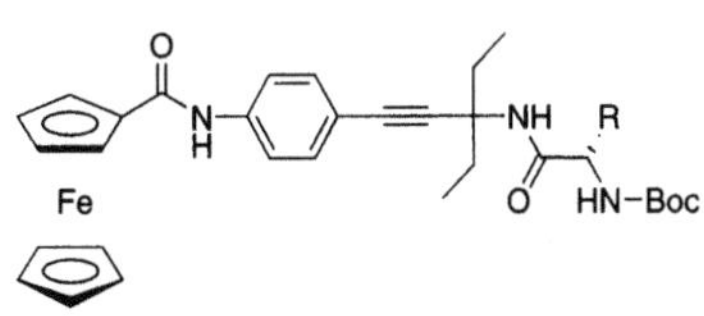

21a: R = Bn
21b: R = *i*-Bu
21c: R = CH$_2$CH$_2$SCH$_3$
21d: R = CH$_2$OH

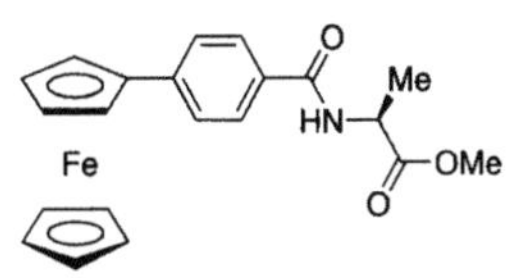

22

an extensive hydrogen bonding network, suggesting that ferrocene-peptide compounds are able to form hydrogen bonds with solvent molecules.

The solid-state IR spectra of the ferrocene **21** indicate that the amide hydrogen atoms are involved in hydrogen bonds in a solid state, although the formation of hydrogen bonds is not detected in solution (CH_2Cl_2) [76]. A hydrogen bonding network is considered to be formed in the solid state. The intermolecular hydrogen bonds between CO (Ala) and NH (Ala of another molecule) to form a hydrogen bonding network is observed in the crystal structure of the ferrocene **22** [77].

The vicinal coupling constant J_{NH} in the NH – CH fragment of Ala is 6.9 Hz for the ferrocene **23a** bearing only one dipeptide chain (–L–Ala–L–Pro–OEt) [66], suggesting a possibility of the C_5 conformation [78]. The ferrocene **23a** exhibits intermolecular hydrogen bonds between C = O (Ala) and N – H (Ala of another molecule), wherein two independent molecules exist in the asymmetric unit and are connected alternately to form an intermolecular hydrogen bonding network without the C_5 conformation observed in the solution, resulting in a left-handed helically ordered arrangement with one turn of pitch height 17.86 Å and a separation (Fe – Fe) of 7.50 Å between the closest ferrocene units (Fig. 8). The ferrocene **23b** bearing only one dipeptide chain (-Gly–L–Pro–OEt) also forms a hydrogen bonding network, in which each molecule is connected to two neighboring molecules through intermolecular hydrogen bonds between CO (Gly) and NH (Gly of another molecule), instead of intramolecular hydrogen bonding [69]. As observed in the ferrocene **23a**, the ferrocene **23b** is packed in a left-handed helically ordered arrangement with one turn of pitch height 15.65 Å through a network of intermolecular hydrogen bonds, within which the distance between the closest ferrocene units is 7.95 Å (Fe – Fe), as depicted in Fig. 9.

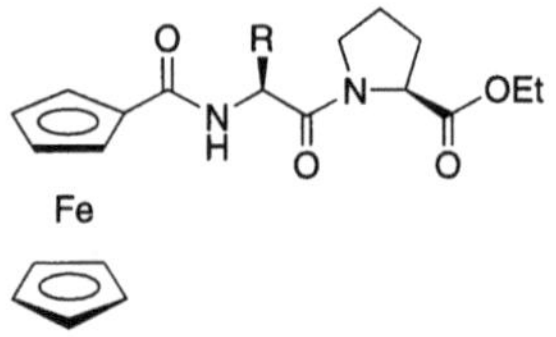

23a: R = Me
23b: R = H

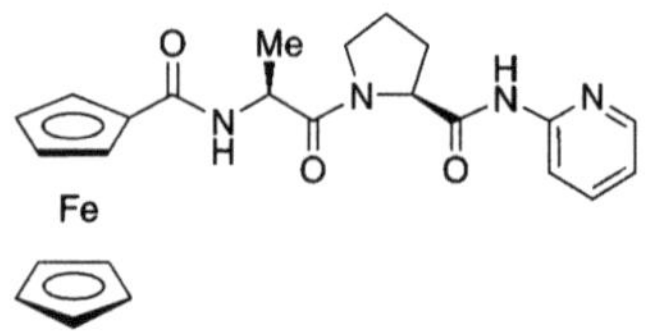

24

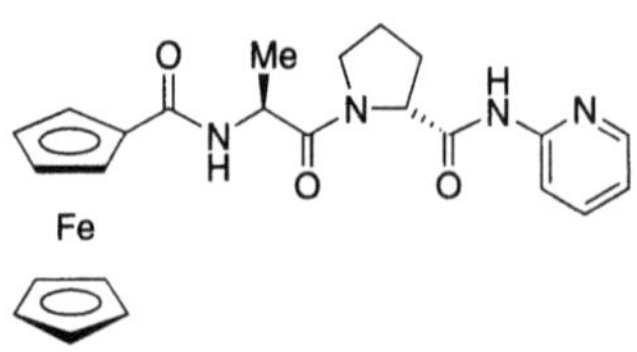

25

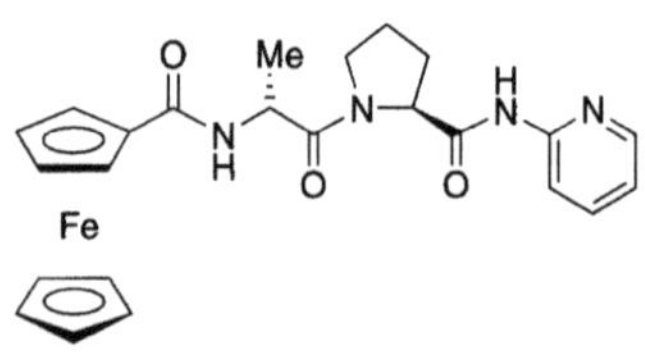

26

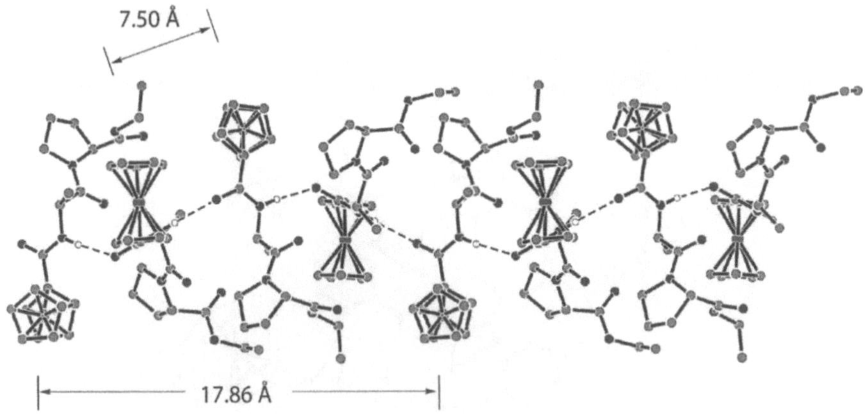

Fig. 8 A portion of a layer containing the helical assembly of crystal packing of **23a**

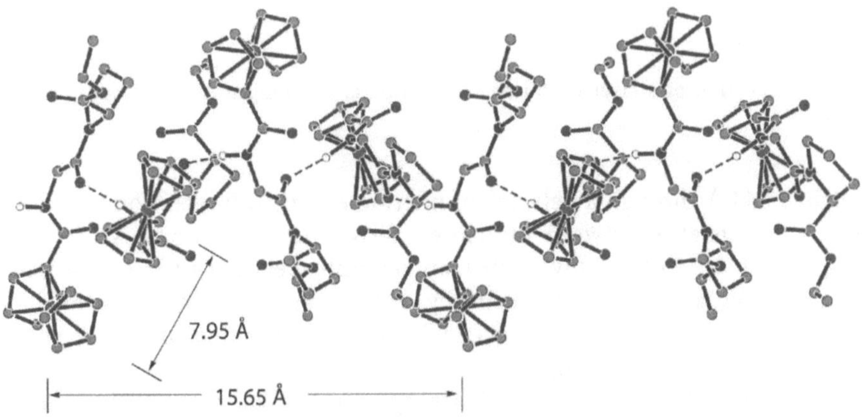

Fig. 9 A portion of a layer containing the helical assembly of crystal packing of **23b**

The ferrocene **24** bearing only one dipeptide chain (-L-Ala-L-Pro-NHPy) is indicated to be present in a non-hydrogen-bonded state by ^{1}H NMR ($CDCl_3$) and FT-IR (CH_2Cl_2) solution analyses [79]. Instead of intramolecular hydrogen bonding, an antiparallel hydrogen-bonded network is formed in the solid state to create a highly organized assembly, wherein each molecule is connected to two neighboring molecules through N – H (Ala)/N (pyridine of another molecule) and N – H (adjacent to pyridine unit of another molecule)/ O (Ala) intermolecular hydrogen bonds to form a seven-membered intermolecularly hydrogen-bonded ring (Fig. 10). In contrast, the ferrocene **25** bearing only one dipeptide chain (-L-Ala-D-Pro-NHPy) exhibits intermolecular hydrogen bonds, wherein two independent molecules exist in the asymmetric unit and are connected alternately to be packed in a left-handed helically ordered arrangement with one turn of pitch

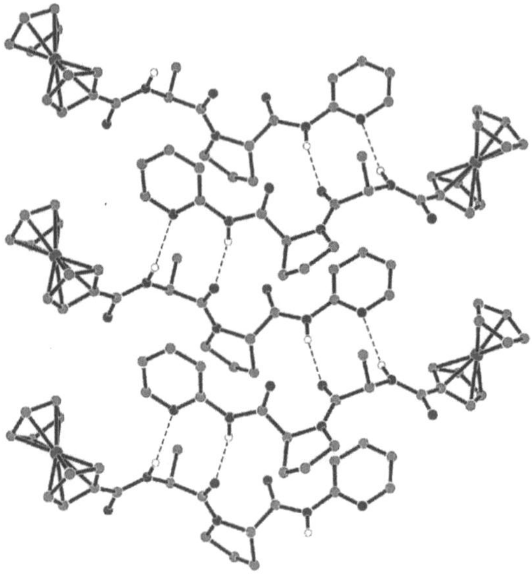

Fig. 10 A portion of a layer containing the antiparallel arrangement of crystal packing of **24**

height 19.44 Å through a network of intermolecular hydrogen bonds, within which the distance between the closest ferrocene units is 6.95 Å (Fe – Fe) (Fig. 11a) [70]. Noteworthy is that an opposite helically ordered molecular assembly, a right-handed helically ordered arrangement, is formed in the crystal packing of the ferrocene **26** bearing only one dipeptide chain (–D–Ala–L–Pro–NHPy) (Fig. 11b).

In contrast to the antiparallel arrangement of **24**, the two independent molecules of the ferrocene **27** bearing only one dipeptide chain (–Gly–Gly–OMe) are held together in a parallel manner through intermolecular hydrogen bonding [80]. Parallel β-sheet layers are held together by intermolecular hydrogen bonding to an interstitial water molecule, and two β-sheets form a tail-to-tail bilayer with the peptide substituents pointing toward each other, in which the bilayer thickness is 20.5 Å. Although the hydrogen bonding pattern of the ferrocene **28a** bearing only one dipeptide chain (–L–Leu–L–Phe–OMe) is similar to that of **27**, the intermolecular hydrogen bonding interaction of **28a** creates supramolecular helicates with the ferrocenyl groups on the outside of the helix and the isobutyl groups on the inside of the helix [80]. The molecules of the ferrocene **28b** bearing only one dipeptide chain (–L–Ala–L–Phe–OMe) are also arranged in a helical fashion by intermolecular hydrogen bonds [63].

X-ray crystallographic analyses of the ferrocenes **29** and **30b–d** bearing only one oligoprolyl chain reveal that these ferrocenes adopt a left-handed polyproline II helix with all prolines in a mutually *trans*-conformation [81].

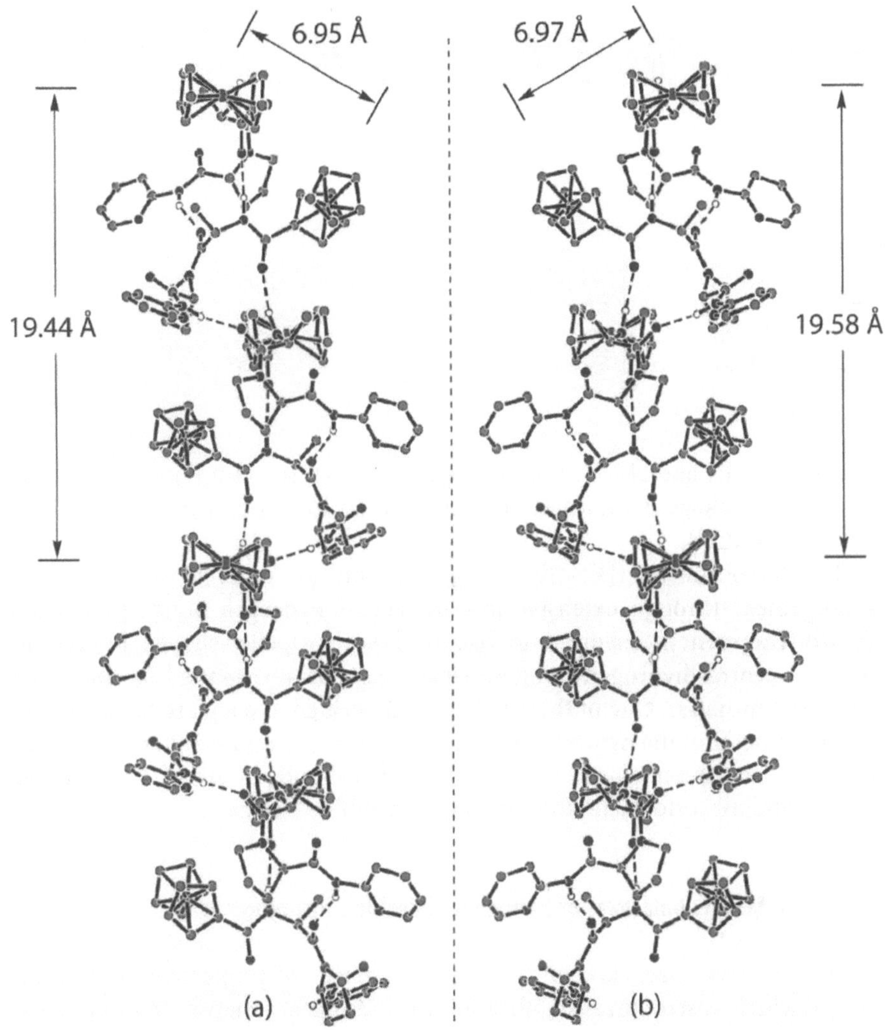

Fig. 11 A portion of a layer containing the helical assembly of crystal packing of **a** **25** and **b** **26**

These crystal structures can be recognized as snapshots of a growing left-handed polyproline II helix. This structure is considered to be maintained in CD_3CN and $CDCl_3$ by NMR studies. On increasing the peptide chain length of **30**, the ferrocene moiety becomes more easly oxidized. The redox potential of ferrocene-peptide bioconjugates is also reported to be highly dependent on the hydrogen bonding ability of the solvent and correlates with the hydrogen donor ability α of the Kamlet–Taft formation [75]. On the other hand, the ferrocene **31** bearing only one tripeptide chain (-L-Pro-L-Pro-L-Phe-OH) forms a strong intramolecular hydrogen bond between $C=O$ (adjacent to the

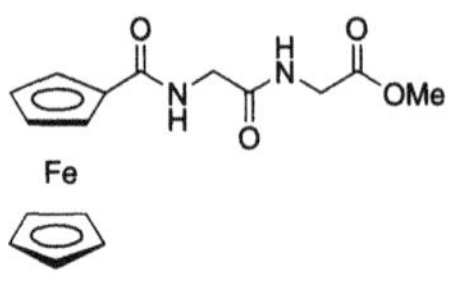

27

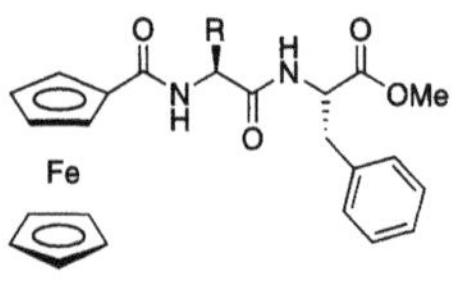

28a: R = *i*-Bu
28b: R = Me

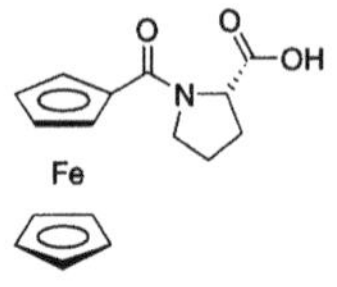

29

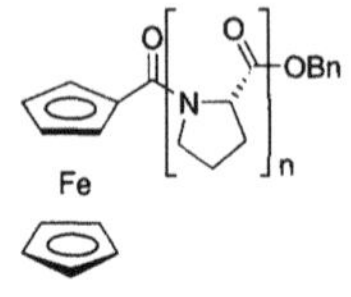

30a: n = 1
30b: n = 2
30c: n = 3
30d: n = 4

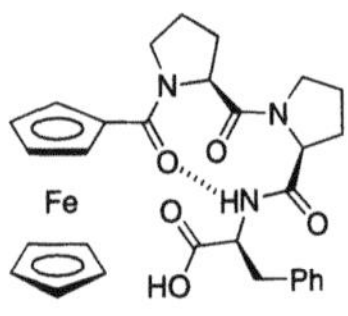

31

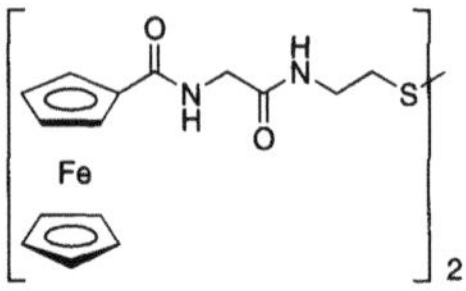

32

ferrocene unit) and N – H (Phe) to create a β-turn structural motif [81]. The *cis*-proline linkage is indicated to be preserved in solution by NOESY NMR studies (CD_3CN).

The ferrocene **32** ([Fc–Gly–CSA]$_2$) exhibits an ordered double helical arrangement through extensive intermolecular hydrogen bonds [82]. Each Fc–Gly fragment is involved in two different propeller-shaped helices, in which a central hydrogen-bonded peptide core is surrounded by the redox ferrocene moieties. One of these helices is described as a square helix with an inner diameter of the hydrogen-bonded core of 3.8 Å, and another is a twisted helix with an inner diameter of the hydrogen-bonded core of 4.1 Å. These helices are linked to each other through a disulfide bridge.

2.3
Transition Metal Complexes of Ferrocene–Peptide Bioconjugates

Metal ions have been known to exhibit a variety of properties in proteins, one of which is structural stabilization for biological function [83, 84]. Metal ions also play a crucial role in the redox processes of proteins [83, 84]. The incorporation of metal coordination sites into peptides has been investigated for the stabilization of secondary structures [85–89] and catalytic activities [90, 91]. Phosphine-containing β-turn ligands are used in asymmetric catalysis [91].

The ferrocene **33** bearing the podand dipeptide chains (–L–Ala–L–Pro–NHPy) is also characterized by chirality organization through intramolecular conformational regulation by the formation of two intramolecular interchain hydrogen bonds, as observed with **4a** (Fig. 12a) [92]. The NH adjacent to the pyridyl moiety of **33** is available for participating in intermolecular hydrogen bonding with the CO adjacent to the ferrocene unit to create the highly organized self-assembly, wherein each molecule is connected to four neighboring molecules (Fig. 13). The ferrocene **33** forms the 1 : 1 *trans* palladium complex

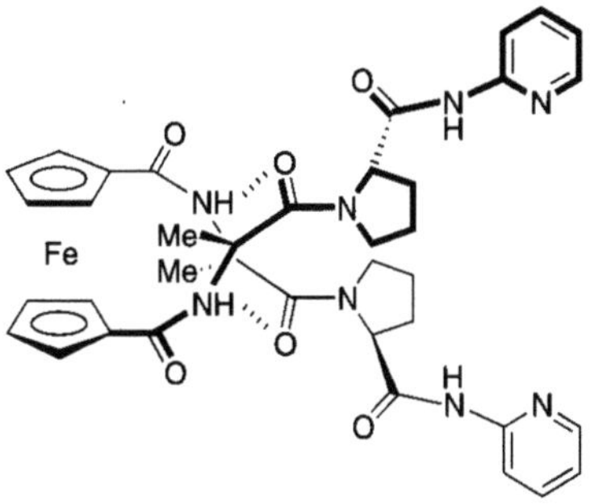

33

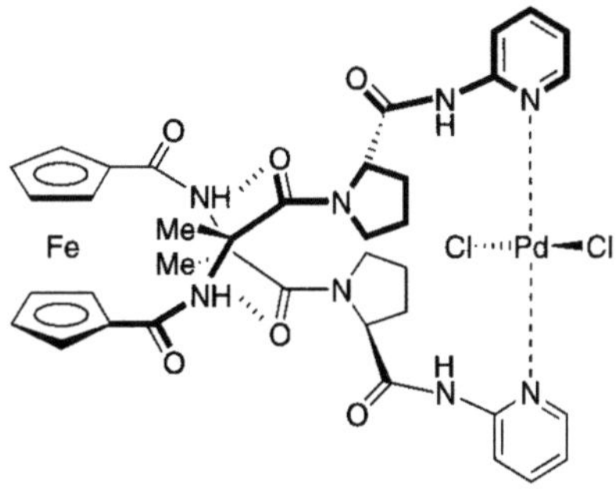

34

34 with $PdCl_2(MeCN)_2$ to stabilize the chirality conformational regulation in both solution and solid states. A greater downfield shifting of the Ala N – H resonance is observed in the ^1H NMR spectrum of 34 in $CDCl_3$ as compared with that of 33, indicating that complexation strengthens the intramolecular hydrogen bonds. The single-crystal X-ray structure determination of 34 confirms the pseudo-helical conformation through palladium coordination and chirality organization based on the preservation of intramolecular interchain hydrogen bonds, as depicted in Fig. 12b [92]. In contrast to the crystal packing of 33, the NH adjacent to the pyridyl moiety cannot participate in intermolecular hydrogen bonding.

Complexation of the ferrocene 24 bearing only one dipeptide chain (-L-Ala-L-Pro-NHPy) with $PdCl_2(MeCN)_2$ affords the 2 : 1 *trans* palladium complex 35 [79]. Two ferrocenyl dipeptide strands of complex 35 are able to rotate with respect to each other about the palladium center by the ball-

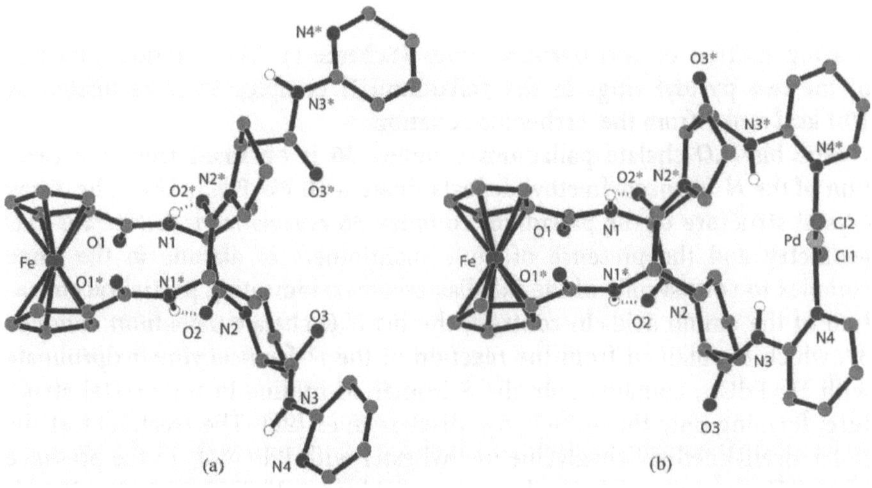

Fig. 12 Molecular structures of **a** 33 and **b** 34

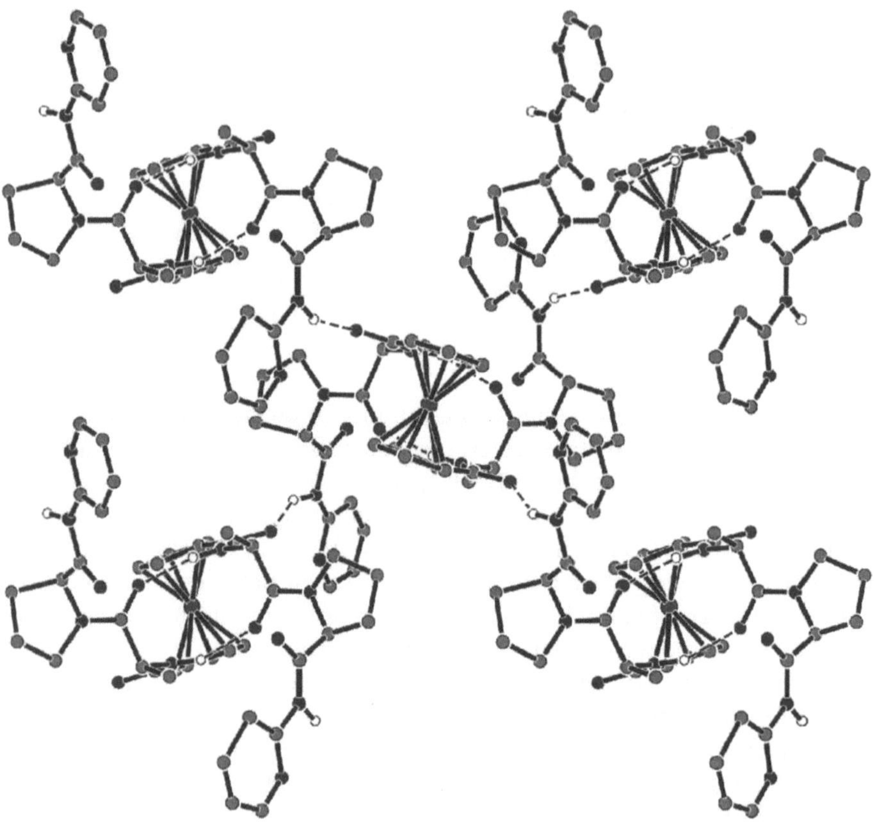

Fig. 13 The crystal packing of **33**. Each molecule is connected to four neighboring molecules by continuous intermolecular hydrogen bonds

bearing motion of two pyridine rings (Scheme 1). The rotational barrier of the two pyridyl rings in the palladium(II) complex **35** is calculated as 10.1 kcal mol^{-1} from the Arrhenius equation.

The bis-*N*,*O* chelate palladium complex **36** is obtained from the reaction of the *N*-ferrocenylmethylidenealaninate with Na_2PdCl_4 [93]. The X-ray crystal structure of the palladium complex **36** reveals *trans-N,N*/*trans-O,O* geometry and the presence of both enantiomers of alanine in the same complex to consist only of the *R*,*S* diastereomer, indicating partial racemization of the amino acid. In contrast, the bis-*N*,*O* chelate palladium complex **37**, which is obtained from the reaction of the *N*-ferrocenylmethylprolinate with Na_2PdCl_4, contains only the *S* isomer of proline in the crystal structure, forming only the $S_C R_N S_{C'} R_{N'}$ diastereomer [93]. The treatment of the *N*-ferrocenylmethylideneglycine methyl ester with $Pd(OAc)_2$ in the presence of MeOH and KOAc affords the acetate-bridged palladium complex **38**. Although the palladium complex **38** has no chiral center, orthopalladation is

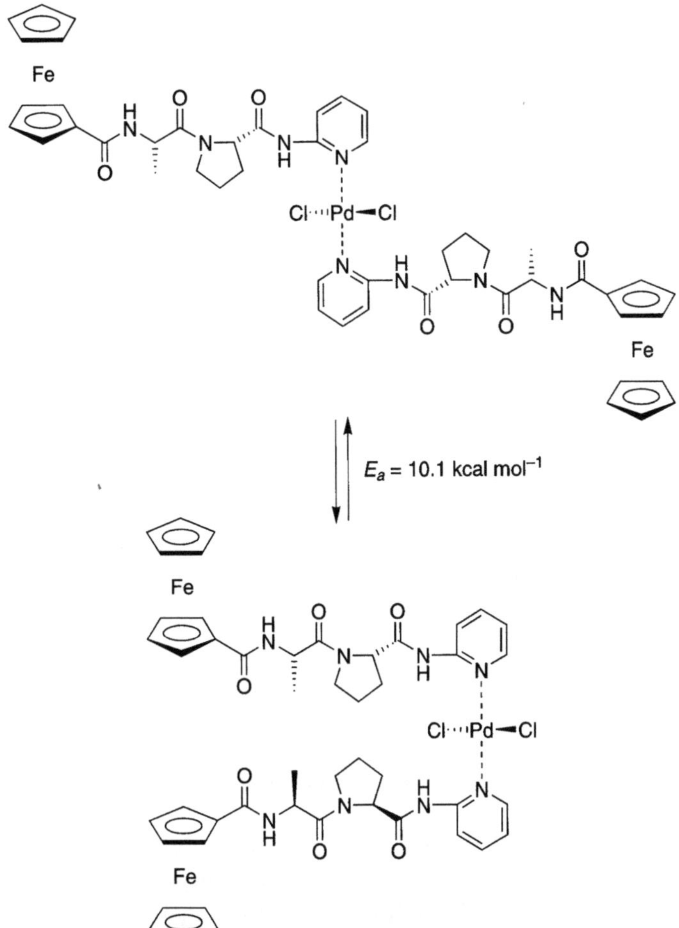

35

Scheme 1 Rotation of complex **35** about the palladium center

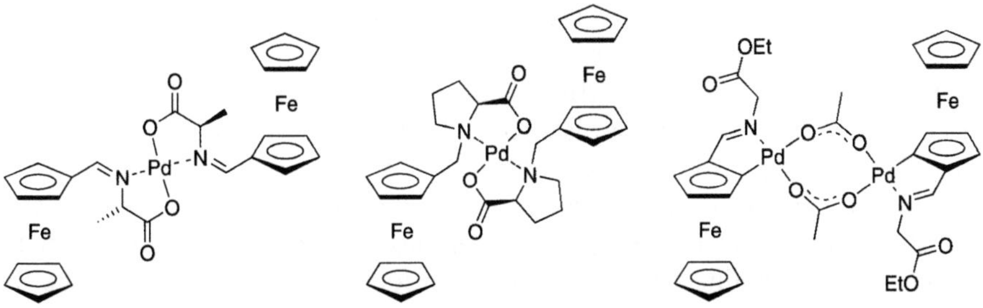

36 **37** **38**

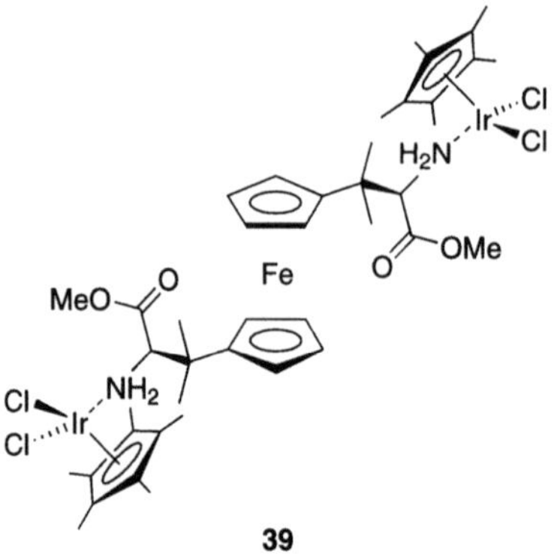

39

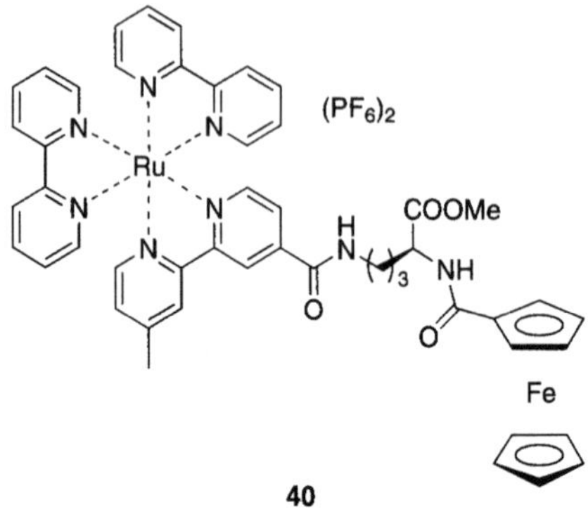

40

expected to induce planar chirality. However, the crystal structure of **38** shows the same configuration of both ferrocene moieties, giving only the $R_{Fc}R_{Fc'}/S_{Fc}S_{Fc'}$ diastereoisomeric pair [93].

The iridium complex **39** is also obtained by the reaction of the 1,1′-bis-(valine methyl ester) ferrocene derivative with $(Cp^*IrCl_2)_2$ [94]. [Fc-Orn {Ru(bipy)$_2$m}-OMe]$(PF_6)_2$ (**40**) shows efficient luminescence quenching by energy transfer [95].

3 Applications of Ferrocene–Peptide Bioconjugates

The organization of host molecules by self-assembly is a useful strategy to form active receptors [96–98]. Metal-templated organization has been exploited to provide oriented binding sites, resulting in the construction of artificial receptors for selective recognition [99–112]. Utilization of self-assembling properties of amino acids as observed in proteins, which are organized into well-defined three-dimensional structures, is considered to be a convenient approach to desired molecular receptors. On the other hand, ferrocenes have been focused on as an organometallic scaffold for molecular receptors based on redox properties and two rotatory coplanar cyclopentadienyl (Cp) rings with ca. 3.3 Å separation [113–120]. Ferrocenylboronic acid [121] and α-ferrocenylalkylamine [122] derivatives have been reported as chiral redox-active receptors.

Electron transfer (ET) reactions along a sequential redox potential field occur effectively over a long distance in biological systems [123]. These redox centers are surrounded by highly ordered polypeptides, which are considered to control the three-dimensional environment of redox centers and the direction of the ET [124]. Self-assembled monolayers (SAMs) of helical ferrocene-contained peptides have been focused on to investigate the ET reaction.

Peptide nucleic acid (PNA) is one of the attracted DNA analogues and considerable effort has been devoted to designing functional PNA derivatives [125–141]. The introduction of redox-active ferrocene derivatives into PNA is expected to provide new properties.

3.1 Ferrocene–Peptide Bioconjugates for Molecular Receptors

The ferrocene receptor **41** bearing the podand dipeptide chains (–L–Ala–L–Pro–NHPyMe) is designed to incorporate L-alanyl-L-proline as a dipeptide capable of hydrogen bonding and imposing conformational constraint on the peptide backbone [142]. The single-crystal X-ray structure determination of **41** confirms the chirality-organized structure, as observed with **4a** (Fig. 14). The formation of a chirality-organized structure even in solution is supported by CD spectrometry. ^{1}H NMR ($CDCl_3$ and $CDCl_3$/DMSO-d_6 (9 : 1)) and FT-IR (CH_2Cl_2) analyses indicate that the additional NH adjacent to the pyridyl moiety is not hydrogen bonded in the solution. The NH adjacent to the pyridyl moiety is expected for hydrogen bonding with dicarboxylic acids. In the ferrocene receptor **41**, the two amido pyridyl moieties as hydrogen bond-

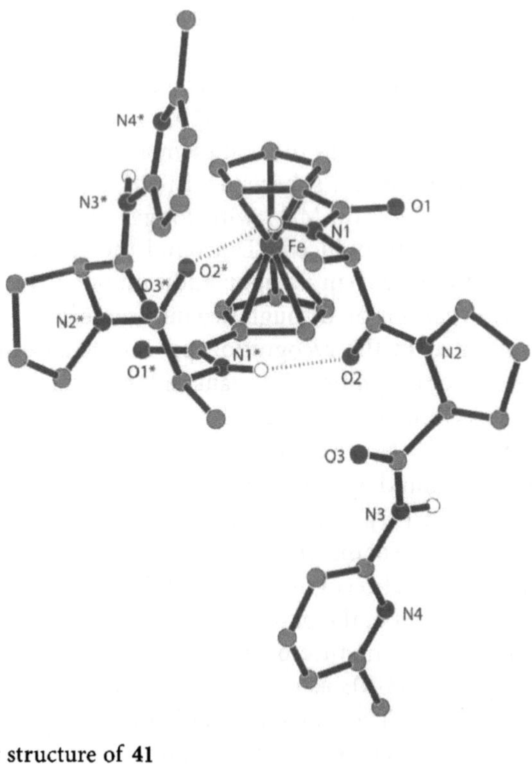

Fig. 14 Molecular structure of **41**

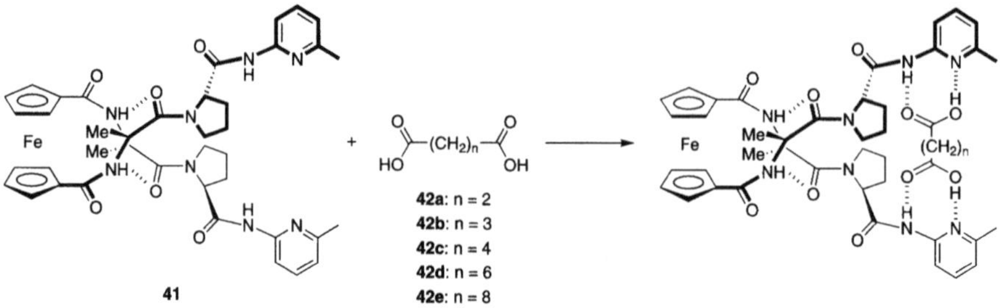

Scheme 2 Binding of dicarboxylic acid to the ferrocene receptor **41**

ing sites are well arranged for dicarboxylic acids by the chirality organization through two intramolecular hydrogen bonds (Scheme 2).

The dicarboxylic acid binding properties of the ferrocene receptor **41** are investigated using ^{1}H NMR spectroscopy and association constants are measured for dicarboxylic acids **42** possessing various chain lengths. ^{1}H NMR studies in $CDCl_3$/acetone-d_6 (5 : 1) reveal the considerable downfield shift of the pyridine amide protons of **41** ($\Delta\delta$ 1.04 ppm) upon addition of one molar equivalent of **42c**, suggesting that amido pyridyl moieties serve as binding sites for dicarboxylic acid. The dicarboxylic acid–receptor stoichiometry is confirmed to be 1 : 1 by Job plots. Titration of **41** in $CDCl_3$/acetone-d_6 (5 : 1) with a series of dicarboxylic acids **42** shows appreciable association constants. It should be noted that a higher association constant for adipic acid (**42c**, $K_a = 2.1 \times 10^4\ M^{-1}$) is observed as compared with succinic acid (**42a**, $K_a = 2.8 \times 10^2\ M^{-1}$) and glutaric acid (**42b**, $K_a = 3.7 \times 10^3\ M^{-1}$). This difference is probably attributable to the complementary binding space size of **41** for adipic acid. Furthermore, suberic acid (**42d**, $K_a = 7.7 \times 10^3\ M^{-1}$) and sebacic acid (**42e**, $K_a = 5.9 \times 10^3\ M^{-1}$) with longer chain lengths exhibited a smaller constant. Since an induced CD around the absorbance of the ferrocene function of **41** hardly changes upon addition of five molar equivalents of **42c**, the chirality organization through two intramolecular hydrogen bonds seems to be maintained in the recognition process to afford a rigid binding site for the selective recognition. To evaluate the chiral recognition ability of **41**, its chiral discrimination properties are examined with biologically important glutamic acids. Noteworthy is that benzoyl-L-glutamic acid (**42f**, $K_a = 5.5 \times 10^3\ M^{-1}$) is bound approximately 15 times more tightly to **41** than benzoyl-D-glutamic acid (**42g**, $K_a = 3.7 \times 10^2\ M^{-1}$). The chirality-organized binding site of **41** is capable of discriminating the chirality of guest molecules. The size-selective and chiral recognition of dicarboxylic acids is achieved by multipoint hydrogen bonds of the binding sites.

Crystallization of a 1 : 2 mixture of **41** and (1*R*,3*S*)-camphoric acid (**CA**) gives the 1 : 2 complex **41**·(**CA**)$_2$ as orange crystals by slow diffusion of hexane into chloroform [143]. The single-crystal X-ray structure determination of **41**·(**CA**)$_2$ reveals a polymeric cocrystal composed of alternating units of **41** and two molecules of **CA**, which are connected by continuous intermolecular hydrogen bonds to form a double-helix-like hydrogen-bonded molecular arrangement (Fig. 15). Each **CA** is found to serve as a hydrogen bonding bridge. One carboxyl moiety of **CA** binds to the amide pyridyl binding site of **41** while another carboxyl moiety of **CA** interacts with the carbonyl group adjacent to the ferrocene unit of another molecule of **41**. As a result, each molecule of **41** is bridged by two molecules of **CA**, as depicted in Fig. 16. Another interesting feature is that the chirality-organized structure of **41** is preserved in spite of complexation with **CA**.

The ferrocenes **19** bearing only one amino acid chain (–Gly–OR) serve as electrochemical anion sensors for HSO_4^-, Cl^-, $H_2PO_4^-$, and BF_4^- [74]. The dihydrogen phosphate anion $H_2PO_4^-$ causes the largest negative perturbation of the Fc/Fc^+ redox couple of 120 mV with **19a**. The ferrocenes **27** and **43a–e** bearing only one dipeptide chain also show the same electrochemical anion sensing behavior [144]. The ferrocene **27** bearing only one dipeptide chain

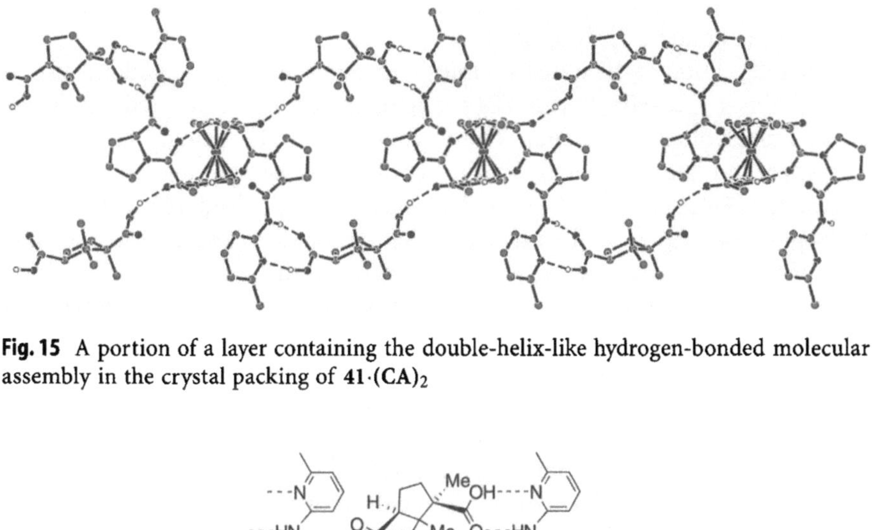

Fig. 15 A portion of a layer containing the double-helix-like hydrogen-bonded molecular assembly in the crystal packing of $\mathbf{41}\cdot(\mathbf{CA})_2$

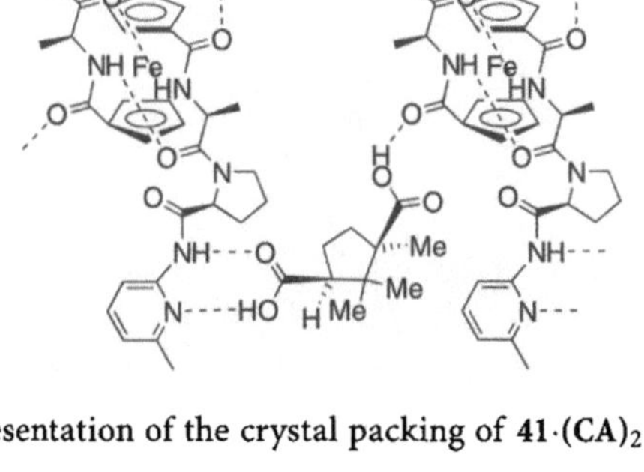

Fig. 16 Schematic representation of the crystal packing of $\mathbf{41}\cdot(\mathbf{CA})_2$

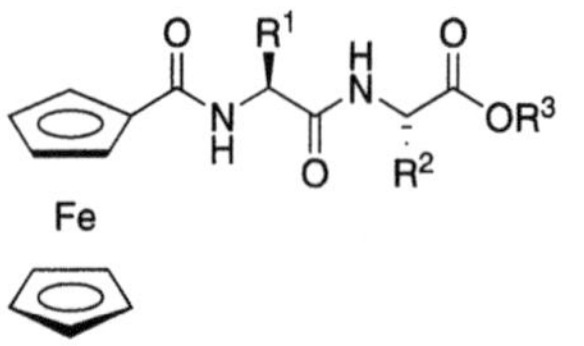

43a: R^1 = *i*-Bu, R^2 = *i*-Bu, R^3 = Et
43b: R^1 = Me, R^2 = Bn, R^3 = Et
43c: R^1 = Bn, R^2 = *i*-Bu, R^3 = Bn
43d: R^1 = Bn, R^2 = CH_2OH, R^3 = Et
43e: R^1 = Bn, R^2 = Bn, R^3 = Me
43f: R^1 = *i*-Pr, R^2 = Bn, R^3 = Me
43g: R^1 = *i*-Pr, R^2 = *i*-Pr, R^3 = Me
43h: R^1 = Me, R^2 = Me, R^3 = Bn
43i: R^1 = Me, R^2 = Me, R^3 = Me
43j: R^1 = *i*-Bu, R^2 = *i*-Bu, R^3 = Me

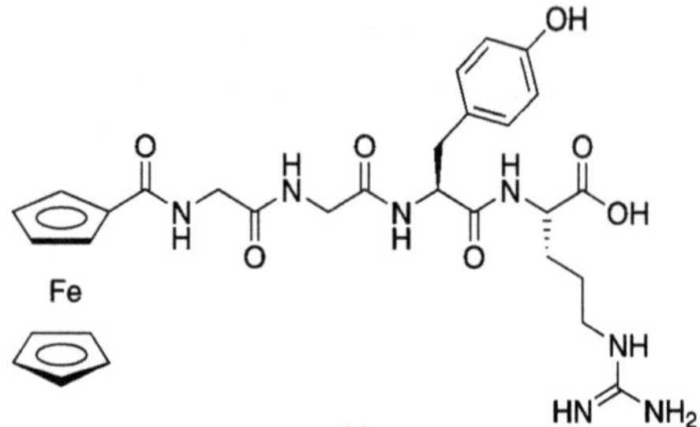

44

(-Gly–Gly–OMe) is the most suitable among the six peptide receptors **27** and **43a–e** because it is less sterically hindered to interact more closely with an anion at the amide CONH moiety.

The interaction of the water-soluble ferrocene **2** bearing the podand amino acid chains (-Gly–NH_2) with DNA oligomer can be monitored by differential pulse voltammetry [62]. The ferrocene **2** binds strongly to the DNA oligomer and the calculated binding constant is comparable to those reported for $[Fe(phen)_3]^{2+/3+}$ [145]. This similarity is considered to be due to hydrogen bonding as well as hydrophobic interaction. The Cp ring separation of ferrocene is ca. 3.3 Å, which is similar to the distance between base pairs in DNA. Each ferrocene molecule is bound to every two base pairs as expected.

The saturation titration experiments of the ferrocenes **27**, **28a**, and **43e–j** bearing only one dipeptide chain with 3-aminopyrazole (3-Apzl) derivatives indicate that these derivatives interact with the top face of the ferrocene–dipeptides by hydrogen bonding to form a 1 : 1 complex as shown in Scheme 3, although binding constants in chloroform are low [80, 146]. The redox potentials are sensitive to complex formation, and the addition of a large

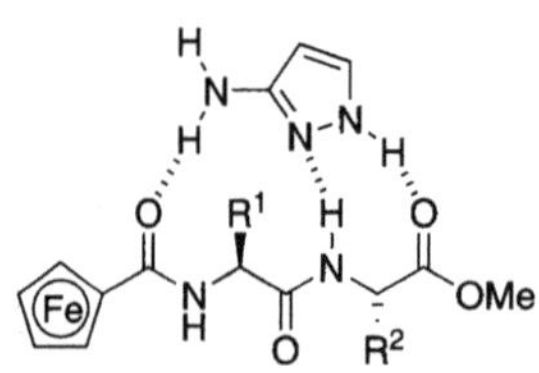

Scheme 3 Interaction of 3-aminopyrazole derivatives with ferrocene–dipeptides to form a 1 : 1 complex

excess of 3-Apzl to the solution of ferrocenes **28a** and **43e–h** results in a cathodic shift of the redox potentials.

The ferrocene **44** bearing only one tetrapeptide chain (–Gly–Gly–L–Tyr–L–Arg–OH) is designed to bind to papain [147]. The ferrocene **44** acts as an efficient competitive papain inhibitor for *N*-benzoylarginine ethyl ester hydrolysis, with an inhibition constant K_i of 9 μM at pH 6.2. Binding of papain to the ferrocene receptor **44** causes an electrochemical response, resulting in a small cathodic shift of the redox potential of the ferrocene moiety of **44**.

3.2 Ferrocene–Peptide Bioconjugates for Biomaterials

The alamethicin derivatives **45** and **46** bearing the *C*-terminal redox-active ferrocene moiety form voltage-dependent ion channels in planar lipid bilayers, wherein the conductance properties are similar to those of alamethicin [148]. The in situ oxidation of **45** by CAN in the planar lipid bilayer apparatus causes a time-dependent elimination of channel openings. The channel activity of the oxidized **45** can be restored by increasing the bilayer potential. In contrast, oxidation of **46** results in the formation of shorter-lived channels.

The *N*-ferrocenoyl-labeled oligoproline cystamines **47** form stable monolayers on gold, in which hydrogen bonds between adjacent molecules are considered to play an important role in the packing and stability of the monolayers [149]. Cystamines **47e–g** adopt the helical all-*trans* polyproline II conformation in solution, although **47b** and **47c** show greater flexibility and potential to undergo *cis–trans* isomerization in solution. The ferrocene moiety of **47** becomes more easily oxidized with increasing length of the oligoproline chain. The significant deviation from Marcus-type behavior implicates

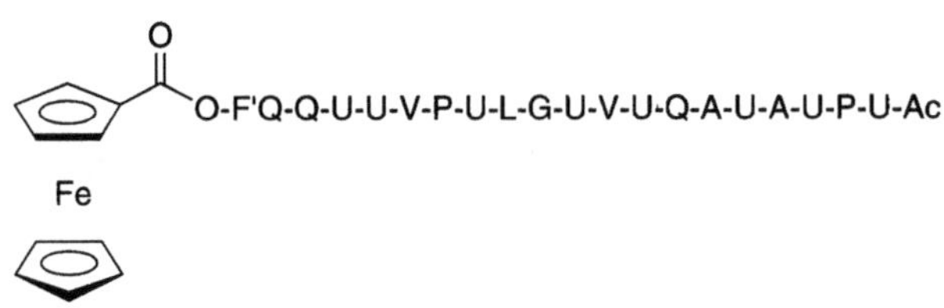

45

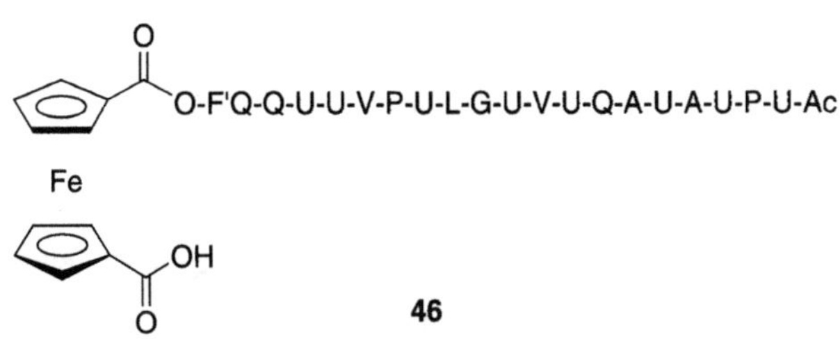

46

a through-bond mechanism in the ET process between the ferrocenyl redox probe and the gold microelectrode surface across the oligoproline spacer, although the k_{ET} value is distance-dependent.

Well-ordered SAMs are also formed from helical peptides **48** and **49** carrying a ferrocene moiety at the *N*- or *C*-terminal end [150]. Electrochemical measurements reveal that a long-range ET reaction over 4 nm occurs with the inelastic hopping mechanism over the superexchange mechanism in the SAMs. The accelerating effect of the helix dipole on the ET rate is also observed, probably due to the lowering of the barrier height between the gold surface and the peptide layer. The ET rate in **49** is about threefold faster than that in the **48** SAM.

The ferrocene derivative of PNA monomer **50** is prepared by the reaction of ferrocene carboxylic acid chloride with thymine-PNA methyl es-

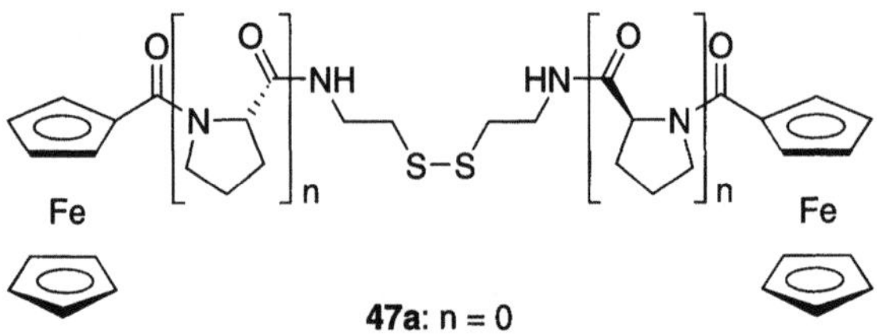

47a: n = 0
47b: n = 1
47c: n = 2
47d: n = 3
47e: n = 4
47f: n = 5
47g: n = 6

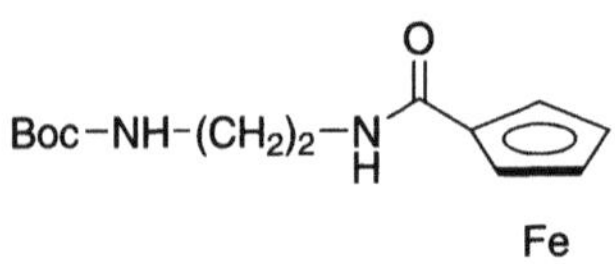

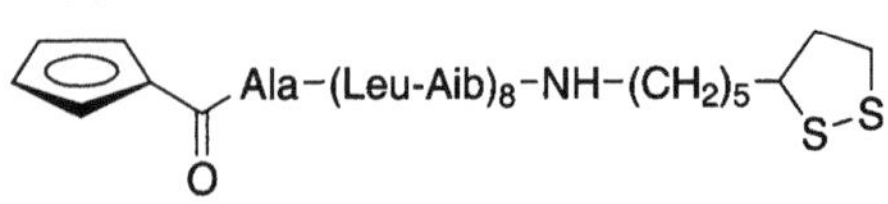

48

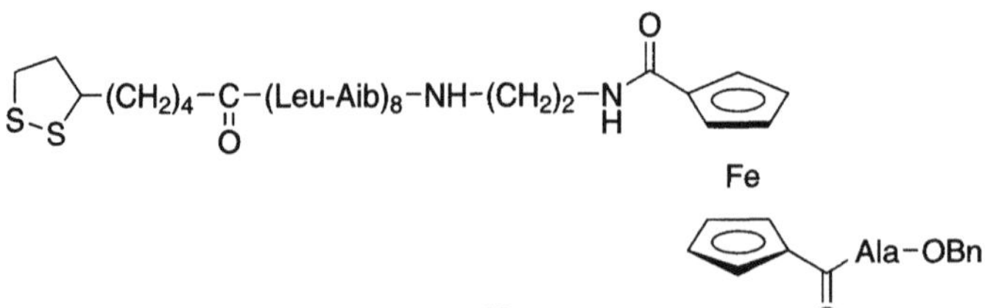

49

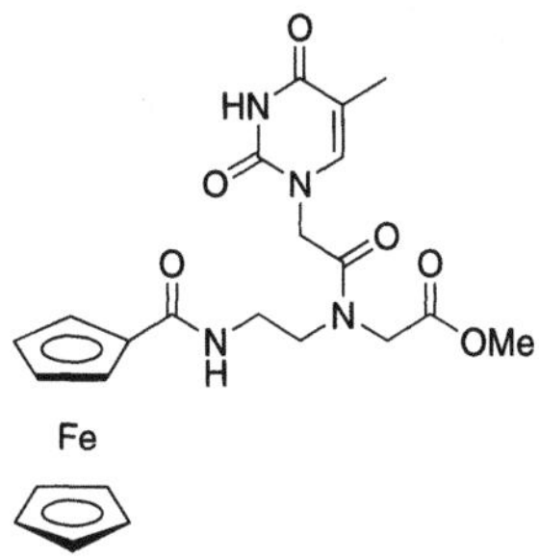

50

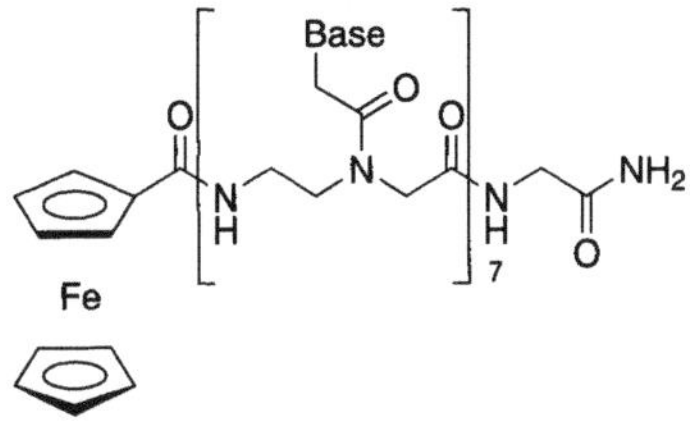

51

ter [151]. Two isomers exist in **50**, as observed in PNA monomers [152, 153]. Variable-temperature ^{1}H NMR studies of **50** reveal a rotation barrier of 75 ± 0.5 kJ mol^{-1}, which is consistent with rotation about the tertiary amide bond [152, 153]. T–T self-association is considered to occur over a concentration range of 0.375–0.0027 mol L^{-1} from ^{1}H NMR studies in $CDCl_3$. Also, the ferrocene derivative of PNA oligomer **51** is prepared by standard solid support peptide synthesis methods [154].

4
Conclusion

Ferrocenes are recognized as an organometallic scaffold in a central reverse-turn unit based on the inter-ring spacing of ferrocene of about 3.3 Å, which is appropriate for hydrogen bonding of the attached peptide strands as observed in *β*-sheets. Another advantage in the use of ferrocenes as a scaffold depends on the electrochemical reversibility of a redox couple, which is able to be available as a redox-switching center and an electrophore. A variety of ferrocene–peptide bioconjugates have been designed to induce highly ordered structures of peptides and develop new biomaterials. The ferrocene serves as a reliable organometallic scaffold for the construction of an ordered structure via intramolecular hydrogen bonding, wherein the attached peptide strands are regulated in the appropriate dimensions. These chemical models of protein secondary structures afford fundamental insight into the factors affecting protein structure and stability. A further noteworthy feature of ferrocene–peptide bioconjugates is their strong tendency to self-assemble through the contribution of all available hydrogen bonding donors in the solid state. The architectural control of molecular assemblies utilizing peptide chains, which possess chiral centers and hydrogen bonding sites, is envisioned to be a useful approach to artificial highly ordered systems. In addition, the ferrocenyl moiety of ferrocene–peptide bioconjugates can act as a redox-switching center and an electrophore in the receptors and biomaterials.

This bioorganometallic chemistry is envisioned to provide not only a peptidomimetic basis for protein folding, but also pharmacologically useful compounds, artificial receptors, asymmetric catalysts, and new materials with functional properties.

References

1. Jaouen G, Vessiéres A, Butler IS (1993) Acc Chem Res 26:361
2. Severin K, Bergs R, Beck W (1998) Angew Chem Int Ed 37:1634
3. Jaouen G (ed) (1999) Bioorganometallic chemistry (special issue). J Organomet Chem 589:1–126
4. Braga D, Grepioni F, Desiraju GR (1998) Chem Rev 98:1375
5. Balzani V, Credi A, Raymo FM, Stoddart JF (2000) Angew Chem Int Ed 39:3348
6. Swiegers GF, Malefetse TJ (2000) Chem Rev 100:3483
7. Jeffrey GA (1997) An introduction to hydrogen bonding, 1st edn. Oxford University Press, New York
8. Conn MM, Rebek J Jr (1997) Chem Rev 97:1647
9. Archer EA, Gong H, Krische MJ (2001) Tetrahedron 57:1139
10. Prins LJ, Reinhoudt DN, Timmerman P (2001) Angew Chem Int Ed 40:2382
11. Kyte J (1995) Structure in protein chemistry. Garland, New York
12. Branden C, Tooze J (1998) Introduction to protein structure, 2nd edn. Garland, New York
13. Togni A, Hayashi T (1995) Ferrocenes. Wiley, Weinheim
14. Schlögl K (1957) Monatsh Chem 88:601
15. Hauser CR, Lindsay JK (1957) J Org Chem 22:1246
16. Osgerby JM, Pauson PL (1958) J Chem Soc 656
17. Urban R, Ugi I (1975) Angew Chem Int Ed Engl 14:61
18. Urban R, Eberle G, Marquarding D, Rehn D, Rehn H, Ugi I (1976) Angew Chem Int Ed Engl 15:627
19. Cuingnet E, Sergheraert C, Tartar A (1980) J Organomet Chem 195:325
20. Carlström A-S, Frejd T (1990) J Org Chem 55:4175
21. Brunner H, König W, Nuber B (1993) Tetrahedron Asymmetry 4:699
22. Jackson RFW, Turner D, Block MH (1996) Synlett 862
23. Kira M, Matsubara T, Shinohara H, Sisido M (1997) Chem Lett 89
24. Basu B, Chattopadhyay SK, Ritzén A, Frejd T (1997) Tetrahedron Asymmetry 8:1841
25. Kayser B, Altman J, Nöth H, Knizek J, Beck W (1998) Eur J Inorg Chem 1791
26. Hess A, Brosch O, Weyhermüller T, Metzler-Nolte N (1999) J Organomet Chem 589:75
27. Brosch O, Weyhermüller T, Metzler-Nolte N (2000) Eur J Inorg Chem 323
28. Hess A, Sehnert J, Weyhermüller T, Metzler-Nolte N (2000) Inorg Chem 39:5437
29. Xu Y, Kraatz H-B (2001) Tetrahedron Lett 42:2601
30. Maricic S, Frejd T (2002) J Org Chem 67:7600
31. Mandal HS, Kraatz H-B (2003) J Organomet Chem 674:32
32. Feigel M (1986) J Am Chem Soc 108:181
33. Brandmeier V, Feigel M, Bremer M (1989) Angew Chem Int Ed Engl 28:486
34. Brandmeier V, Feigel M (1989) Tetrahedron 45:1365
35. Brandmeier V, Sauer WHB, Feigel M (1994) Helv Chim Acta 77:70
36. Kemp DS, Bowen BR (1988) Tetrahedron Lett 29:5077

37. Kemp DS, Bowen BR (1988) Tetrahedron Lett 29:5081
38. Kemp DS (1990) Trends Biotechnol 8:249
39. Kemp DS, Muendel CC, Blanchard DE, Bowen BR (1990) In: Rivier JE, Marshall GR (eds) Peptides: chemistry, structure, and biology. Proceedings of the 11th American peptide symposium. ESCOM, Leiden, p 674
40. Kemp DS, Bowen BR, Muendel CC (1990) J Org Chem 55:4650
41. Díaz H, Kelly JW (1991) Tetrahedron Lett 32:5725
42. Díaz H, Espina JR, Kelly JW (1992) J Am Chem Soc 114:8316
43. Díaz H, Tsang KY, Choo D, Espina JR, Kelly JW (1993) J Am Chem Soc 115:3790
44. Díaz H, Tsang KY, Choo D, Kelly JW (1993) Tetrahedron 49:3533
45. Graciani NR, Tsang KY, McCutchen SL, Kelly JW (1994) Bioorg Med Chem 2:999
46. Tsang KY, Díaz H, Graciani N, Kelly JW (1994) J Am Chem Soc 116:3988
47. Nowick JS, Powell NA, Martinez EJ, Smith EM, Noronha G (1992) J Org Chem 57:3763
48. Nowick JS, Abdi M, Bellamo KA, Love JA, Martinez EJ, Noronha G, Smith EM, Ziller JW (1995) J Am Chem Soc 117:89
49. Nowick JS, Smith EM, Noronha G (1995) J Org Chem 60:7386
50. Nowick JS, Mahrus S, Smith EM, Ziller JW (1996) J Am Chem Soc 118:1066
51. Nowick JS, Holmes DL, Mackin G, Noronha G, Shaka AJ, Smith EM (1996) J Am Chem Soc 118:2764
52. Nowick JS, Smith EM, Pairish M (1996) Chem Soc Rev 25:401
53. Nowick JS, Pairish M, Lee IQ, Holmes DL, Ziller JW (1997) J Am Chem Soc 119:5413
54. Nowick JS (1999) Acc Chem Res 32:287
55. Nowick JS, Tsai JH, Bui Q-CD, Maitra S (1999) J Am Chem Soc 121:8409
56. Moriuchi T, Tamura T, Hirao T (2002) J Am Chem Soc 124:9356
57. Jones IG, Jones W, North M (1998) J Org Chem 63:1505
58. Ranganathan D, Haridas V, Kurur S, Thomas A, Madhusudanan KP, Nagaraj R, Kunwar AC, Sarma AVS, Karle IL (1998) J Am Chem Soc 120:8448
59. Hackenberger CPR, Schiffers I, Runsink J, Bolm C (2004) J Org Chem 69:739
60. Herrick RS, Jarret RM, Curran TP, Dragoli DR, Flaherty MB, Lindyberg SE, Slate RA, Thornton LC (1996) Tetrahedron Lett 37:5289
61. Oberhoff M, Duda L, Karl J, Mohr R, Erker G, Fröhlich R, Grehl M (1996) Organometallics 15:4005
62. Georgopoulou AS, Mingos DMP, White AJP, Williams DJ, Horrocks BR, Houlton A (2000) J Chem Soc Dalton Trans 2969
63. van Staveren DR, Weyhermüller T, Metzler-Nolte N (2003) Dalton Trans 210
64. Cerichelli G, Floris B, Ortaggi G (1974) J Organomet Chem 76:73
65. Nomoto A, Moriuchi T, Yamazaki S, Ogawa A, Hirao T (1998) Chem Commun 1963
66. Moriuchi T, Nomoto A, Yoshida K, Hirao T (1999) J Organomet Chem 589:50
67. Moriuchi T, Nomoto A, Yoshida K, Ogawa A, Hirao T (2001) J Am Chem Soc 123:68
68. Bauer W, Polborn K, Beck W (1999) J Organomet Chem 579:269
69. Moriuchi T, Nomoto A, Yoshida K, Hirao T (2001) Organometallics 20:1008
70. Moriuchi T, Nagai T, Hirao T (2005) Org Lett 7:5265
71. Xu Y, Saweczko P, Kraatz H-B (2001) J Organomet Chem 637–639:335
72. Maricic S, Berg U, Frejd T (2002) Tetrahedron 58:3085
73. Kraatz H-B, Lusztyk J, Enright GD (1997) Inorg Chem 36:2400
74. Gallagher JF, Kenny PTM, Sheehy MJ (1999) Inorg Chem Commun 2:200
75. Baker MV, Kraatz H-B, Quail JW (2001) New J Chem 25:427
76. Brosch O, Weyhermüller T, Metzler-Nolte N (1999) Inorg Chem 38:5308
77. Savage D, Gallagher JF, Ida Y, Kenny PTM (2002) Inorg Chem Commun 5:1034

78. Cung MT, Marraud M, Néel J, Aubry A (1976) J Chim Phys 73:213
79. Moriuchi T, Yoshida K, Hirao T (2001) J Organomet Chem 637–639:75
80. Saweczko P, Enright GD, Kraatz H-B (2001) Inorg Chem 40:4409
81. Kraatz H-B, Leek DM, Houmam A, Enright GD, Lusztyk J, Wayner DDM (1999) J Organomet Chem 589:38
82. Bediako-Amoa I, Silerova R, Kraatz H-B (2002) Chem Commun 2430
83. Anfinsen CB, Edsall JT, Richards FM, Eisenberg DS (1991) Advances in protein chemistry. Academic, New York
84. Kaim W, Schwederski B (1994) Bioinorganic chemistry: inorganic elements in the chemistry of life. Wiley, New York
85. Ghadiri MR, Choi C (1990) J Am Chem Soc 112:1630
86. Handel T, DeGrado WF (1990) J Am Chem Soc 112:6710
87. Ruan F, Chen Y, Hopkins PB (1990) J Am Chem Soc 112:9403
88. Schneider JP, Kelly JW (1995) J Am Chem Soc 117:2533
89. Gilbertson SR, Pawlick RV (1996) Angew Chem Int Ed Engl 35:902
90. Merkle DL, Schmidt MH, Berg JM (1991) J Am Chem Soc 113:5450
91. Gilbertson SR, Collibee SE, Agarkov A (2000) J Am Chem Soc 122:6522
92. Moriuchi T, Yoshida K, Hirao T (2001) Organometallics 20:3101
93. Freiesleben D, Polborn K, Robl C, Sünkel K, Beck W (1995) Can J Chem 73:1164
94. Dialer H, Polborn K, Ponikwar W, Sünkel K, Beck W (2002) Chem Eur J 8:691
95. Geißer B, Alsfasser R (2003) Inorg Chim Acta 344:102
96. Kim M, Gokel GW (1987) J Chem Soc Chem Commun 1686
97. Wyler R, de Mendoza J, Rebek J Jr (1993) Angew Chem Int Ed Engl 32:1699
98. Cha X, Ariga K, Onda M, Kunitake T (1995) J Am Chem Soc 117:11833
99. Schneider H-J, Ruf D (1990) Angew Chem Int Ed Engl 29:1159
100. Kobuke Y, Sumida Y, Hayashi M, Ogoshi H (1991) Angew Chem Int Ed Engl 30:1496
101. Schwabacher AW, Lee J, Lei H (1992) J Am Chem Soc 114:7597
102. Hunter CA, Sarson LD (1994) Angew Chem Int Ed Engl 33:2313
103. Goodman MS, Weiss J, Hamilton AD (1994) Tetrahedron Lett 35:8943
104. Prévot-Halter I, Smith TJ, Weiss J (1996) Tetrahedron Lett 37:1201
105. Habata Y, Bradshaw JS, Zhang XX, Izatt RM (1997) J Am Chem Soc 119:7145
106. Linton B, Hamilton AD (1997) Chem Rev 97:1669
107. Goodman MS, Jubian V, Linton B, Hamilton AD (1995) J Am Chem Soc 117:11610
108. Huc I, Krische MJ, Funeriu DP, Lehn J-M (1999) Eur J Inorg Chem 1415
109. Haino T, Yamanaka Y, Araki H, Fukazawa Y (2002) Chem Commun 402
110. Ayabe M, Ikeda A, Shinkai S, Sakamoto S, Yamaguchi K (2002) Chem Commun 1032
111. Crowley JD, Gosche AJ, Bosnich B (2003) Chem Commun 392
112. Tobey SL, Jones BD, Anslyn EV (2003) J Am Chem Soc 125:4026
113. Saji T, Kinoshita I (1986) J Chem Soc Chem Commun 716
114. Medina JC, Li C, Bott SG, Atwood JL, Gokel GW (1991) J Am Chem Soc 113:366
115. Medina JC, Goodnow TT, Rojas MT, Atwood JL, Lynn BC, Kaifer AE, Gokel GW (1992) J Am Chem Soc 114:10583
116. Beer PD (1996) J Chem Soc Chem Commun 689
117. Beer PD, Graydon AR, Johnson AOM, Smith DK (1997) Inorg Chem 36:2112
118. Li C, Medina JC, Maguire GEM, Abel E, Atwood JL, Gokel GW (1997) J Am Chem Soc 119:1609
119. Carr JD, Lambert L, Hibbs DE, Hursthouse MB, Malik KMA, Tucker JHR (1997) Chem Commun 1649
120. Carr JD, Coles SJ, Hursthouse MB, Light ME, Tucker JHR, Westwood J (2000) Angew Chem Int Ed 39:3296

121. Ori A, Shinkai S (1995) J Chem Soc Chem Commun 1771
122. Laurent P, Miyaji H, Collinson SR, Prokeš I, Moody CJ, Tucker JHR, Slawin AMZ (2002) Org Lett 4:4037
123. Page CC, Moser CC, Chen XX, Dutton PL (1999) Nature 402:47
124. Beratan DN, Onuchic JN, Winkler JR, Gray HB (1992) Science 258:1740
125. Nielsen PE, Egholm M, Berg RH, Buchardt O (1991) Science 254:1497
126. Egholm M, Buchardt O, Christensen L, Behrens C, Freier SM, Driver DA, Berg RH, Kim SK, Norden B, Nielsen PE (1993) Nature 365:566
127. Wittung P, Nielsen PE, Buchardt O, Egholm M, Nordén B (1994) Nature 368:561
128. Nielsen PE, Haaima G (1997) Chem Soc Rev 96:73
129. Kuwahara M, Arimitsu M, Sisido M (1999) J Am Chem Soc 121:256
130. Nielsen PE (1999) Acc Chem Res 32:624
131. Seitz O, Bergmann F, Heindl D (1999) Angew Chem Int Ed 38:2203
132. Schütz R, Cantin M, Roberts C, Greiner B, Uhlmann E, Leumann C (2000) Angew Chem Int Ed 39:1250
133. Kozlov IA, Orgel LE, Nielsen PE (2000) Angew Chem Int Ed 39:4292
134. Wilhelmsson LM, Holmén A, Lincoln P, Nielsen PE, Nordén B (2001) J Am Chem Soc 123:2434
135. Vilaivan T, Lowe G (2002) J Am Chem Soc 124:9326
136. Okamoto A, Tanabe K, Saito I (2002) J Am Chem Soc 124:10262
137. Cichon MK, Haas CH, Grolle F, Mees A, Carell T (2002) J Am Chem Soc 124:13984
138. Popescu D-L, Parolin TJ, Achim C (2003) J Am Chem Soc 125:6354
139. Zelder FH, Brunner J, Krämer R (2004) Chem Commun 902
140. Krishnan-Ghosh Y, Stephens E, Balasubramanian S (2004) J Am Chem Soc 126:5944
141. Mokhir A, Krämer R, Wolf H (2004) J Am Chem Soc 126:6208
142. Moriuchi T, Yoshida K, Hirao T (2003) Org Lett 5:4285
143. Moriuchi T, Yoshida K, Hirao T (2003) J Organomet Chem 668:31
144. Gallagher JF, Kenny PTM, Sheehy MJ (1999) Inorg Chem Commun 2:327
145. Carter JC, Rodriguez M, Bard AJ (1989) J Am Chem Soc 111:8901
146. Saweczko P, Kraatz H-B (1999) Coord Chem Rev 190–192:185
147. Plumb K, Kraatz H-B (2003) Bioconjug Chem 14:601
148. Schmitt JD, Sansom MSP, Kerr ID, Lunt GG, Eisenthal R (1997) Biochemistry 36:1115
149. Galka MM, Kraatz H-B (2002) ChemPhysChem 4:356
150. Morita T, Kimura S (2003) J Am Chem Soc 125:8732
151. Hess A, Metzler-Nolte N (1999) Chem Commun 885
152. Dueholm KL, Engholm M, Behrens C, Christensen L, Hansen HF, Vulpius T, Petersen KH, Berg RH, Nielsen PE, Buchardt O (1994) J Org Chem 59:5767
153. Chen S-M, Mohan V, Kiely JS, Griffith MC, Griffey RH (1994) Tetrahedron Lett 35:5105
154. Verheijen JC, van der Marel GA, van Boom JH, Metzler-Nolte N (2000) Bioconjug Chem 11:741

Top Organomet Chem (2006) 17: 177–210
DOI 10.1007/3418_001

Published online: 4 February 2006

Medicinal Properties of Organometallic Compounds

Claire S. Allardyce · Paul J. Dyson (✉)

Laboratoire de Chimie Organometallique et Medicinale,
Institut des Sciences et Ingénierie Chimiques,
Ecole Polytechnique Fédérale de Lausanne (EPFL), 1015 Lausanne, Switzerland
Paul.Dyson@epfl.ch

Abstract Although organometallic compounds are ubiquitous in nature, synthetic organometallic compounds are generally considered to be toxic or non-compatible with biological systems. Despite this perception the medicinal properties of organometallic compounds, in particular organo-transition metal compounds, have been probed for a long time and in the last few years the area has grown considerably. In this chapter the medicinal properties of organometallic compounds are reviewed, with notable applications in the treatment and diagnosis of cancer and in the treatment of viral, fungal, bacterial and parasitic infections.

Keywords Antibacterial · Anticancer · Antifungal · Antiviral · Bioorganometallic · Parasitic infections · Pharmaceutical

1 Introduction

Inorganic compounds (metal complexes) have been used to treat various diseases and ailments for many centuries. Around 5000 years ago the Egyptians used copper metal to sterilise water, and gold was used in a variety of medicines in Arabia and China, but the practise emanated from the value of the pure metal rather than from known therapeutic effects.

In contrast, the introduction of inorganic complexes as therapies was usually based on an observed medicinal effect. Two of the earliest inorganic remedies involved the use of mercurous chloride as a diuretic, and iron complexes as mineral supplements, introduced about 500 years ago. More recently, gold complexes have been used as antibacterials, in particular for the treatment of tuberculosis at the beginning of the twentieth century. Other traditional inorganic drugs include the use of arsenic complexes, such as arsephenamaine, to treat syphilis [1] and antimony compounds for the treatment of leishmaniasis.

2 Cisplatin and Beyond

Inorganic therapies came of age with the discovery of *cis*-$Pt(NH_3)_2Cl_2$, cisplatin (Fig. 1a), by Rosenberg in 1965 [2]. Cisplatin is a truly remarkable drug in that, for the last thirty years, it has been used to treat more than 70% of all cancer patients. It is still particularly useful for treatment of testicu-

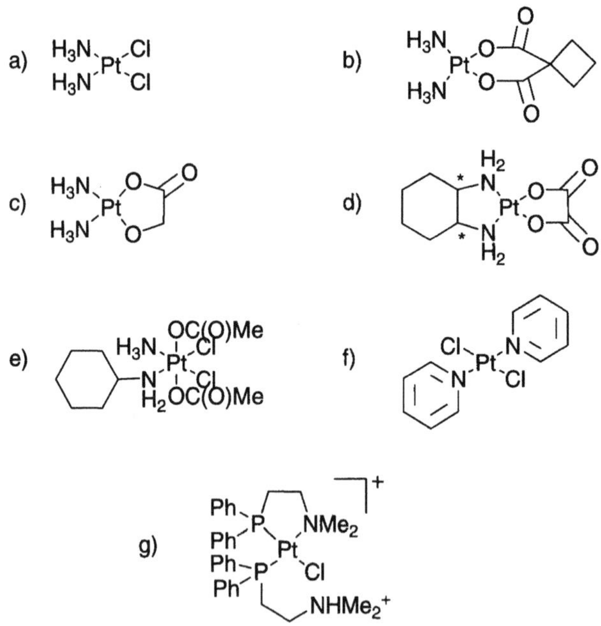

Fig. 1 Selected platinum-based anticancer compounds: **a** cisplatin; second generation anticancer drugs **b** carboplatin, **c** nedaplatin and **d** oxaliplatin; and novel complexes with interesting clinical properties such as the orally administered **e** satraplatin and complexes designed to overcome cisplatin resistance **f** *trans*-dipyridine dichloroplatinum and **g** a platinum amino phosphine complex

lar, ovarian, oropharyngeal, bronchogenic, cervical and bladder carcinomas, lymphoma, osteosarcoma, melanoma and neuroblastoma [3, 4].

Although effective, there are a number of problems associated with cisplatin, such as its high general toxicity that results in many unwanted side effects and the fact that it is inactive against several common malignancies, including lung carcinomas and adenocarcinomas of the colon and rectum [5]. Acquired drug resistance is also a problem, and cisplatin resistant cancers are often responsible for cancer mortality. The problems associated with the use of cisplatin have driven the development of new inorganic anticancer therapies.

Despite a massive research effort, the exact details of the mechanism of cisplatin activity is a matter of debate, but DNA is widely accepted to be the target of this drug and platinated DNA has been isolated from tumour biopsies of cisplatin treated patients. In vitro experiments show that cisplatin has high base specificity, binding to the N7 of adjacent guanine residues. Both chloride ligands are lost, with the cisplatin forming mainly intrastrand cross links that prevent DNA replication. The crystal structure of cisplatin bound in such a manner to an oligonucleotide was obtained by Lippard and co-workers and showed that bending of the DNA helix also takes place [6].

In addition to the affinity of cisplatin to bind to DNA, this drug also reversibly binds to thiol groups of proteins. Indeed, this type of interaction is likely to occur in the blood, where proteins like transferrin and albumin can transport the drug in a soluble form. On release in the cell, the active drug species is hydrolysed to $[cis\text{-}Pt(NH_3)_2Cl(H_2O)]^+$, a crucial step with respect to the activity of the drug. Hydrolysis is promoted inside cells, where the chloride concentration is significantly lower than that of the extracellular environment, and such a phenomenon might also be important for many organometallic based anticancer agents (see below).

As cisplatin has been such a successful drug, many researchers used this molecule as a starting point to design improved therapies. Consequently, the first alternative remedies to cisplatin closely resembled the molecule, with coordinated amine groups in a *cis* geometry and the platinum in the +II oxidation state, giving rise to the second generation cisplatin analogues now in the clinic, which include carboplatin, nedaplatin and oxaliplatin (Fig. 1b–d, respectively). These early empirical rules have since been broken in order to try to develop drugs that are effective on cisplatin resistant cancer cells. The platinum(IV) complex satraplatin (Fig. 1e), was designed to be orally administered as well as having both improved anticancer activity and lower toxicity than cisplatin. Complexes with a *trans* geometry, such as *trans*-$Pt(Py)_2Cl_2$ (Fig. 1f), may be useful against cisplatin resistant cell lines [7]. This complex is considerably more active than transplatin, possibly because the bulkier pyridine ligands alter the biological properties of the drug, promoting DNA interstrand cross-linking [8]. Other ligand substitutions also affect biological activity, and therefore may be used to overcome drug resistance. For example, coordinating an amino phosphine ligand to the Pt centre (Fig. 1g) leads to a preference for thymine base specificity, thereby overcoming cisplatin drug resistance that involves accelerated repair of the CG cross-links that are characteristic of this drug.

The discovery of new cancer therapies has been driven by the requirements to overcome the problems of drug resistance and to reduce the high toxicity of the existing treatments. Similar driving forces apply to other areas of inorganic medicinal chemistry. For example, resistance of the parasite that causes malaria, *Plasmodium falciparum*, to chloroquine led to the development of inorganic-chloroquine derivatives, including the triphenylphosphinegold(I) chloroquine complex shown in Fig. 2, which has a nine-fold higher therapeutic activity against chloroquine-resistant human *Plasmodium* strains than chloroquine diphosphate [9]. Similar strategies have been used to restore the activity of other antiparasitic and antibiotic drugs where resistance is now a problem (see below).

Some therapeutic or diagnostic applications of metal complexes, thus excluding the mechanical use of metals in biomedical applications, are listed in Table 1.

It is worth noting that none of the complexes based on the metals listed in Table 1 are organometallic compounds, i.e., with a direct metal-carbon bond. And as far as we are aware no organo-transition metal compounds are currently used in the clinic; however, there is considerable and growing interest in the medicinal properties of organometallic compounds and several show promising results in clinical trials. Thimerosal (sodium ethylmercurithiosalicylate), an organometallic mercury-containing antimicrobial preservative, has been used in cosmetics, skin cleansers, antiseptic sprays and vaccines since 1930s [110]. However, with the increased number of vaccines now given to children, a link between this preservative and autism has been highlighted and this additive has now been removed in many countries. It is also likely that tiny quantities of organometallic compounds are present as contaminants in many pharmaceuticals. For example, palladium catalysed reactions are widely used in the manufacture of pharmaceutical products, and while residual palladium must be very low, typically less than 5 ppm, traces are often present, some of which are likely to be organo-palladium intermediates [111]. However, these impurities are unwanted and in general the perception of

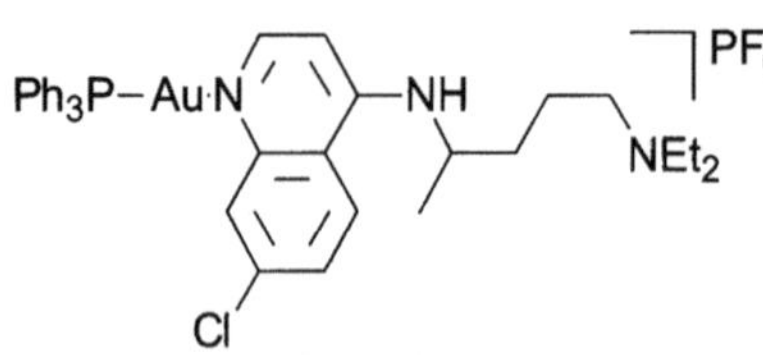

Fig. 2 The triphenylphosphinegold(I)-chloroquine complex that restores chloroquine activity in resistant *Plasmodium falciparum*

Table 1 Metals with biomedicinal applications (or potential applications) [10–13]

Metal	Applications	Refs.
Arsenic	Syphilis, ulcers, parasitic disease, acute promyelocytic leukaemia	[1, 14, 15]
Aluminium	Antacid, dermatology, hyperphosphataemia	[16–18]
Antimony	Leishmaniasis	[19, 20]
Barium	Diagnostic agents	[21, 22]
Bismuth	Gastrointestinal disorders, syphilis, angina, adenotonsillectomy	[23–26]
Copper	Diagnostic and imaging agents, radiopharmaceuticals, photodynamic therapy, Menkes disease	[27–30]
Chromium	Diagnostic agent, diabetes	[31, 32]
Calcium	Hyperphosphataemia	[18]
Cobalt	Diagnostic and imaging agents, photodynamic therapy	[33–37]
Iron	Photodynamic therapy, hypotensive, hyperphosphatamia	[38–42]
Gadolinium	Diagnostic and imaging agents	[43–45]
Gallium	Diagnostic and imaging agents, cancer	[46, 47]
Germanium	Cancer	[48]
Gold	Rheumatoid arthritis, bronchial asthma, malaria, bacterial infections, cancer, viral infections including AIDS	[9, 49–58]
Holmium	Radiopharmaceuticals	[59]
Indium	Diagnostic and imaging agents	[60, 61]
Lead	Ulcer treatment	[62]
Lithium	Manic depressive psychoses and viral infections including AIDS	[63]
Magnesium	Antacid, laxative, hyperparathyroidism	[64–67]
Manganese	Photodynamic therapy	[68, 69]
Mercury	Diuretic, microbial infections, dermatology (syphilis), heart failure	[70–75]
Molybdenum	Menkes disease	[27]
Palladium	Photodynamic therapy, cancer, HIV	[76–78]
Platinum	Cancer, photodynamic therapy, microbial infections, viral infections including AIDS	[79–82]
Rhodium	Leishmaniasis, radiotherapy, bacterial infections	[83–85]
Ruthenium	Malaria, cancer, Chagas' disease, bacterial infections, septic shock, HIV	[86–91]
Rhenium	Diagnostic and imaging agents, radiotherapy	[92]

Table 1 (continued)

Metal	Applications	Refs.
Silver	Microbial infections, fungal infections, dermatology, diagnosis and imaging agents	[93–96]
Strontium, Radium, Plutonium, Samarium	Radiopharmaceuticals	[97, 98]
Tin	Radiopharmaceuticals, photodynamic therapy	[99, 100]
Thallium	Diagnostic and imaging agents	[101]
Technetium	Diagnostic and imaging agents	[31, 102, 103]
Vanadium	Insulin mimics	[104]
Yttrium	Diagnostic and imaging agents, radio-immunotherapy	[105, 106]
Zinc	Photodynamic therapy, Menkes disease, dermatology, HIV	[27, 107–109]

organometallic compounds is that they are toxic and do not have useful medicinal properties. Despite this misconception, organometallic species are often formed during the biological processing involving heavy metals. In the case of many mercury, lead and tin species, the organometallic species are more toxic than inorganic forms, whereas for arsenic and selenium, organometallic species are more readily eliminated from the body [112]. Hence organometallic drugs can be less toxic and more readily processed by the body than inorganic coordination complexes.

3 Organometallic Pharmaceuticals

Over the last four decades thousands of inorganic drugs have been screened for their medicinal activity in a wide range of diseases, but only a handful have made it into the clinic. During the screening process it has been shown that small structural differences can markedly alter the medicinal properties of a putative drug. The reason for such differences is often poorly understood, although much progress has been made in recent years. It has also become clear that organometallic ligands present opportunities not always possible with traditional coordination complex based drugs. In particular, organometallic complexes have shown promise in overcoming several different types of drug resistance, as well as allowing improved specificity and drug targeting, thereby reducing the side effects associated with chemotherapy.

3.1
Overcoming Drug Resistance

One way of restoring the activity of organic drugs for which resistance has emerged is to modify the structure to contain a metal, and some of these compounds are organometallic. As early as 1975, it was found that substituting the aromatic groups in the antibiotics penicillin and cephalosporine with ferrocenyl moieties (Fig. 3a and b) produced compounds with altered antibacterial activity compared to the starting materials [113]. Against various strains of *Staphlococcus aureus*, ferrocenyl-penicillin showed comparable activity to benzyl-penicillin and also inhibited β-lactamase, which is one of the enzymes responsible for bacterial resistance to penicillin-type antibiotics.

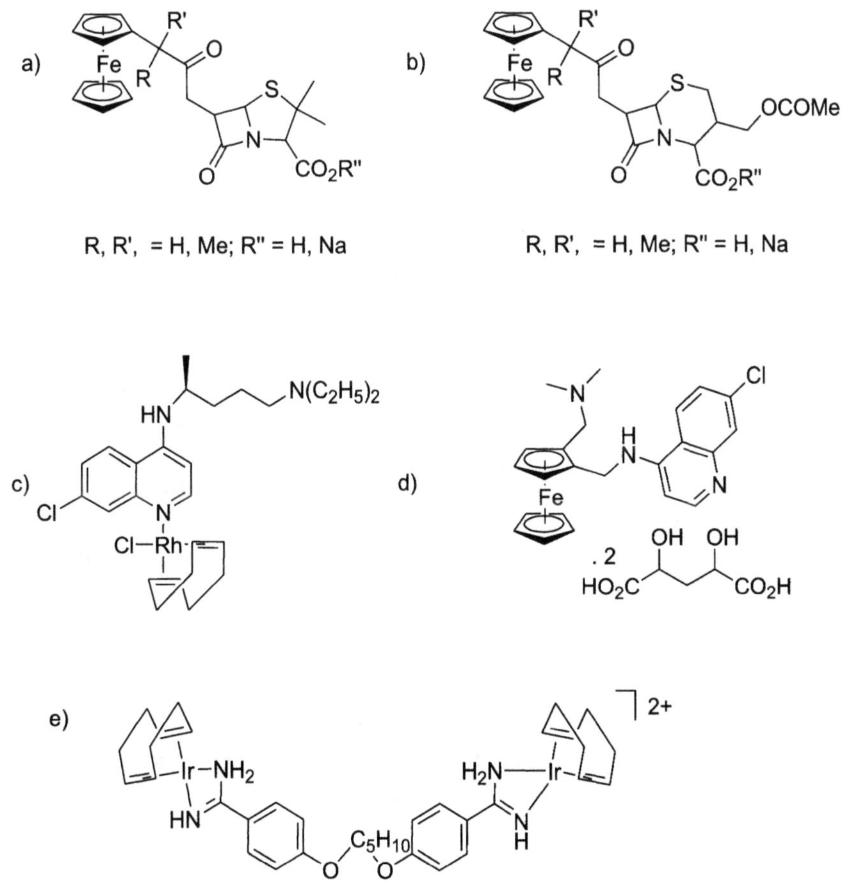

Fig. 3 Organometallic complexes of known organic antiproliferation agents designed to overcome drug resistance: **a** ferrocenyl-penicillin derivatives; **b** ferrocenyl-cephalosporin derivatives; **c** rhodium chloroquine complex; **d** ferrocenyl-chloroquine derivative; **e** iridium-COD-pentamidine complex

Organometallic derivatives of organic antiproliferation drugs have been used to overcome drug resistance of antiparasitic drugs, with antimalarials being particularly well studied. Nearly half the world's population lives under the continuous threat of malaria [114]. Malaria is a parasitic disease caused by one of four species of *plasmodium* protazoa, the most deadly of which is the *plasmodium falciparum*. There are two major groups of therapies. The first appears to target folinic acid (folate) synthesis which includes compounds such as chloroguanide, pyrimethamine, and sulfonamides. However, these drugs require long periods to show any effect and drug resistance, largely due to mutation in the target enzymes, is common. The second group of compounds include chloroquine, primaquine, quinine and their analogues [115]. These drugs show a more rapid response compared to the first group and are thought to accumulate in the parasite [116, 117], interfering with the way it metabolises the host's haemoglobin, blocking the polymerisation of heme into hemozin. Inhibition of this process results in the accumulation of toxic heme monomers, which are thought to poison the parasite [118–122]. These drugs may also alter the pH in the digestive vacuole, disrupting enzymatic functions [123, 124].

Organic drugs such as chloroquine offer fast acting relief against the malaria parasite. However, drug resistance to chloroquine, which involves both drug uptake [125] and efflux [126] mechanisms, is becoming increasingly common. The rate of chloroquine efflux is 40-fold higher in resistant parasites compared to sensitive strains [126]. Many of the conventional treatments of malaria are now ineffective driving forward research for new therapies.

Attaching chloroquine to metal centres, either via direct coordination or by covalent attachment, has been shown to overcome drug resistance, with inorganic complexes of chloroquine based on gold (shown in Fig. 2) and ruthenium, and organo-rhodium (Fig. 3c) [86] and organo-iron (Fig. 3d) [127, 128] complexes, demonstrating enhanced antimalarial activity. However, only the organo-iron compound, a ferrocene derivative, totally restores chloroquine activity on drug resistant strains [127]. Ferrocene alone shows no antimalarial activity [128], but substituting the carbon chain of chloroquine with a hydrophobic ferrocenyl-group while maintaining the positions of the two exocyclic chloroquine nitrogen atoms produces a powerful antimalarial drug.

Similarly, many organic drugs used to treat the parasitic diseases leishmaniasis and Chagas' disease, as well as those used to treat helminth worm infections, are becoming increasingly ineffective due to drug resistance. *Leishmania donovani*, which causes leishmaniasis [129–131], transferred via the bite of a sandfly, infects approximately 10–15 million people worldwide. The disease may be fatal if not treated and the effectivity of traditional organic drugs such as the pentamidine, amphotericin B, aminosidine and antimonials is declining due to drug resistance [132]. In the case of Chagas' disease, drug therapy has never been particularly successful, and it remains an incurable parasitic disease that affects millions of people in Latin America, caused by

Trypanosoma cruzi infection. The current treatment involves noxious organic drugs, the dose being limited by the toxicity of the therapy. Helminth worms include flukes, tapeworm and roundworms, which cause a diverse range of infections all over the world. For example, the annual incidence of infection from threadworm exceeds 200 million cases, including high levels of infection in Europe, Canada and the US.

In a similar manner to the metallation of chloroquine, a series of iridium, platinum, rhodium, palladium, antimony and osmium complexes of various organic drugs [133–135] has been prepared and evaluated for activity against *L. donovani*, *T. cruzi*, the helminth worms and other parasites. Pentamidine, one of the organic antiparasitic drugs, has been complexed to several different metal centres, and its activity has been evaluated against different parasite species. Of these, an organo-osmium derivative was found to have a 7.5-fold higher therapeutic index than pentamidine alone in treating leishmaniasis, curing infected mice in a single dose. Organo-platinum [136] and organo-iridium [137] derivatives were used against *T. brucei* in rodent models, curing mice infected with the parasite in a single dose. The organo-iridium compound, shown in Fig. 3e, was also shown to have antifilarial activity with the infective larvae of human helminth parasite models of *Molinema dessetae* and *Brugia pahangi* helminth parasites [138], and exhibits slight activity against Leishmaniasis [139]. In this latter study the iridium complex was found to be considerably less toxic compared to pentamidine isethionate alone, allowing administration of the drug at fatal concentrations of pentamidine isethionate in the mouse models. The organo-iridium derivative of pentamidine, accumulates in the *L. donovani* promastigotes, binding to ribosomal subunits in vitro, but not effecting macromolecular synthesis [140]. In contrast, other organometallic complexes have different mechanisms of action, for example, osmium(II) complexes incorporating amino acids and peptides as ligands inhibit *L. donovani* growth [141], mediated through the inhibition of macromolecular (protein, RNA and DNA) synthesis.

A series of organometallic benzimidazole derivatives have also been evaluated for antiparasitic activity. Of the various compounds screened, a ruthenium-aminobenzothiazole complex was found to be the most potent compound of those tested for antifilarial activity using models of infective larvae of *Molinema dessetea* and adult females of *Brugia pahangi*. Although some heavy metals are known to have antifilarial activity, only the organo-ruthenium complex showed enhanced activity compared to the organic moieties used to prepare the organometallic drugs tested [138].

The reasons for the success of organometallic derivatives overcoming resistance to organic therapies depend on the mechanism of drug resistance. Where drug resistance is due to altered uptake of the drug, the organometallic derivative may provide a new way of entering the cell. For example, organo-iron complexes may exploit the transferrin-mediated uptake of nutrient iron, and as iron is an essential nutrient, drug resistance to this type of drug delivery system is unlikely to occur. Coordination of known agents such as sulfamethoxydiazine, dithiocarbamate or diphenyl-dithiocarbamate to the rhodium-dicarbonyl fragment "$Rh(CO)_2$" has resulted in compounds that overcome drug resistance due to altered uptake mechanisms [142, 143].

Whereas changes in drug effectiveness due to alterations in the uptake mechanisms affect the drug in question, together with a small range of structurally similar compounds, multidrug resistance can make diseased cells more resistant to a broad spectrum of compounds. Multidrug resistance usually involves accelerated efflux of drugs from the cell, amongst other changes in protein activity. The increased rate of efflux can result from altered patterns of protein expression involving accelerated glutathione molecular tagging and increased numbers of drug transport proteins. Glutathione serves as a molecular tag for processing and excretion of compounds. This thiol-containing tripeptide either reacts spontaneously with metal complexes or is catalytically conjugated to metals and electrophilic centres by the glutathione S-transferase family of enzymes. Both elevated levels of glutathione and glutathione S-transferase enzymes have been linked with mulitdrug resistance phenotypes. Glutathione conjugates are usually more soluble and less toxic than the unconjugated molecules, and are therefore more readily eliminated from the body. Often, some processing occurs before excretion, with the glutathione tag signalling for further molecular modifications and efflux from the cells, usually via the multidrug resistance protein.

The multidrug resistance protein is a general efflux pump, and its overexpression has been linked to drug resistance. Other similar pumps exist, such as *P*-glycoprotein, which also accelerate the rate of drug efflux from cells, thereby reducing the therapeutic activity of drugs in corresponding cell lines. There are several agents known to block these pumps, such as verapamil and daunorubicin, which restore the activity of a range of drugs including those used to combat cancer, leishmaniasis and malaria.

Multi drug efflux pumps have also been shown to actively transport imaging agents as well as drugs, interfering with diagnosis. For example, several organo-technetium complexes have enhanced accumulation in drug-resistant tumour cell lines when co-administered with agents that block these efflux pumps [144, 145].

More specific drug resistance occurs with certain drugs when the biomolecular target or repair mechanisms in the cells alter so that the drug is no longer effective. In the case of protein targets, the alteration may be a mutation that blocks the interaction between the drug and the target. Such changes can be responsible for resistance in protease-targeting anti-HIV drugs. Organometallic derivatives of these drugs may overcome resistance by providing alternative enzyme binding or inactivation by the organic drug. Metal centres in organometallic complexes may coordinate to thiol, amine or hydroxyl functional groups of the protein in the drug binding site, resulting in a stronger interaction than the non-covalent association of the organic drug alone. Metal centres may also help to displace metal ions from the protein that are important for function and stability, causing a more potent inactivation of the protein function.

Where DNA is the biomolecular target, drug resistance is usually due to alterations in the proteins that recognise and repair DNA lesions. Although there is some cross-reactivity between the recognition and processing of structurally similar damage caused by some drugs, other compounds cause significantly different lesions such that drug resistance to one compound does not diminish the activity of other compounds. In the case of platinum-based anticancer compounds, drug resistance to cisplatin can be due to accelerated excision and repair of the intramolecular DNA cross-links that are preferentially formed by this drug. Alternatively, some ruthenium-arene compounds form intermolecular cross-links that are not recognised in the same way as the platinum-induced intramolecular cross-links and so the activity of the ruthenium compounds towards cisplatin resistance cancers are not necessarily affected.

3.2 Drug Targeting

In order to be effective, drugs must have some selectivity for the diseased cells, exploiting differences between the biomolecules and metabolic pathway that are present in the diseased and healthy cells, thereby causing minimal damage to the healthy cells.

Designing antibacterial and antibiotic agents has the advantages that bacterial cells are sufficiently evolutionary distinct compared to human cells and there are many properties that can be used to allow drugs to specifically accumulate in bacterial cells or to interfere selectively with bacterial cell growth. For example, bacterial cells are contained within a wall which is essential for their function whereas human cells are not. Drugs that disrupt cell wall synthesis can, therefore, be specifically used to control bacterial cell growth without effecting human cell growth. Penicillin is the classic example of a drug with this type of selectivity, causing selective cell death by disrupting bacterial cell wall synthesis. Unfortunately, as with many of the drugs that were once highly effective, many bacterial strains are now emerging with drug resistance to penicillin; hence, the search for alternative agents is needed. Organometallic derivatives of penicillin and other antibiotics have been shown to retain their antibiotic selectivity, whilst providing a means of overcoming drug resistance and thereby enhancing drug activity.

Antiviral and anticancer therapies possibly offer the greatest challenge in terms of selectivity. A virus hijacks the host's cellular machinery in order to make more viruses, thus there are few targets that are unique to virally infected cells for antiviral therapy. Viruses sometimes have unique proteins that are essential for their function. Organometallic complexes that have been shown to have interesting antiviral properties include vanadocene complexes [146, 147], and a highly water soluble tetraruthenium cluster, $[H_4Ru_4(\eta^6-C_6H_6)_4]^{2+}$ [188].

The vanadocene complex shown in Fig. 4a might also have applications in preventing HIV transmission [146, 147]. It is stable at physiological pH and has no detectable toxicity, but has anti-HIV and spermicidal activity, making it an interesting candidate for novel contraceptives. The mechanism of the vanadocene compound is thought to involve its ability to intercalate with the membrane surrounding either the virus or the sperm, resulting in immobilisation, without damaging the epithelial layer of the vagina. In addition to showing direct antiviral activity, metallocenes have been shown to pose interesting anti-HIV properties when used in combination with existing drugs such as 5-bromo-6-methoxy-5,6-dihydro-3′-azidiothyamine-5′-(*p*-bromophenyl-methoxy alanyl) phosphate (WHI-07) [146].

The ruthenium cluster shown in Fig. 4 is active against polio virus type 1, which is still a cause of major physical disabilities in the world. This compound showed little toxicity in healthy cells, yet promising antiviral activity [147]. A number of other cluster compounds have been shown to interact with DNA in vitro [148], and while they may prove to have activity as anticancer agents, the fact that they bind DNA indicates that they may also have general toxicity.

Cancer cells originate from the patients' own bodies, so initially share the same genes, proteins and other cellular machinery, providing few differences between healthy and diseased cells that can be used as drug targets to selectively kill cancer cells. As the disease progresses the cells undergo changes that can give rise to new drug targets, including the expression of new enzymes, changes in the carbohydrate moieties on the cell surface, and changes to the physiological conditions inside the cell and in the immediate surroundings. These changes are aimed at improving the proliferation of the diseased cells or arise as a consequence of the uncontrolled growth.

Within each cell there are thousands of different proteins, often with overlapping substrate specificities; hence, it is difficult to design drugs that block

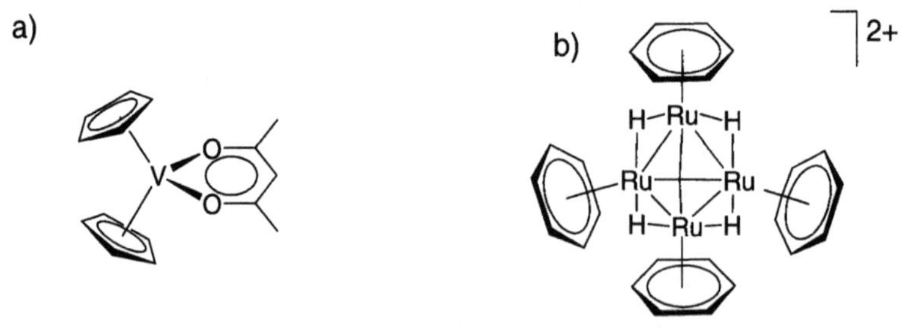

Fig. 4 Organometallic compounds with antiviral properties: **a** a vanadocene spermicides/antiHIV compound; **b** the tetraruthenium cluster $[H_4Ru_4(\eta^6-C_6H_6)_4]^{2+}$

specific proteins that do not have side effects through interacting with other protein species. Most protein targets for antiviral and anticancer therapy are involved in DNA synthesis, repair and manipulation. Putative new therapies for HIV infections commonly target the reverse transcriptase and the protease, which are expressed in infected cells and not in healthy cells. The reverse transcriptase is involved in the early stages of HIV infection, converting the RNA genome of the virus into a DNA sequence that can then integrate in to the host's genome, lying dormant until it is reactivated. The HIV-1 protease is an enzyme that is essential for viral maturation, dividing the polypeptide that is produced when the host cell translates the viral genome into distinct protein units for viral assembly.

DNA and DNA-processing enzymes are common targets because DNA transcription and replication are integral to rapid cell growth. In addition, diseased cells often undergo changes that make them more susceptible to cell death through DNA damage. Healthy cells have a proof reading mechanism that checks DNA replication to ensure integrity, however this process is slow and limits their replication rate. Rapidly dividing, diseased cells often loose this proof reading mechanism, making mutations and errors much more common. These errors give rise to a genetically heterogeneous population of cells. In some cases these mutations can help the cells develop drug resistance and other undesirable properties, but if considerable DNA damage occurs, for example the lesions caused by cisplatin, often the damage is too severe for the cells to overcome, resulting in their death.

Although the differences in DNA replication are sufficient to give drugs that cause DNA lesions a slightly higher toxicity in diseased compared to healthy cells, the difference is not sufficient enough to completely eliminate toxicity in healthy cells, which results in unwanted side effects. As such, compounds that cause DNA damage alone are often too toxic to use in the clinic. For example, alkyltin(IV) derivatives have been known to have antiproliferative activity against specific cancers for more than twenty years [149–153]. While DNA is believed to be their main target, these compounds seem to have a number of targets in the cell, eventually leading to the induction of apoptosis in sensitive cell lines [154, 155]. However, these compounds have limited applications in the clinical because of their high general toxicity. Many other organometallic compounds have been shown to damage DNA and some of these have promising anticancer properties, including organo-gold(III) complexes [156, 157, 231], alkylcobalt(III) complexes [158, 159], $[Co(cyclam)(H_2O)(Me)]^{2+}$ [160, 161], spirogermanium [162], mefloquine complexes [163], and a range of metallocenes, $[M(\eta^5-C_5H_5)_2X_2]$, based on iron, molybdenum, niobium, titanium, vanadium and rhenium [164–180]. However, for entry into the clinic they must be shown to have superior activity to currently available drugs or demonstrate new mechanisms of delivery or action. The general toxicity of anticancer drugs is one feature that researchers are trying to improve with new therapies, using drug targeting and activation mechanisms, with the aim of increasing the toxicity of the drug at the site of the disease and decreasing it elsewhere.

Transferrin mediated drug delivery may provide specific targeting to diseased cells, in addition to providing a mechanism of drug delivery that is unlikely to lead to drug resistance. Transferrin is a natural transport protein, binding iron(III) in the blood and delivering it to cells according to the number of transferrin-receptors expressed on the cell surface. Rapidly dividing cells, such as those infected by parasites or cancer cells, have a higher iron requirement than healthy cells and one way of satisfying this requirement is to increase the expression of transferrin-receptors. Thus, drugs that exploit transferrin-mediated transport and delivery have been shown to accumulate in diseased cells.

In addition to binding and transporting iron, transferrin can bind and deliver other metals to the diseased cells. For example, the anticancer compound titanocene dichloride can interact with transferrin. The drug seems to bind to the protein in a similar way to iron, resulting in the cyclopentadienyl ligands being released [181] and allowing the titanium metal to be delivered to the cancer cells. Titanocene dichloride (Fig. 5a) was first recognised as an anticancer agent in 1979 [182] and until recently being evaluated for activity against cisplatin resistant ovarian and metastatic renal-cell carcinomas [183]. Titanocene dichloride demonstrates general antiproliferation activity and has also been shown to be effective against more than five types of cancer cells [174]. Other metallocenes have more specific activity, for example ferrocifen, illustrated in Fig. 5b. In this example, oestrogen mimics are tethered to the ferrocene unit, which specifically targets the drug to cells that carry oestrogen receptors [167, 168]. Ferrocifen is described in detail together with related compounds in Chapter by Jaouen, this volume.

Combining metallocenes with other organometallic centres, such as gold (see the compound shown in Fig. 5c) provides another means of modifying drug activity and specificity. In this case the complex shows promising activity against bladder and colon cancers [170].

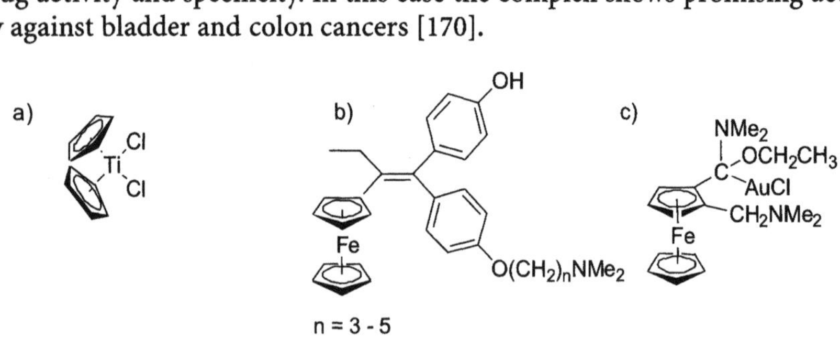

Fig. 5 Selected examples of anticancer compounds with metallocene fragments: **a** titanocene dichloride; **b** ferrocenyl oestrogen mimic complexes – see Chapter by Jaouen (this volume), for more examples; **c** a ferrocenyl gold complex

In the same way that titanocene dichloride can exploit transferrin mediated delivery to diseased cells, ruthenium(III) has been shown to be able to mimic iron binding to this transport protein. Therefore organo-ruthenium drugs that retain this feature could potentially exploit the iron delivery system to gain entry into diseased cells. Once at the disease site, the ruthenium(III) complexes are released from transferrin into the interior of the diseased cell where they are reduced to ruthenium(II) complexes, which in general are more toxic to biological systems. This mechanism of transport and activation has been suggested as an explanation for the remarkably low toxicity of the ruthenium(III) salt, ImH[Ru(Im)(Me_2SO)Cl_4] (Im = Imidazole), called NAMI-A, which shows promising antimetastatic activity in clinical trials [184, 185]. Metastasis is the term used to describe the spread of cancer from one part of the body to another. Once metastasis is in progress, the survival rate of the patient falls dramatically, in part because there are few effective drugs against metastisizing cancers.

As the mechanism of NAMI-A begins to unfold, it has been shown that the drug may not need to enter the cell in order to have a therapeutic effect; hence transferrin delivery to the cell interior may not be part of the low toxicity of this drug. NAMI-A causes changes in the extracellular matrix around the cancer cells such that they are contained and suffocated. Further studies have shown that certain ruthenium(II)-arene complexes, such as those in Fig. 6a and b, also demonstrate remarkably low toxicity and anticancer/antimetastasis activity, despite the fact that these complexes enter the

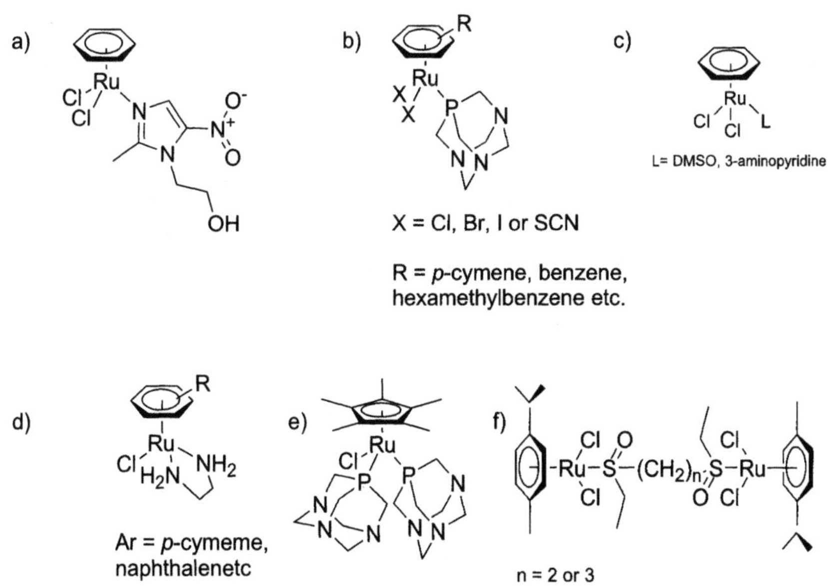

Fig. 6 Ruthenium(II)-arene anticancer complexes: **a** from [186] **b** from [187, 188]; **c** from [189, 190]; **d** from [191]; **e** from [192]; **f** from [193]

physiological environment in the more biologically active ruthenium(II) form, and therefore are unlikely to be transported in the same way as ruthenium(III) compounds. In contrast, it is postulated that the specific cytotoxicity that these ruthenium(II) compounds demonstrated in cancer cells may be due to the altered ligand exchange kinetics induced by the physiological changes that accompany rapid cell proliferation without an adequate nutrient supply.

During rapid proliferation, the environment in and around the cell changes and these changes can be exploited to specifically activate drugs at the site of infections and disease. When cells grow normally, they are fed by an adequate blood supply that delivers sufficient nutrients and oxygen and removes waste products at such a rate that the physiological conditions for growth are optimised. However, during periods of rapid cell growth, oxygen can be consumed and waste products can accumulate faster than can be buffered by the blood supply. In addition, during tumour cell growth, clusters of cells form, with the outer cells suffocating the inner cells. Once the diameter of the tumour has reached 2 mm or above the rate of diffusion is not sufficient to feed and remove waste products from the centre of the tumour and the cells on the interior begin to die, triggering the angiogenesis process which involves the creation of new blood vessels to supply the growing tumour.

When cells are not provided with enough oxygen, waste products including carbon dioxide begins to accumulate and their physiological environment changes. The build up of carbon dioxide reduces the pH of the cytosol and extracellular fluid and with the low oxygen levels creates a hypoxic environment, favouring reduction. The hypoxic nature of rapidly dividing cells was also postulated to contribute to the low toxicity of ruthenium(III) drugs, whereas the biologically less active ruthenium(III) complex is delivered to the diseased cells it becomes rapidly reduced to the more cytotoxic ruthenium(II) species, resulting in drug activation. Similar activation in hypoxic cells has been observed with alkylcobalt(III) complexes, such as the one shown in Fig. 7, which undergo pH dependent degradation three times faster at pH 6.5 than 7.5, producing alkyl radicals that can damage tumours [194, 195]. Thus, these differences between healthy and diseased cells can be exploited to activate drugs at the target site.

A hypoxic cell environment not only effects the metal, but can also result in modifications to the ligands, causing protonation or affecting the kinetics of ligand exchange. In the case of the Ru(arene)Cl_2(pta) (pta = 1,3,5-triaza-

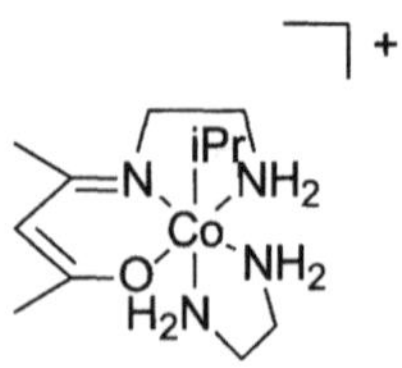

Fig. 7 An alkylcobalt complex that undergoes rapid degradation at the pH characteristic of many cancer cells producing alkyl radicals that kill the cells

7-phosphaadamantane) series of compounds, (termed RAPTA, Fig. 6b), pH dependent drug activation has been observed both against the DNA target and in whole cell systems; however, this cannot be due to the reduction of the metal, since it is administered in the ruthenium(II) oxidation state. In contrast, it is thought that the pH dependent activity of these compounds, which show promising antimetastatic activity in preclinical trials, could to be due to the protonation of the pta ligand. It is postulated that the RAPTA compounds carry a neutral change in normal physiological conditions, allowing them to more readily cross cell membranes. However, once in the hypoxic cell environment, the pta becomes protonated, trapping it in the cell. In addition, the lower chloride concentration inside cells promotes hydrolysis of the chloride ligands, producing the active drug species.

In addition to forming adducts with DNA, organo-ruthenium(II) complexes have been shown to poison DNA processing enzymes. The ruthenium(II) arene compound shown in Fig. 6c poisons topoisomerase II, an important enzyme involved in the processing of DNA to allow the code to be read by the protein making machinery in the cell. These compounds have also been shown to be active against in vitro breast and colon carcinoma cells.

Recently, antiproliferation agents based on the 11-vertex ferratricarbaborane [196] (Fig. 8a) have been demonstrated. These complexes may be viewed as ferrocene analogues, but they are more resistant to degradation and possibly exhibit lower toxicity. Ferratricarbaboranes show promising activity against several tumour cell lines. Similar broad antiproliferation activity is demonstrated by $M(C_5H_5)(MeC_3B_7H_9)$ complexes (see Fig. 8b) [196], which display cytotoxicity in a variety of tumour cell lines.

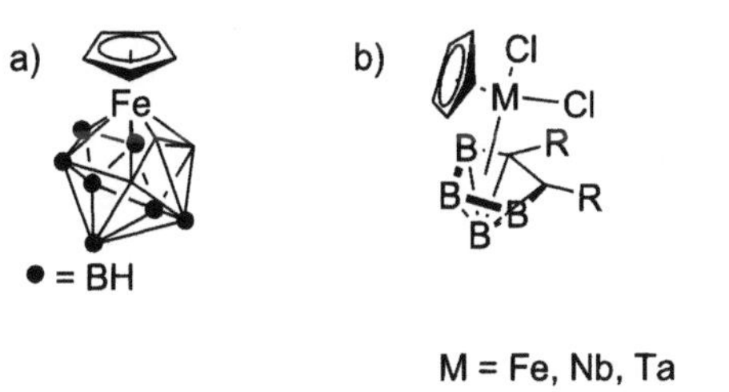

Fig. 8 Ferrocene analogue anticancer complexes: **a** an 11-vertex ferratricarbaborane; **b** $M(C_5H_5)(MeC_3B_7H_9)$ complexes

3.3 Ligand Substitutions

As discussed above, coordinating biologically active organic ligands to metals can produce drugs with novel biological activities that can be used to overcome drug resistance or enhance the properties of the organic molecules. However, it is not surprising that nearly all ligand substitutions can affect the biological properties of putative drugs. Fine tuning of drug activity and specificity by ligand modifications have been observed with rhodium, iridium ruthenium and gold complexes [197–200]. Rhodium and ruthenium-carbene complexes (Fig. 9b) [199], and rhodium, iridium and ruthenium tripodal phosphine complexes (Fig. 9c) [197], showed a greater selectivity towards gram-positive bacteria compared to gram-negative bacteria, possibly due to facilitated uptake in these organisms, inhibiting the growth at micromolar concentrations.

Ligand dependent specificity was shown with a series of gold complexes based on the [1,3-(dimesitylmethylimidazolinium)] chloride ligand (Fig. 9d) [200]. For example, when $R = CH_2C_6H_4NMe_2$-*p*, the drug inhib-

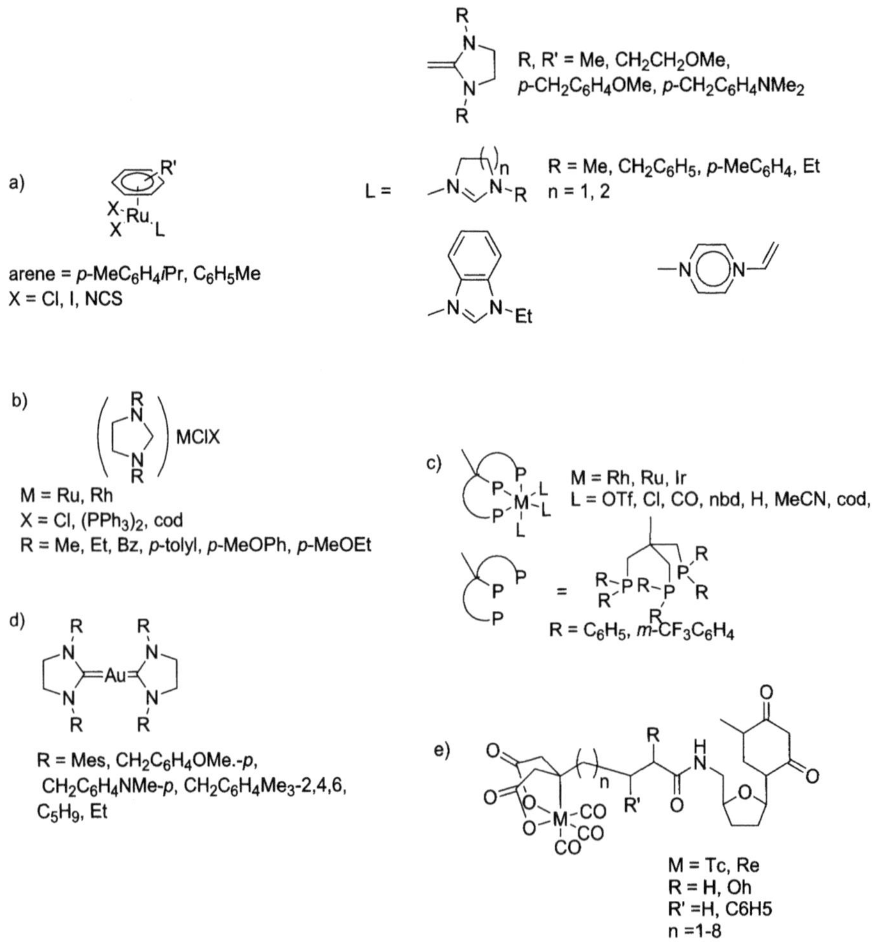

Fig. 9 **a** ruthenium(II)-arene complexes; **b** rhodium and ruthenium carbene complexes; **c** rhodium, iridium and ruthenium tripodal phosphine complexes; **d** gold [1,3-(dimesitylmethylimidazolinium)] chloride complexes; **e** metal carboxamide derivatives

ited the growth of *Escherichia coli* (ATCC 25922) at the lowest concentration tested, with little effect on the growth of the other gram negative bacteria, gram positive bacteria or yeast cells. In contrast, when R = C_5H_9, the complex inhibits the growth of *Enterobacter cloacae* and *Staphylococcus aureus*, with little effect on the other systems tested.

The [Ru(arene)(L)X_2] complexes (Fig. 9a) have been shown to have antibacterial and anticancer properties, with the exact activity and specificity of the drugs being linked to the type of ligands coordinated to the metal centre. For example, [Ru(η^6-*p*-cymene)Cl_2(4-vinylpyridine)] demonstrates only moderate antifungal activity, but high selectivity for fungal growth inhibition. In the [Ru(arene)(pta)X_2] (pta = 1,3,5-triaza-7-phosphatricyclo[3.3.1.1]decane) series, the X ligand determines selectivity between bacterial, fungal and yeast cells, with chloride ligands dictating selectivity for *Cladosporium resinae*, iodide ligands for *Bacillus subtilis* and NCS ligands selectivity for *Trichrophyton mentagrophytes* [188].

Nucleoside kinases have been shown to be important targets for antiviral and anticancer drugs. Expression of thymidine kinase is increased in proliferating cells compared to quiescent cells, and in addition to acting as a direct target, could be used as a reporter gene for diagnosis of tumours. In the case of the metal carboxamide derivatives of aminothymidine shown in Fig. 9e, the inhibition capacity of the compound with respect to human thymidine kinase increased with spacer length and the compounds showed only slight (if any) inhibition of the herpes simplex equivalent [201].

A more extensive study on the effect of ligand modifications on biological activity has been performed with metallocene compounds. Ferrocene complexes have diverse biomedical applications, partially because of their stability, relatively non-toxic nature and, importantly, because they are amenable to extensive derivatisation. Several examples of ferrocene derivatives of organic drugs were shown to overcome drug resistance (see above).

In addition to providing an organometallic scaffold onto which organic drugs with proven activity can be grafted, some novel ferrocenyl compounds have also been shown to possess interesting clinical activity. The bis-1,1′-disubstituted-ferrocenyl carbohydrazone complexes (Fig. 10a) inhibit the growth of five types of bacteria [202], showing about half the activity of the control, imipenum. Chelation of the complexes with metal ions giving the structure shown in Fig. 10b enhances activity by up to fifty percent. Similar results were obtained following the evaluation of the antifungal properties of these compounds; when tested against six fungal strains, the antifungal activity was enhanced by up to 42% on chelation of a metal ion within the macrocyclic cavity. In the case of the antifungal activity of these drugs, the activity of the chelate complexes closely approached that of the control drug miconazole.

Ligand substitutions on metallocenes can also produce compounds with antiviral and contraceptive applications. Vanadocene dithiocarbamate [203]

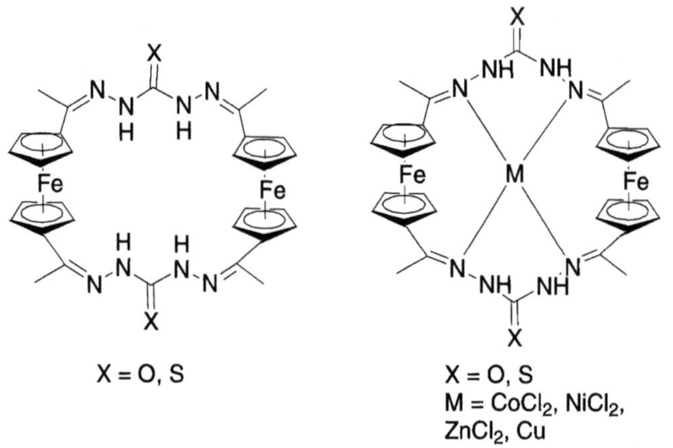

Fig. 10 Structures of (ferrocenyl)carbohydrazone complexes: **a** bis-1,1′-disubstituted ferrocenylcarbohydrazone complexes; **b** and subsequent chelation derivatives

was shown to be the most stable and potent of the various compounds tested as contraceptives. Female mice were rendered reversibly infertile after intravaginal administration. During these studies, there were no observable side effects on the mice and no long term effects on fertility.

Silver nitrate has been used for decades to prevent the infection of burn wounds. The key feature of drug activity is the release of silver ions around the site of infection. The problem with using silver nitrate itself is that much of the drug reacts with the chloride ions that are present in the physiological environment, forming an insoluble (and inactive) precipitate, and thereby reducing the amount of therapeutic silver ions that are available to fight infection. Silver(I) *n*-pincer-type heterocyclic bis-carbene complexes, such as the one shown in Fig. 11 exhibit slightly improved antimicrobial activity against *Escherichia coli, Staphylococcus aureus* and *Pseudomonas aeruginosa* compared to silver nitrate itself. The pincer ligands were shown to be inactive and it was postulated that these ligands protected the metal centre from reactions with chloride, stabilising the compounds in biological solutions and enabling the slow release of silver ions to fight infection. In particular these properties

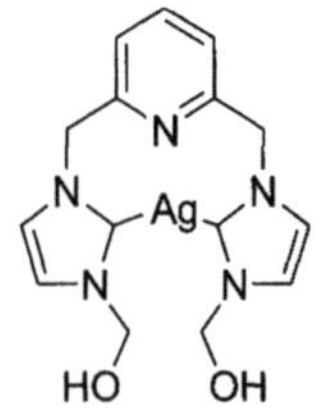

Fig. 11 A pincer silver(I)-carbene complex designed for the controlled release of Ag^+ ions to fight bund wounds

are attributed to the slow breakdown of the Ag – C bond, leading to a more stable complex.

Inclusion of appropriate co-ligands in metal-carbonyl compounds has led to the fine tuning of organometallic carbon monoxide releasing molecules for medical applications. Carbon-monoxide releasing molecules can induce blood vessel dilation, and acts as cardio- and cyto-protectants. Carbon monoxide is naturally generated in living organisms and has been shown to induce vasodilation, and possess anti-inflammatory and antiapoptotic properties. Research into the biological role of carbon monoxide has been hampered by the lack of known molecules that can carry and deliver carbon monoxide in biological systems. Recently, however, several carbon monoxide-releasing molecules have been studied, based on metals such as manganese, iron, cobalt, nickel and ruthenium. $Mn_2(CO)_{10}$ and $[Ru(CO)_3Cl_2]_2$ (Fig. 12a and b, respectively) release carbon monoxide under physiological conditions to cause long-lasting vasodilation, with no detectable cytotoxicity [204] and the ruthenium compound shown in Fig. 12c has demonstrable cardio-protective action [205]. Carbon monoxide release can be promoted by using alternative ligands for carbon monoxide [206–208] and photodissociation has been reported for some compounds [209]. Further, ligands can be introduced that give rise to additional biological properties, for example, the complex containing the $Fe(CO)_3$ unit, $Fe(CO)_3(\eta^4 - C_4H_6)$, can be used to generate iron carbonyl nucleoside compounds (Fig. 12d) that induce apoptosis and therefore have potential applications in cancer therapy.

Carbon monoxide releasing compounds are also being evaluated as potential hypertensive and anti-inflammatory drugs, including allergic in-

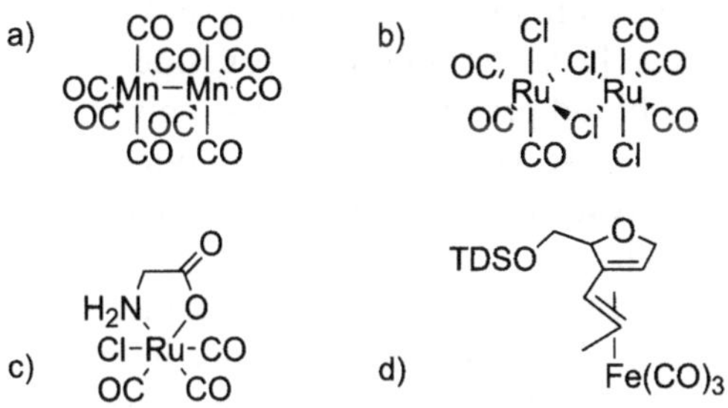

Fig. 12 Carbon monoxide releasing compounds: **a** manganese decacarbonyl; **b** tricarbonylruthenium chloride dimer; **c** a ruthenium-glycinate complex; **d** iron carbonyl nucleoside analogues, TDSO = thexyldimethylsilyloxy

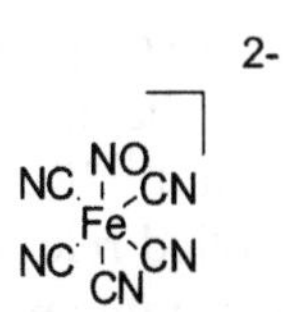

Fig. 13 The anion in sodium nitroprusside, a hypotensive agent

flammation, vasodilators and protectors against vascular injury, suppressing transplant-associated vascular arteriosclerosis. Carbon monoxide gas is known to protect against organ graft rejection, oxidative tissue damage, ischemia/reperfusion injury, septic shock, as well as being involved in apoptosis regulation and blood pressure; hence, there are many potential applications of these carbon monoxide releasing molecules in the clinic.

Whereas the discovery of carbon monoxide signalling properties is relatively new, the role of NO in biological signalling and in particular vasodilation, has been known for some years. NO is produced intracellularly by a ubiquitous class of enzymes known as the NO synthases, and a variety of compounds that spontaneously release NO in biological systems have been developed, reproducing the physiological or pathophysiological function of NO. Hypertension can be a long term illness, and it is becoming increasingly common, severely limiting the quality of life of the sufferer. The main symptom comprises elevated systolic and diastolic blood pressure, but other changes also occur in the body during the progression of the disease. For example, elevated blood pressure flow to the heart can cause major physiological changes in the vasculature and hypertrophy of the left ventricle, which in turn can lead to stroke or sudden cardiac death. Because of the long term nature of the condition and therefore the long term nature of medication, there is a strong incentive to develop drugs that can regulate blood pressure without causing side effects or becoming resistant over the course of the treatment. Many of the current leading therapies are based on organic drugs, although some inorganic compounds also exhibit excellent activity.

Metal-cyanide compounds possess interesting biological properties and are worth mentioning. Sodium nitroprusside, the anion is shown in Fig. 13, can be used to release NO in biological systems and has been investigated as a potential hypotensive [210]. Sodium nitroprusside can be administered by infusion and reduces blood pressure within two minutes, the effect depending on the rate of NO release. Other similar compounds have been studied for potential applications as vasodilators, including vanadium, cobalt and molybdenum analogues [211].

In addition to the release of carbon monoxide, transition metal carbonyl complexes have also found applications in anticancer chemotherapy [212], radiolabelling [213–216], and as photosensitisers. The cobalt complex shown in Fig. 14 a demonstrates a higher anticancer activity than cisplatin in certain mammary tumor cells lines [217,218]. Complexes such as $Tc-99m(I)(CO)_3(OH_2)_3$ along with hydroxymethyl phosphine derivatives [219] and the cyclopentadienyl complex shown in Fig. 14b have useful radiopharmaceutical applications, in diagnosis, using ^{99m}Tc, as the metal centre, and in therapy based on ^{188}Re and ^{186}Re isotopes [220].

In the same way that the controlled release of silver ions is important in the antibacterial activity of silver nitrate and silver(I) *n*-pincer-type heterocyclic carbene complexes, the release of carbon monoxide in a controlled

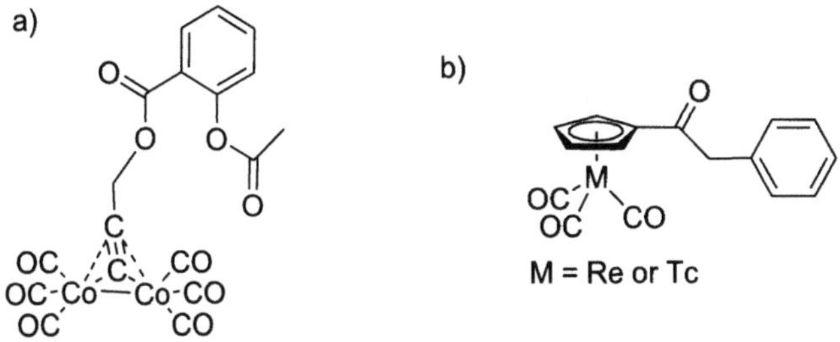

Fig. 14 Metal carbonyl anticancer and diagnosis compounds: **a** a dicobalthexacarbonyl-alkyne agent; **b** a metal tricarbonyl-cyclopentadienyl species

fashion is important for blood vessel dilation and cardio- and cyto-protectant activity of carbon monoxide-releasing molecules. Ligand exchange in different environments can also be critical for the activity and selectivity of other drugs. In the case of many putative anticancer compounds, such as the ruthenium(II)-arene compounds, hydrolysis of the chloride ligand(s) is a key step in determining cytotoxicity. In blood plasma the level of free chloride ions is relatively high, typically 100 mM, whereas in cells the chloride concentration falls to around 4 nM. This difference in chloride concentration can be used to ensure that the drugs are activated only when they enter cells.

The methyl complex $[Co(cyclam)(H_2O)(Me)]^{2+}$ has been investigated as a photosensitiser in anticancer therapy [221]. The complex is stable in water and air, but in the presence of photons produces methyl radicals via homolysis of the Co – C bond. The methyl radicals produced bring about the cleavage of the DNA without base specificity [222, 223].

4 Drug Cross Reactivity

Although cancer cells originate from mutations within the host, the aim of cancer treatment is to stop these diseased cells growing, that is, to kill the cancer cells without killing the surrounding healthy cells. Treatment of bacterial, parasitic, fungal or viral infections has essentially the same aim, to kill the invading organism without causing harm to the host. Thus, there are similarities in effective drug design for the treatment of cancer and infections, the key difference being that, in order to be effective, the drugs must have some selectivity for the diseased cells, exploiting differences in the biomolecules and metabolic pathway that are present in the diseased and healthy cells, thereby causing minimum damage to the healthy cells. Cancer and infections, irrespective of whether they are parasitic, bacterial, fungal or viral, are related in that cells or, in the case of viral infections, virons, proliferate in an uncontrolled fashion and in a manner that damages their "host". Consequently, some drugs that have activity against cancer may also be active against infections and vice versa. Indeed, cisplatin, which is arguably the most successful anticancer drug of all time, was first observed to inhibit the growth of bacterial cells. Subsequently, Rosenberg had the insight to investigate whether the same drug would inhibit cancer cell growth, and the outcome of these experiments is now well known.

In addition to specific differences between healthy cells and certain types of diseased cells, there are general changes that result because of the infection or mutations, which cause the disease. Diseased cells commonly grow much more quickly than healthy cells, the exception being with diseases that go through a stage of dormancy, such as the early stages of HIV infection, but even then, when the virus revives, the proliferation rate increases. Rapid cell proliferation can cause subtle changes, for example, in nutrient requirements and cytosolic pH, giving a handle by which to target or activate drugs specifically where they are needed. Because of the similarities between all rapidly proliferating cells, nutrient transport as a targeting mechanism can be used as a tool to ensure drugs specifically accumulate in bacterial, fugal, parasitically and virally infected cells as well as cancer cells. Consequently, many drugs that are active against one of these diseases may also be active against others.

Promising anticancer agents have been developed by attaching classical antiparasitic drugs such as *benznidazole*, nifurtimox and niridazole to rhodium-cycloocta-1,5-diene fragments [224, 225], demonstrating the cross reactivity of antiproliferation drugs between different types of disease. The bidentate cyclooctadiene ligand has been proposed to act as a non-labile group [226]. Several square planar rhodium-cycloocta-1,5-diene complexes [227–229] have demonstratable anticancer properties. For example, the rhodium complex in Fig. 15 shows antimetastatic activity, reducing the number and size of metastatic tumours as well as acting on the primary tumour. Similarly, rhodium(I) carbonyl complexes of dithiocarbamate and xanthate derivatives (Fig. 15b) have been assayed as cytostatic and antitumour agents as well as against trypanosome strains.

Drugs can be active against more than one type of disease because blocking the same biochemical pathways serve as targets for chemotherapy. Folate synthesis is a drug target in treating both malaria and cancer. Traditionally drugs such as chloroguanide, pyrimethamine, and sulfonamides are used

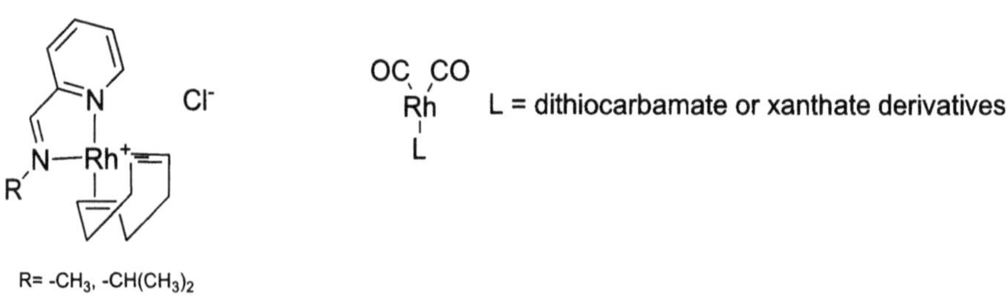

Fig. 15 Rhodium(I) anticancer compounds

as antimalarials, and methotrexate as the anticancer equivalent, although all these compounds target the same biological pathway.

The same drugs can even be used to treat different types of diseases. For example, several complexes known to have antiproliferic activity in tumour cells have been tested against parasitic diseases [135], and certain drugs identified for their antibacterial activity show promising anticancer activity [230, 231]. The first reported medicinal properties of cationic rhodium-2-hydroxybenzothiazole complexes and antimony-*N*,*N*′-piperazine-dithiocarbamate complexes were as anticancer agents. However, these compounds have since been shown to completely inhibit the growth of *L. donovani*, which causes Leishmaniasis. Similarly, drug resistance mechanisms in different diseases can involve the same mechanism and can therefore be treated using the same drug. For example, co-administering multi-drug resistance protein or *P*-glycoprotein blocking agents, such as verapamil or daunorubicin, can restore chloroquine activity in resistant cells, suggesting that in both cancer and malaria there are common drug resistance mechanisms.

The ruthenium(II)-arene RAPTA complexes show interesting antiproliferation activity against bacteria, fungi, and yeast, and also show promising antitumour activity [188]. Similarly, some organo-gold complexes that are active against *Staphylococus aureus* and *Enterococus faecalis* and more cytotoxic to Chinese hamster ovary cells than cisplatin; equally against HT1376 bladder tumor and SK-OV-3 ovarian tumor cells [230, 231].

The cross reactivity of drugs between different diseases is not just reserved to those within the antiproliferation catagory. Certain gold complexes such as those used in the treatment of rheumatoid arthritis, an inflammatory, autoimmune disease, also exhibit antimalarial activity and can enhance other antimalarial drugs when used in combination therapy [232]. A recent clinical report also showed that an AIDS patient, not being treated with anti-HIV drugs, but being treated with the gold-based drug auranofin used for psoriatic arthritis had an increased CD4+ T cell count, the decrease of which is a characteristic symptom of the disease [233]. While auranofin is not an organometallic compound, it is processed to auranocyanide in the body, which in turn is more readily taken up into cells compared to auranofin. Aurocyanide inhibits proliferation of HIV in cultured T-9 cells [234]. The activation of the drug via cyanation explains why smokers treated with auranofin suffer from more side effects than non-smokers, since they have greater levels of cyanide in their blood.

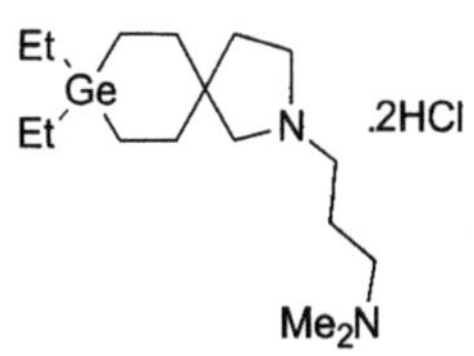

Fig. 16 Spirogermanium

Spirogermanium (Fig. 16) was shown to have activity against psoriasis [235], arthritis [236], as well as anticancer activity [237]. This azaspirane containing compound was first reported to have antitumour activity; however, further studies indicate that it normalises interleukin-1 production acting as an immunoregulator and reducing inflammation caused by arthritis via a different mechanism compared to the commonly used antiarthritis drugs auranofin and indomethacin.

5 Imaging and Diagnosis

Targeting is not just an important feature of drugs, but also of imaging agents and probes used in the diagnosis of diseases. Organometallic analogues of drugs have been used to evaluate drug activity and as metallotracers to label drugs and probe their metabolism [238]. For example, cobalt or iron analogues of the antiepileptic drugs phenobarbital and phenytoin have been used in drug evaluation and tricarbonyl η^5-cyclopentadienyl manganese as a metal tracer to investigate drug clearance and activity in vivo [239]. In this latter example the manganese tracer did not affect the inhibition or affinity constant of the drug. Co-57-cyanocobalamin, which is based on one of the few naturally occurring organometallic compounds, cobalamin, has been used in the diagnosis of pernicious anaemia and in intestinal imaging, to measure tissue permeability, for example, in acute-pancreatitis [240] and in investigating the biological functions of cobalamin uptake and transport [241]. Co-57-cyanocobalamin is also used to measure cobalamin levels in patients on long-term chemotherapy, to evaluate the potential toxicity and side effects of the drugs [242].

The technecium-isonitrile complexes (Fig. 17) were designed as imaging agents. For example, cardiolite (Fig. 17, R $=$ $CH_2C(OMe)Me_2$) is a useful myocardial perfusion imaging agent that behaves as a potassium mimic, and is taken up by the myocardium [243]. However, these compounds were also shown to have other medicinal applications. These compounds were the first compounds shown to be *P*-glycoprotein transport substrates and have since

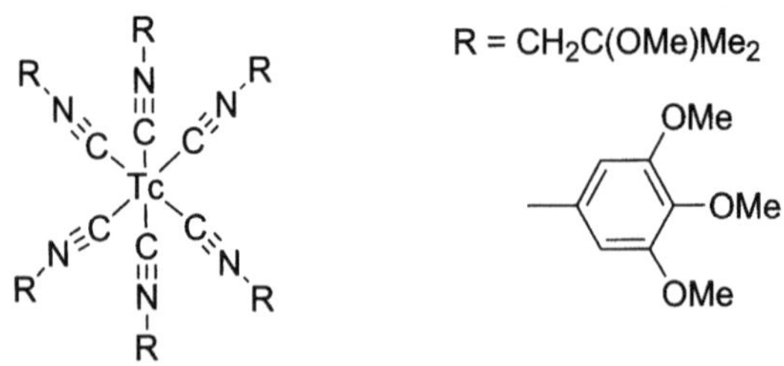

Fig. 17 Technecium-isonitrile imaging agents

proven to be useful for monitoring multidrug resistance. The complexes are non-metabolisable and so accumulate in cells unless the *P*-gp multidrug resistance protein has been expressed.

Changing the substituents attached to the isonitrile ligands in cardiolite alters the properties and the specificity of the compound. For example, Tc-TMPI also accumulates in cells with inverse proportionally to *P*-glycoprotein expression, inhibiting cardiolite transport and displacing iodoarylazidoprazosin, a *P*-glycoprotein-specific label [145].

6 Concluding Remarks

In general, organometallic compounds are considered as highly toxic, air sensitive and incompatible with an aqueous environment. Despite these conceptions, many organometallic compounds are proving to be well suited to the physiological environment and are finding applications in medicine and diagnosis. In particular, it would appear that derivatisation of known organic drugs with organometallic fragments can improve the activity of the drug and, in some cases, restore activity where resistance has developed for the organic compound. Understanding the molecular basis for such observations will require considerable research efforts at the interface of organometallic chemistry, biochemistry and physiology and should ultimately lead to superior drugs to those currently in use.

References

1. Albert A (1985) Selective Toxicity, 7th Ed. Chapman and Hall, New York, p 206
2. Rosenberg B, VanCamp L, Krigas T (1965) Nature 205:698
3. Reedijk J (1996) Chem Commun :801
4. Wong E, Giandomenico CM (1999) Chem Rev 99:2451
5. Sava G, Alessio E, Bergamo A, Mestroni G (1999) In: Clarke MJ, Sadler PJ (eds) Metallopharmaceuticals I, DNA Interactions. Springer-Verlag, Berlin Heidelberg, p 143
6. Takahara PM, Frederick CA, Lippard SJ (1996) J Am Chem Soc 118:12309
7. Farrell N, Kelland LR, Roberts JD, Van Beusichem M (1992) Cancer Res 52:5065
8. Zou Y, Houten B, Farrell N (1993) Biochemistry 32:547
9. Navarro M, Pérez H, Sánchez-Delgado RA (1997) J Med Chem 40:1937
10. Martell AE, Hancock RD (1996) In: Fackler JP (ed) Modern Inorganic Chemistry. Plenum, New York
11. Howard-Lock HH, Lock CJL (1987) In: Wilkinson G (ed) Comprehensive Coordination Chemistry. Pergamon, Oxford
12. Hughes MN (1987) In: Wilkinson G (ed) Comprehensive Coordination Chemistry. Pergamon, Oxford
13. Bertini I, Gray HB, Lippard SJ, Valentine JS (1994) Bioinorganic Chemistry. University Science Books, Mill Valley, CA, p 505
14. Dhubhghaill OMN, Sadler PJ (1991) Struct Bond 78:129

15. Shen ZX, Chen GQ, Ni JH, Li XS, Xiong SM, Giu QY, Zhu J, Tang W, Sun GL, Yang KQ, Chen Y, Zhou L, Fang ZW, Wang YT, Ma J, Zhang P, Zhang TD, Chen SJ, Chen Z, Wang ZY (1997) Blood 89:3354
16. Clemow PR, Lowry JW (1954) J Pharm Pharmacol 6:610
17. Perrenoud D, Gallezot D, Van Melle G (2001) Contact Dermatology 45:134
18. Bleyer AJ (2003) Expert Opin Pharmacol 4:941
19. Sengupta P, Mukherjee AM, Chatterjee A (1961) J Ind Med Assoc 37:585
20. Gellhorn A, Vandyke HB (1946) J Pharmacol Exp Ther 88:162
21. Vanson JAM, Julsurd PR, Hagler DJ, Sim EKW, Puga FJ, Schaff HV, Danielson GK (1994) Ann Thorac Surg 57:604
22. Shorvon P, Lamb G (1994) Curr Opin Gastroenterol 10:163
23. Briand GG, Burford N (1999) Chem Rev 99:2601
24. Paris J, Merveill P (1968) Lille Med 13:1038
25. Degros R (1977) Int J Dermatol 16:391
26. Lin CY, Shen YH, Wu SH, Lin CH, Hwang SM, Tsai YC (2004) Biochem Biophys Res Commun 315:830
27. Green MA (1987) Nucl Med Biol 14:58
28. Allan CC, Archibald SJ, Silversides JD, Lewis EA (2003) J Inorg Biochem 96:91
29. Benites PJ, Holmberg RC, Rawat DS, Kraft BJ, Klein LJ, Peters DG, Thorp HH, Zaleski JM (2003) J Am Chem Soc 125:6434
30. Sarkar B (1999) Chem Rev 99:2535
31. Macke H (1991) Schweiz Med Wochenschr 121:299
32. Schachter S, Nelson RW, Kirk CA (2001) J Vet Intern Med 15:379
33. Pertuiset B, Visot A, Metzger J (1976) Acta Neurochir 34:151
34. Vaillant G, Bertrand A, Nouel JP (1975) J Radiol Electrol Med Nucl 56:380
35. Truhaut R, Renault H, Claude JR, Rudler M (1975) Ann Biol Clin 33:201
36. Jin L, Yang, P (1997) Polyhedron 16:3395
37. Bellet M, Labat JP, Barthele L, Bellet M (1974) J Radiol Electrol Med Nucl 55:535
38. Hamblin MR, Newman EL (1994) J Photochem Photobiol B-Biol 26:45
39. Bates JN, Baker MT, Guerra R, Harrison DG (1991) Biochem Pharmac 42:S157
40. Manukhina EB, Malyshev IY, Malenyuk EB, Zenina TA, Podkidyshev DA, Mikoyan VD, Kubrina LN, Vanin AF (1998) Bull Exp Biol Med 125:23
41. Vanin AF, Kleshchev AL, Mordvintsev PI, Sedov KR (1985) Dokl Akad Nauk SSSR 281:742
42. Hergesell O, Ritz E (1999) Nephrol Dial Transplant 14:863
43. Caravan P, Ellison JJ, McMurry TJ, Lauffer RB (1999) Chem Rev 99:2293
44. Ledermann HP, Morrison WB, Schweitzer ME (2003) Radiology 227:305
45. Bozzao A, Floris R, Fasoli F, Fantozzi LM, Colonnese C, Simonetti G (2003) Eur Radiol 13:592
46. Assassa GS, Siegel ME, Chen DCP, Ansari A (1994) Clin Nucl Med 19:614
47. Collery P, Keppler BK, Madoulet C, Desoize B (2002) Crit Rev Oncol Hematol 42:283
48. Jao SW, Lee W, Ho YS (1990) Dis Colon Rectum 33:99
49. Howard-Lock HH, Lock CGL (1987) In: Wilkinson G (ed) Comprehensive Coordination Chemistry. Pergamon, Oxford, p 755
50. Okatani Y (1980) J Asthma Res 17:165
51. Van Arsdel PP (1981) J Allergy Clin Immunol 67:348
52. Muranaka M, Nakajima K, Suzuki S (1981) J Allergy Clin Immunol 67:350
53. Shaw CF (1999) Chem Rev 99:2589
54. Agrawal KC, Bears KB, Marcus D, Jonassen HB (1978) Proc Am Assoc Cancer Res 69:110

55. Simon TM, Kunishima DH, Vibert GJ, Lorber A (1979) Cancer 44:1965
56. Okada T, Patterson BK, Ye S-Q, Gurney ME (1993) Virology 192:631
57. Tepperman K, Shinju Y, Roy PW, Floyd R, Zhao Z, Dorsey J, Elder RC (1994) Met Based Drugs 1:433
58. Shapio DL, Masci JR (1996) J Rheumatol 23:1818
59. Louw WKA, Dormehl IC, van Rensburg AJ, Hugo N, Alberts AS, Forsyth OE, Beverley G, Sweetlove MA, Marais J, Lotter MG, van Aswegen A (1996) Nucl Med Biol 23:935
60. Schettino CJ, Kramer EL, Noz ME, Taneja S, Padmanabhan P, Lepor H (2004) Am J Roentgenol 183:519
61. Nakajima S, Yamauchi H, Sakata I, Hayashi H, Yamazaki K, Maeda T, Kubo Y, Samejima N, Takemura T (1993) Nucl Med Biol 29:231
62. Moder KP (1996) Expert Opin Ther Pat 6:29
63. Birch NJ (1999) Chem Rev 99:2659
64. Armstrong J, Beidas AS, Craik DJ, Stephenson WH (1955) J Pharm Pharmacol 7:862
65. Morniere P, Maurouard C, Boudailliez B, Westeel PF, Achard JM, Boitte F, Elesper N, Compagnon M, Maurel G, Bouillon R, Pamphile R, Fournier A (1992) Nephron 60:154
66. Kinnunen O, Salokannel J (1987) Ann Clin Res 19:321
67. Kinnunen O, Karppanen H, Salokannel J (1989) Magnesium Bull 11:68
68. Takehara Y, Sakahara H, Masunaga H, Isogai S, Kodaira N, Sugiyama M, Takeda H, Saga T, Nakajima S, Sakata I (2002) Magn Reson Med 47:549
69. Zhou R, Balasubramanian SV, Kahl SB, Straubinger RM (1999) J Pharm Sci 88:912
70. Pitts RF, Kessler RH, Lozano R (1957) Science 125:750
71. Milnor JP (1950) Proc Soc Exp Biol Med 75:63
72. Overman WJ, Gordon WH, Burch GE (1950) Circulation 1:496
73. Porzi G, Minellibertazzoni E (1975) Boll Soc Ital Biol Sper 51:1591
74. Sartin JS, Perry HO (1995) J Am Acad Dermatol 32:255
75. Ventura HO, Mehra MR, Young JB (2001) J Card Fail 7:277
76. Borle F, Radu A, Fontolliet C, van den Bergh H, Monnier P, Wagnieres G (2003) Br J Cancer 89:2320
77. Peschel RE, Colberg JW, Chen Z, Nath R, Wilson LD (2004) Cancer J 10:170
78. Bruno G, Cusumano M, Giannetto A, Zappala M, Nicolo F, Romeo R (1997) J Inorg Biochem 65:97
79. Brunner H, Obermeier H, Szeimies RM (1995) Chem Berichte 128:173
80. Desoize B, Madoulet C (2002) Crit Rev Oncol Hematol 42:317
81. Gumus F, Pamuk I, Ozden T, Yildiz S, Diril N, Oksuzoglu E, Gur S, Ozkul A (2003) J Inorg Biochem 94:255
82. Margiotta N, Fanizzi FP, Kobe J, Natile G (2001) Eur J Inorg Chem 1303
83. Rodriguez-Cabezas MN, Mesa-Valle CM, Azzouz S, Moraleda-Lindez V, Craciunescu D, Gutierrez-Rios MT, De Frutos MI, Osuna A (2001) Pharmacology 63:112
84. Powell NA, Hill AM, Levason W, Webster M (1996) J Chem Soc Dalton 467
85. Bien M, Pruchnik FP, Seniuk A, Lachowicz TM, Jakimowicz P (1999) J Inorg Biochem 73:49
86. Sanchez-Delgado RA, Navarro M, Perez H, Urbina JA (1996) J Med Chem 39:1095
87. Sava G, Bergamo A (2000) Int J Oncol 17:353
88. Otero L, Noblia P, Gambino D, Cerecetto H, Gonzalez M, Ellena JA, Piro OE (2003) Inorg Chim Acta 344:85
89. Anderberg PI, Harding MM, Luck IJ, Turner P (2002) Inorg Chem 41:1365
90. Morbidelli L, Donnini S, Filippi S, Messori L, Piccioli F, Orioli P, Sava G, Ziche M (2003) Brit J Cancer 88:1484

91. Maurya MR, Unny IR, Alekar NA, Gopinathan S, Gopinathan C (1997) Indian J Chem Sect A – Inorg Bio-Inorg Phys Theor Anal Chem 36:1058
92. Kohlickova M, Jedinakova-Krizova V, Melichar F (2000) Chem Listy 94:151
93. Silver S (2003) FEMS Microbiol Rev 27:369
94. Ford MM (2004) Vet Clin N Am-Small 34:669
95. Singal A, Thami GP (2003) J Dermatol 30:644
96. Piva JR, Canal AM, Piva CE, Reus V, Salvetti NR, Ortega HH (2003) J Histotechnol 26:31
97. Bijwaard H, Brugmans MJP, Leenhouts HR (2004) Radiat Res 162:171
98. Wang RF, Zhang CL, Zhu SL, Zhu M (2003) Med Princ Pract 12:97
99. Vanlic-Razumenic N (1999) J Radioanal Nucl Chem 242:235
100. McCaughan JS (1999) Drug Aging 15:49
101. Tomura N, Watanabe O, Omachi K, Sakuma I, Takahashi S, Otani T, Kidani H, Watarai J (2004) Eur Radiol 14:1249
102. Liu S, Edwards DS (1999) Chem Rev 99:2235
103. Arano Y (2002) Ann Nucl Med 16:79
104. Thompson KH, McNeill JH, Orvig C (1999) Chem Rev 99:2562
105. Milowsky MI, Nanus DM, Kostakoglu L, Vallabhajosula S, Goldsmith SJ, Bander NH (2004) J Clin Oncol 22:2522
106. Tsimberidou AM, Murray JL, O'Brien S, Wierda WG, Keating MJ (2004) Cancer 100:2195
107. Konan YN, Gurny R, Allemann E (2002) J Photochem Photobiol B 66:89
108. Sparsa A, Bonnetblanc JM (2004) Ann Dermatol Vener 131:255
109. Bhowon MG, Laulloo BSJ (2004) Indian J Chem Sect A – Inorg Bio-Inorg Phys Theor Anal Chem 43:1131
110. www.ncirs.usyd.edu.au/facts/f-thiomersal.html
111. Garrett CE, Prasad K (2004) Adv Synth Catal 346:889
112. Templeton DM, Fresen J (1999) Anal Chem 363:505
113. Edwards EI, Epton R, Marr G (1975) J Organometal Chem 85:C23
114. Sturchler D (1989) Parasitol Today 5:39
115. Muller R, Baker JR (1990) Medical Parasitology. Gower Medical Publishing, London
116. Veignie E, Moreau S (1991) Ann Trop Med Parasitol 85:229
117. Krogstad DJ, Gluyman IZ, Herwaldt BL, Schlesinger PH, Wellems TE (1987) Am J Trop Med Hyg 36:213
118. Slater A, Swiggard W, Orton B, Flitter W, Goldberg D, Cerami A, Henderson G (1991) Proc Natl Acad Sci USA 88:325
119. Ling V, Thompson LH (1974) Proc Natl Acad Sci Cell Physiol 83:103
120. Slater AF, Cerami A (1992) Nature 355:167
121. Bendrat K, Berger B, Cerami A (1995) Nature 378:138
122. Ridley R, Dorn A, Matile H, Kansy M (1995) Nature 378:138
123. Wernsdorfer W, McGregor I (eds) (1988) Malaria: Principles and Practice of Malariology. Churchill Livingstone New York, p 1818
124. Fitch C, Chevli R, Banyal H, Phillips G, Pfallar M, Krogstad D (1982) Antimicrob Agents Chemother 21:819
125. Sanchez C, Wunsch S, Lanzer M (1997) J Biol Chem 272:2652
126. Krogstad DJ, Gluzman IY, Kyle DE, Oduola AM, Martin SK, Milhous WK, Schlesinger PH (1987) Science 283:1283
127. Biot C, Glorian G, Maciejewski L, Brocard J (1997) J Med Chem 40:3715
128. Domarle O, Blampain G, Agnaneit H, Nzadiyabi T, Lebibi J, Brocard J, Maciejewski L, Biot C, Georges A, Millet P (1998) Antimicrob Agents Chemother 42:540

129. Croft S (1988) Trends Pharmacol Sci 9:376
130. Berman J (1988) J Infect Dis 10:560
131. Ashford R, Desjeux P, de Raadt P (1992) Parasitol Today 8:104
132. Ouellette M, Papadopoulou B (1993) Parasitol Today 9:150
133. Loiseau PM, Craciunescu DG, Doadriovillarejo JC, Certadfombona G, Gayral P (1992) Tropical medicine and parasitology 43:110
134. Zinsstag J, Brun R, Craciunescu DG, Iglesias EP (1991) Tropical medicine and parasitology 42:41
135. Mesa Valle CM, Moraleda Lindez V, Craciunescu D, Alonso M, Osuna A (1993) Arzneimittel Forsch/Drug Res 43:1010
136. Loiseau PM, Dreyfuss G, Doadrio L, Parrondo E, Craciunescu DG (2001) J Chemother 13:59
137. Loiseau PM, Reyfuss G, Daulouède S, Lachâtre G, Vincendaeu P, Craciunescu DG (1997) Trop Med Int Health 2:19
138. Loiseau PM, Jaffe JJ, Craciunescu DG (1998) Int J Parasitol 28:1279
139. Loiseau PM, Mbongo N, Bories C, Boulard Y, Craciunescu DG (2000) Parasite 7:103
140. Mbongo N, Loiseau PM, Lawerence F, Craciunescu DG, Iglesias EP (1991) Tropical medicine and parasitology 42:41
141. Castilla JJ, Mesa Valle CM, Sanchez Moreno M, Arnedo T, Rosales MJ, Mascaro C, Crraciunescu D, Osuna A (1996) Arzneimittel-forschung-drug research 46:990
142. Craciunescu DG, Scarcia V, Furlani A, Papaioannou A, Parrondo Iglesias E, Alonso MP (1991) In Vivo 5:329
143. Craciunescu G, Scarcia V, Furlani A, Parrondo Iglesias E, Ghirvu C, Papaioannou A (1989) Anticancer Res 9:781
144. Piwnicaworms D, Rao VV, Kronauge JF, Croop JM (1995) Biochemistry 34:12210
145. Rao VV, Herman LW, Kronauge JF, Piwnica-Worms D (1998) Nucl Med Biol 25:225
146. D'Cruz OJ, Waurzyniak B, Uckum FM (2004) Antiomicrobial agents and chemotherapy 48:1082
147. D'Cruz OJ, Waurzyniak B, Uckum FM (2001) Contraception 64:177
148. Allardyce CS, Dyson PJ (2001) J Cluster Sci 12:563
149. Crowe AJ, Smith PJ, Atassi G (1980) Chem-Biol Interact 32:171
150. Crowe AJ, Smith PJ, Atassi G (1984) Inorg Chim Acta 93:179
151. Han GY, Yang P (2002) J Inorg Biochem 91:230
152. Triolo F, Pellerito C, Stocco GC, Fiore T, Aggio F, Pellerito L, Triolo R (1999) Appl Organomet Chem 13:733
153. Barbieri R, Silvestri A (1991) J Inorg Biochem 41:31
154. Bollo E, Ceppa L, Cornaglia E, Nebbia C, Biolatti B, Dacasto M (1996) Hum Exp Toxicol 15:219
155. Boualam M, Willem R, Biesemans M, Gielen M (1991) Appl Organomet Chem 5:497
156. Buckley RG, Elsome AM, Fricker SP, Henderson GR, Theobald BR, Parish RV, Howe BP, Kelland LR (1996) J Med Chem 39:5208
157. Marcon G, Carotti S, Coronnello M, Messori L, Mini E, Orioli P, Mazzei T, Agostina Cinellu M, Minghetti G (2002) J Med Chem 45:1672
158. Pratviel G, Bernadou J, Meunier B (1995) Angew Chem Int Ed 34:746
159. Vol'pin M, Levitin I, Osinsky S (1996) Angew Chem Int Ed 35:2395
160. Bakac A, Espenson JH (1989) Inorg Chem 28:3901
161. Riodan CG, Wei P (1994) J Am Chem Soc 116:2189
162. Keppler BK, Vogel EA (1993) In: Pinedo HM, Schronagel JH (eds) Platinum and other metal coordination compounds in cancer chemotherapy 2. Plenum, New York, p 253

163. Hubel R, Polborn K, Knizek J, Noth H, Beck W (2000) Zeitschrift für anorganische und allgemeine Chemie 626:1701
164. Popova LV, Babin VN, Belousov YA, Nekrasov YS, Snegireva AE, Borodina NP, Shaposhnikova GM, Bychenko OB, Raevskii PM (1993) Appl Organomet Chem 7:85
165. Koepf-Maier P, Koepf H, Neuse EW (1984) J Cancer Res Clin 108:336
166. Henderson W, Alley SR (2001) Inorg Chim Acta 322:106
167. Top S, Vessières A, Cabestaing C, Laios I, Leclercq G, Provot C, Jaouen G (2001) J Organomet Chem 637:500
168. Jaouen G, Top S, Vessieres A, Leclercq G, Quivy J, Jin L, Croisy A (2000) CR Acad Sci Ic 3:89
169. Rosenfeld A, Blum J, Gibson D, Ramu A (1992) Inorg Chim Acta 201:219
170. Viotte M, Gautheron B, Kubicki MM, Nifant'ev IE, Fricker SP (1995) Met Based Drugs 2:311
171. Liu R-C, Ma Y-Q, Yu L, Li J-S, Cui J-R, Wang R-Q (2003) Appl Organomet Chem 17:662
172. Koepf-Maier P, Klapoetke T (1992) Cancer Res Clin 118:216
173. Koepf-Maier P, Klapoetke T (1992) Cancer Chemoth Pharm 29:361
174. Koepf-Maier P, Koepf H (1987) Chem Rev 87:1137
175. Koepf-Maier P, Neuse E, Klapoetke T, Koepf H (1989) Cancer Chemoth Pharm 24:23
176. Ghosh P, D'Cruz OJ, Narla RK, Uckun FM (2000) Clin Cancer Res 6:1536
177. Caruso F, Rossi M (2004) Mini Revs Med Chem 4:159
178. Bernard J, Ortner K, Spingler B, Pietzsch H-J, Alberto R (2003) Inorg Chem 42:1014
179. Moksi G, Harding MM (2001) J Inorg Biochem 86:611
180. Waern JB, Harding MM (2004) Inorg Chem 43:206
181. Guo ML, Sun HZ, McArdle HJ, Gambling L, Sadler PJ (2000) Biochemistry 39:10023
182. Köpf H, Köpf-Maier P (1979) Angew Chemie 91:509
183. Clarke MJ, Zhu FC, Frasca DR (1999) Chem Rev 99:2511
184. Sava G, Bergamo A (2000) Int J Oncol 17:353
185. Clarke MJ, Galang RD, Rodriguez VM, Kumar R, Pell S, Bryan DM (1988) In: Nicolini M (ed) Platinum and Other Metal Coordination Compounds in Cancer Chemotherapy. Martinus Nijhoff, Boston, p 582
186. Dale LD, Tocher JH, Dyson TMD, Edwards I, Tocher DA (1992) Anti-Cancer Drug Design 7:3
187. Allardyce CS, Dyson PJ, Ellis DJ, Heath SL (2001) Chem Commun 1396
188. Allardyce CS, Dyson PJ, Ellis DJ, Salter PA, Scopelliti R (2003) J Organomet Chem 668:35
189. Gopal YNV, Jayaraju D, Kondapi AK (1999) Biochemistry 38:4382
190. Gopal YNV, Konuru N, Kondapi AK (2002) Arch Biochem Biophys 401:53
191. Morris RE, Aird RE, Murdoch PDS, Chen H, Cummings J, Hughes ND, Parsons S, Parkin A, Boyd G, Jodrell DI, Sadler PJ (2001) J Med Chem 44:3616
192. Akbayeva DN, Gonsalvi L, Oberhauser W, Peruzzini M, Vizza F, Brueggeller P, Romerosa A, Sava G, Bergamo A (2003) Chem Commun 264
193. Huxham LA, Cheu ELS, Patrick BO, James BR (2003) Inorg Chim Acta 352:238
194. Pratviel G, Bernadou J, Meunier B (1995) Angew Chem Int Ed 34:746
195. Vol'pin M, Levitin I, Osinsky S (1996) Angew Chem Int Ed 35:2395
196. Grimes RN (2000) Coord Chem Rev 200–202:773
197. Sülü M, Küçükbay H, Durmaz R, Günel S (1998) Arzneimittel-forschung-drug research 48:291
198. Çetinkaya B, Özdemir I, Binbaşioğlu B, Durmaz R, Günel S (1999) Arzneimittel-forschung-drug research 49:538

199. Çetinkaya B, Çetinkaya E, Küçükbay H, Durmaz R (1996) Arzneimittel-forschung-drug research 46:821
200. Özdemir I, Denizci A, Öztürk HT, Çetinkaya B (2004) Appl Organometal Chem 18:318
201. Schibli R, Netter M, Scapozza L, Birringer M, Schelling P, Dumas C, Schoch J, Schubiger PA (2003) J Organomet Chem 668:67
202. Chohan ZH, Khan KM, Supran CT (2004) Appl Organometal Chem 18:305
203. D'Cruz OJ, Waurzyniak B, Uckum FM (2004) Antiomicrobial agents and chemotherapy 48:1082
204. Motterlini R, Clark JE, Foresti R, Sarathchandra P, Mann BE, Green CJ (2002) Circ Res 90:e17
205. Clark JE, Naughton P, Shurey S, Green CJ, Johnson TR, Mann BE, Foresti R, Motterlini R (2003) Circ Res 93:178
206. Darensbourg DJ, Klausmeyer KK, Reibenspies JH (1995) Inorg Chem 34:4933
207. Darensbourg DJ, Klausmeyer KK, Reibenspies JH (1996) Inorg Chem 35:1529
208. Pearson J, Cooke J, Takats J, Jordan RB (1998) J Am Chem Soc 120:1434
209. Hughey JL, Bock CR, Meyer TJ (1975) J Am Chem Soc 97:4440
210. Tuzel IH (1974) J Clin Pharmacol 14:494
211. Hayton TW, Legzdins P, Sharp WB (2002) Chem Rev 102:935
212. Craciunescu G, Scarcia V, Furlani A, Parrondo Iglesias E, Ghirvu C, Papaioannou A (1989) Anticancer Res 9:781
213. Kothari KK, Pillarsetty NK, Katti KV, Volkert WA (2003) Radiochimica Acta 91:53
214. Schibili R, Schubiger PA (2002) Euro J Nucl Med Mol Imaging 29:1529
215. Alberto R, Schibli R, Waibel R, Abram U, Schubiger AP (1999) Coord Chem Rev 190–192:901
216. Zobi F, Spingler B, Fox T, Alberto R (2003) Inorg Chem 42:2818
217. Schmidt K, Jung M, Keilitz R, Schnurr B, Gust R (2000) Inorg Chim Acta 306:6
218. Roth T, Eckert C, Fiebig H-H, Jung M (2002) Anticancer Res 22:2281
219. Kothari KK, Pillarsetty NK, Katti KV, Volkert WA (2003) Radiochimica Acta 91:53
220. Bernard J, Ortner K, Spingler B, Pietzsch H-J, Alberto R (2003) Inorg Chem 42:1014
221. DeRosa MC, Crutchley RJ (2002) Coord Chem Rev 233:351
222. Bakac A, Espenson JH (1989) Inorg Chem 28:3901
223. Riodan CG, Wei P (1994) J Am Chem Soc 116:2189
224. Giraldi T, Sava G, Mestroni G, Zassinovich G, Stolfa D (1978) Chem-Biol Interact 22:231
225. Sava G, Giraldi T, Mestroni G, Zassinovich G (1983) Chem-Biol Interact 45:1
226. Zassinovich G, Maestroni G, Camus A (1975) J Organomet Chem 91:379
227. Giraldi T, Zassinovich G, Mestroni G (1974) Chem-Biol Interact 9:389
228. Sava G, Zorzet S, Pacor S, Mestroni G, Zassinovich G (1989) Cancer Chemoth Pharm 24:302
229. Giraldi T, Sava G, Bertoli G, Mestroni G, Zassinovich G (1977) Cancer Res 37:2662
230. Buckley RG, Elsome AM, Fricker SP, Henderson GR, Theobald BR, Parish RV, Howe BP, Kelland LR (1996) J Med Chem 39:5208
231. Parish RV, Howe BP, Wright JP, Mack J, Pritchard RG, Buckley RG, Elsome AM, Fricker SP (1996) Inorg Chem 35:1659
232. Navarro M, Pérez H, Sánchez-Delgado RA (1997) J Med Chem 40:1937
233. Shapiro DL, Masci JR (1996) J Rheumatology 23:1818
234. Tepperman K, Shinju Y, Roy PW, Floyd R, Zhao Z, Dorsey J, Elder RC (1994) Met Based Drugs 1:433
235. Bernstein LR, Reed DE (1998) Pub Int Appl PCT/US97/11220

236. Dimartino MJ, Lee JC, Badger AM, Murhead KA, Mirabelli CK, Hanna N (1986) J Pharma Exp Thera 236:103
237. Slavik M, Elias L, Mrema J, Saiers JH (1982) Drug Exp Clin Res 8:379
238. Atteke C, Ndong JMM, Aubouy A, Naciejewski L, Brocard J, Lebibi J, Deloron P (2003) J Antimicrobial Chemotherapy 51:1021
239. Mariet F, Brosser P (1991) Pharmaceutica Acta Helvetiae 66:60
240. Sweiry JH, Mann GE (1991) Am J Physiol 261:G685
241. Oreshkin AE, Myasishcheva NV (1990) B Exp Biol Med 109:957
242. Schenk BE, Festen HPM, Kuipers EJ, Klinkenberg-Knol EC, Meuwissen SGM (1996) Aliment Pharm Therap 10:541
243. Deutsch E, Bushong W, Glavan KA, Elder RC, Sodd VJ, Scholz KL, Fortman DL, Lukes SJ (1981) Science 214:85

Author Index Volumes 1–17

The volume numbers are printed in italics

Subject Index

MIX
Papier aus verantwortungsvollen Quellen
Paper from responsible sources
FSC® C105338

If you have any concerns about our products,
you can contact us on
ProductSafety@springernature.com

In case Publisher is established outside the EU,
the EU authorized representative is:
Springer Nature Customer Service Center GmbH
Europaplatz 3, 69115 Heidelberg, Germany

Printed by Libri Plureos GmbH
in Hamburg, Germany